Fang · Zhang Generalized Multivariate Analysis

Fang Kai-Tai Zhang Yao-Ting

Generalized Multivariate Analysis

Science Press
Beijing

Springer-Verlag
Berlin Heidelberg New York
London Paris Tokyo Hong Kong

Fang Kai-Tai
Institute of Applied Mathematics, Academia Sinica,
Beijing, China

Zhang Yao-Ting
Wuhan University
Wuhan, China

Translation of the Chinese edition of this book will be published by Science Press Beijing as the 20th volume in the Series in Pure and Applied Mathematics.

Distribution rights throughout the world, excluding The People's Republic of China, granted to Springer–Verlag Berlin Heidelberg New York London Paris Tokyo Hong Kong

Mathematics Subject Classification (1980): 62 H XX, 62 E XX

ISBN 3–540–17651–9 Springer–Verlag Berlin Heidelberg New York
ISBN 0–387–17651–9 Springer–Verlag New York Berlin Heidelberg
ISBN 7–03–000234–2 / O·67 Science Press Beijing

Printed in Hong Kong

Printing and binding: C & C Joint Printing Co., (H. K.) Ltd.
2141/3140-543210

PREFACE

There are hundreds of books on multivariate analysis throughout the world now, of which the overwhelming majority concentrate on multivariate normal distributions, or, in other words, *classical multivariate analysis*. To meet the requirements in theory and practice, we must develop the multivariate analysis of non–normal population. The book "Nonparametric Methods in Multivariate Analysis" by Puri and Sen gives a treatment of multivariate analysis from the nonparametric angle. The multivariate analysis of discrete variables has also been studied by other authors. In the past fifteen years, statisticians have discovered that many properties of the class of *elliptically contoured distributions* are very similar to those of the multivariate normal distributions. This class involves infinite multivariate distributions. In particular, it involves the multivariate normal distribution, the multivariate t–distribution, the multivariate Cauchy distribution, the multivariate Laplace distribution, the multivariate stable law and the multivariate uniform distribution. Many authors have developed a parallel theory and methodology for multivariate analysis based on the elliptically contoured distributions, and this is called *generalized multivariate analysis*.

The establishment and development of the theory of generalized multivariate analysis is a brilliant achievement in the field of multivariate analysis, and is welcomed and praised by many practical workers. The purpose of this book is to summarize the theory of generalized multivariate analysis which has grown up in the past fifteen years. Because mathematical tools used in generalized multivariate analysis are nothing more than those of classical multivariate analysis, this book can be designed as a textbook for the course in multivariate analysis, presenting directly to the students the generalized multivariate analysis with classical multivariate analysis as its special case. Parts of this book were discussed in lectures given by Fang Kai–Tai at Jinan University in China in 1983, and aroused great interest among the audience. While these parts constitute the core of this book new results are incorporated, which were covered by lectures given by the first author at the Stanford University, University of Maryland, George Washington University, Yale University, University of Pennsylvania, Princeton University, Columbia University in U. S. A., Simon Fraser University in Canada, Swiss Federal Institute of Technology, University of Berne in Switzerland, University College London, Cambrige University in UK and in the University of Hong Kong. The second author discussed parts of generalized multivariate analysis in his lectures delivered at the University of Pittsburgh, in U. S. A.

Chapter I gives the elementary knowledge of matrix theory needed for later chapters. The derivative of a matrix and the Jacobian are discussed in detail and the maximal invariants are presented also. Chapter II deals with the fundamentals of generalized multivariate analysis, discussing systematically the properties of the class of elliptically contoured distributions, and the related central distributions and noncentral distributions. In order to expand the sampling theory of generalized multivariate analysis, the concept of elliptically contoured distribution is extended from the case of a vector to the case of a matrix in Chapter III, in which the properties of spherical matrix distributions and elliptically contoured matrix distributions are discussed systematically. The various estimators of parameters of elliptically contoured distributions including the maximum–likelihood estimators, the minimax estimators and the robust M–estimators are studied in Chapter IV. Moreover various tests about mean vector and covariance matrix and other tests are discussed in Chapter V. The last chapter treats the theory of generalized linear model, including the estimation of parameters (the least squares estimators and the shrunken estimators), testing hypotheses and the theory of distributions, discriminant analysis and the relationship between discriminant model and regression model.

The emphasis of this book is laid on the new results of generalized multivariate analysis. In the part though not elematary but similar to that of classical multivariate analysis, we shall give only references or simply results, the reader can easily find the related material in other books on multivariate analysis.

This book has been planned as a textbook for a one–semester course for postgraduate students and as a reference book for lecturers and researchers. There are many exercises attached at the end of each chapter.

The first three chapters were written by the first author. The second author wrote the last chapter. The manuscripts of Chapters IV and V were prepared by Mr. Chen Han–feng under the guidance of the first author, and then the second author checked on them. Finally, the first author made a review for the whole work and the second author prepared a complete index.

We are very grateful to Professor T. W. Anderson for introducing this useful direction of generalized multivariate analysis to the first author of the book. Many results in the book are taken from the theses written by T. W. Anderson and the first author. Anderson's famous book "An Introduction to Multivariate Statistical Analysis" has been very helpful to us. In many places of the book references are made to the works" Aspects of Multivariate Statistical Theory" by R. J. Muirhead and "An Introduction to Multivariate Statistics" by M. S. Srivastava and C. G. Khatri. Contained in this book are some results completed by our students— Chen Han–feng, Fan Jianqing, Li Gang, Quan Hui, Wu Yue–hua, Xu Jing–lun and Zhang Hong–qing. We are greatly indebted to Xu Jing–lun and Chen Han–feng for their excellent typing.

We are also indebted to Professor K. L. Chung who first encouraged us to write this book and to Professor Deng Wei–cai and Professor He Ling who carefully read the manuscript of the book and made many useful comments thus making the substantial improvements possible.

We would greatly appreciate being informed of any errors found or any suggestions concerning our book.

Fang Kai-Tai
Institute of Applied Mathematics,
Academia Sinica, Beijing, China

Zhang Yao-Ting
Wuhan University, Wuhan, China

CONTENTS

CHAPTER I
SOME MATRIX THEORY AND INVARIANCE

CHAPTER II
ELLIPTICALLY CONTOURED DISTRIBUTIONS

CHAPTER III
SPHERICAL MATRIX DISTRIBUTIONS

CHAPTER IV
ESTIMATION OF PARAMETERS

CHAPTER V
TESTING HYPOTHESES

CHAPTER VI
LINEAR MODELS

CHAPTER I
SOME MATRIX THEORY AND INVARIANCE

In this chapter, we summarize some of the well-known definitions and results of matrix algebra that are needed in the rest of the book. Many of the results should be familiar to the reader already; the more basic of these are not proved here. A number of results which are not always found in books on matrix algebra are proved here. In the last section of this chapter, groups are defined and discussed. Many maximal invariants under a group needed in this book are obtained.

1.1. Definitions

1.1.1. Matrices

Definition 1.1.1. An $n \times p$ real matrix $\boldsymbol{A}$ is an ordered rectangular array of elements a_{ij} (reals)

$$(1.1.1) \qquad \boldsymbol{A} = \begin{bmatrix} a_{11} & \cdots & a_{1p} \\ \vdots & & \vdots \\ a_{n1} & \cdots & a_{np} \end{bmatrix},$$

and write $\boldsymbol{A} = (a_{ij})$.

If $n = p$ then $\boldsymbol{A}$ is called a *square matrix of order p*. If $p = 1$ then $\boldsymbol{A}$ is a column vector, and if $n = 1$ then $\boldsymbol{A}$ is a row vector. Two matrices of the same dimension $\boldsymbol{A}(n \times p)$, $\boldsymbol{B}(n \times p)$ are said to be *equal* (written as $\boldsymbol{A} = \boldsymbol{B}$) if $a_{ij} = b_{ij}$ for $i = 1, \ldots, n$, $j = 1, \ldots, p$. If all $a_{ij} = 0$, then $\boldsymbol{A}$ is called a *zero matrix*, written as $\boldsymbol{A} = \boldsymbol{O}$. If $p = n$, $a_{ii} = 1$ for $i = 1, \ldots, p$ and $a_{ij} = 0$ for $i \neq j$ then $\boldsymbol{A}$ is called the *identity matrix of order p*, written as $\boldsymbol{A} = \boldsymbol{I}_p$ or $\boldsymbol{A} = \boldsymbol{I}$. The diagonal elements of a $p \times p$ matrix $\boldsymbol{A}$ are $a_{11}, \ldots, a_{pp}$.

The transpose of an $n \times p$ matrix $\boldsymbol{A}$ is a $p \times n$ matrix $\boldsymbol{A}'$:

$$\boldsymbol{A}' = \begin{bmatrix} a_{11} & \cdots & a_{n1} \\ \vdots & & \vdots \\ a_{1p} & \cdots & a_{np} \end{bmatrix}$$

and is obtained by interchanging the rows and columns of $\boldsymbol{A}$. The matrix $\boldsymbol{A}$ is expressed in terms of elements, columns, and rows as

(1.1.2) $$A = (a_{ij}) = (\boldsymbol{a}_1, \dots, \boldsymbol{a}_p) = \begin{bmatrix} \boldsymbol{a}'_{(1)} \\ \cdot \\ \cdot \\ \cdot \\ \boldsymbol{a}'_{(n)} \end{bmatrix},$$

Let $\boldsymbol{A}$ be a square matrix of order p. It is called *symmetric* if $\boldsymbol{A}' = \boldsymbol{A}$ and *skew–symmetric* if $\boldsymbol{A} = -\boldsymbol{A}'$. Clearly, all diagonal elements of a skew–symmetric matrix are equal to zero.

A square matrix $\boldsymbol{A} = (a_{ij})$ of order p is said to be a *diagonal matrix* if all its off–diagonal elements are zero, written as $\boldsymbol{A} = \operatorname{diag}(a_{11}, \dots, a_{pp})$. A square matrix of order p $\boldsymbol{A} = (a_{ij})$ with $a_{ij} = 0$ for $j < i$ is called an *upper triangular matrix*, written as $\boldsymbol{A} \in \mathrm{UT}(p)$; if $a_{ij} = 0$ for $j > i$, then $\boldsymbol{A}$ is called a *lower triangular matrix*, written as $\boldsymbol{A} \in \mathrm{LT}(p)$. A square matrix of order p is said to be *orthogonal* if $\boldsymbol{A}'\boldsymbol{A} = \boldsymbol{A}\boldsymbol{A}' = \boldsymbol{I}_p$, written as $\boldsymbol{A} \in O(p)$.

The sum of two $n \times p$ matrices $\boldsymbol{A}$ and $\boldsymbol{B}$ is defined by $\boldsymbol{A} + \boldsymbol{B} = (a_{ij} + b_{ij})$. The product of two matrices $\boldsymbol{A}(p \times q)$ and $\boldsymbol{B}(q \times r)$ is the $p \times r$ matrix defined by $\boldsymbol{AB} = \boldsymbol{C} = (c_{ij})$ where

$$c_{ij} = \sum_{k=1}^{q} a_{ik} b_{kj}.$$

The product of a matrix $\boldsymbol{A}$ by a scalar α is defined by $\alpha\boldsymbol{A} = (\alpha a_{ij})$.

It can be verified that the above operations have the following properties (here, if sums or products are involved, it is assumed that these are defined):

$$\left\{ \begin{aligned} \boldsymbol{A} + \boldsymbol{B} &= \boldsymbol{B} + \boldsymbol{A} \\ (\boldsymbol{A} + \boldsymbol{B}) + \boldsymbol{C} &= \boldsymbol{A} + (\boldsymbol{B} + \boldsymbol{C}) \\ \boldsymbol{A} + (-1)\boldsymbol{A} &= \boldsymbol{O} \\ (c + d)\boldsymbol{A} &= c\boldsymbol{A} + d\boldsymbol{A} \\ c(\boldsymbol{A} + \boldsymbol{B}) &= c\boldsymbol{A} + c\boldsymbol{B} \\ (\boldsymbol{AB})' &= \boldsymbol{B}'\boldsymbol{A}' \\ (\boldsymbol{A}')' &= \boldsymbol{A} \\ (\boldsymbol{A} + \boldsymbol{B})' &= \boldsymbol{A}' + \boldsymbol{B}' \\ \boldsymbol{A}(\boldsymbol{BC}) &= (\boldsymbol{AB})\boldsymbol{C} \\ \boldsymbol{A}(\boldsymbol{B} + \boldsymbol{C}) &= \boldsymbol{AB} + \boldsymbol{AC} \\ (\boldsymbol{A} + \boldsymbol{B})\boldsymbol{C} &= \boldsymbol{AC} + \boldsymbol{BC} \\ \boldsymbol{AI} &= \boldsymbol{IA} = \boldsymbol{A} \end{aligned} \right.$$

1.1.2. Determinants

Definition 1.1.2. The determinant of a square matrix $\boldsymbol{A}$ of order p, denoted by $|\boldsymbol{A}|$, is defined as

(1.1.3) $$|\boldsymbol{A}| = \sum_{\pi} \varepsilon_\pi a_{1j_1} a_{2j_2} \dots a_{pj_p},$$

where $\sum_\pi$ denotes the summation over all $p!$ permutations $\pi = (j_1, \dots, j_p)$ of $(1, \dots, p)$ and $\varepsilon_\pi = 1$ or -1 according as the permutation π is even or odd.

A minor is the determinant of a submatrix of $\boldsymbol{A}$. The minor of an element a_{ij} is the determinant of the submatrix of $\boldsymbol{A}$ obtained by deleting the i-th row and j-th column. The cofactor of a_{ij}, say A_{ij}, is $(-1)^{i+j}$ times the minor of a_{ij}. It can be shown that

$$(1.1.4)\qquad |\boldsymbol{A}| = \sum_{j=1}^{p} a_{ij}A_{ij} = \sum_{j=1}^{p} a_{jk}A_{jk}.$$

The following are elementary properties of determinants:

(1) If $\boldsymbol{a}_i = 0$ or $\boldsymbol{a}_{(j)} = 0$ for some i or j, then $|\boldsymbol{A}| = 0$.

(2) $|\boldsymbol{A}| = |\boldsymbol{A}'|$.

(3) $|(\boldsymbol{a}_1, \ldots, \boldsymbol{a}_{i-1}, \alpha\,\boldsymbol{a}_i, \boldsymbol{a}_{i+1}, \ldots, \boldsymbol{a}_p)| = \alpha|\boldsymbol{A}|$.

(4) $|\alpha\boldsymbol{A}| = \alpha^p|\boldsymbol{A}|$.

(5) $|\boldsymbol{AB}| = |\boldsymbol{A}|\,|\boldsymbol{B}|$ and $|\boldsymbol{A}_1 \ldots \boldsymbol{A}_m| = |\boldsymbol{A}_1| \ldots |\boldsymbol{A}_m|$.

(6) If $\boldsymbol{A}$ is $p \times p$, then $|\boldsymbol{AA}'| = |\boldsymbol{A}'\boldsymbol{A}| \geqslant 0$.

(7) $\left|\begin{bmatrix} \boldsymbol{A} & \boldsymbol{C} \\ \boldsymbol{O} & \boldsymbol{B} \end{bmatrix}\right| = \left|\begin{bmatrix} \boldsymbol{A} & \boldsymbol{O} \\ \boldsymbol{D} & \boldsymbol{B} \end{bmatrix}\right| = |\boldsymbol{A}|\cdot|\boldsymbol{B}|$, where $\boldsymbol{A}{:}\,p \times p$, $\boldsymbol{B}{:}\,q \times q$, $\boldsymbol{C}{:}\,p \times q$, and $\boldsymbol{D}{:}\,q \times p$.

(8) $|\boldsymbol{I}_p + \boldsymbol{AB}| = |\boldsymbol{I}_q + \boldsymbol{BA}|$, where $\boldsymbol{A}{:}\,p \times q$ and $\boldsymbol{B}{:}\,q \times p$.

(9) $|\boldsymbol{T}| = \prod_{i=1}^{p} t_{ii}$, where $\boldsymbol{T} = (t_{ij}) \in \mathrm{UT}(p)$.

(10) $|\boldsymbol{H}| = \pm 1$, if $\boldsymbol{H} \in O(p)$.

1.1.3. Inverse of a Matrix

Definition 1.1.3. If $|\boldsymbol{A}| \neq 0$, there exists a unique matrix $\boldsymbol{B}$ such that $\boldsymbol{AB} = \boldsymbol{I}$. $\boldsymbol{B}$ is called the inverse of $\boldsymbol{A}$ and is denoted by $\boldsymbol{A}^{-1}$.

The (i, j) element of $\boldsymbol{B} = \boldsymbol{A}^{-1}$ is given by $b_{ij} = A_{ji}/|\boldsymbol{A}|$, where A_{ji} is the cofactor of a_{ji}. A matrix whose determinant is not zero is called *nonsingular*. The following properties are elementary:

(1) $\boldsymbol{AA}^{-1} = \boldsymbol{A}^{-1}\boldsymbol{A} = \boldsymbol{I}$.

(2) $(\boldsymbol{A}^{-1})' = (\boldsymbol{A}')^{-1}$.

(3) $(\boldsymbol{AB})^{-1} = \boldsymbol{B}^{-1}\boldsymbol{A}^{-1}$, if $\boldsymbol{A}$ and $\boldsymbol{B}$ are nonsingular $p \times p$ matrices.

(4) $|\boldsymbol{A}^{-1}| = |\boldsymbol{A}|^{-1}$.

(5) $\boldsymbol{A}^{-1} = \boldsymbol{A}'$, if $\boldsymbol{A} \in O(p)$.

(6) If $\boldsymbol{A} = \mathrm{diag}(a_{11}, \ldots, a_{pp})$ with $a_{ii} \neq 0$ $(i = 1, \ldots, p)$, then $\boldsymbol{A}^{-1} = \mathrm{diag}(a_{11}^{-1}, \ldots, a_{pp}^{-1})$.

(7) If $\boldsymbol{A} \in \mathrm{UT}(p)$ then $\boldsymbol{A}^{-1} \in \mathrm{UT}(p)$ and its diagonal elements are a_{ii}^{-1}, $i = 1, \ldots, p$.

1.1.4. Partition of Matrices

Definition 1.1.4. An $n \times p$ matrix $\boldsymbol{A} = (a_{ij})$ is said to *be partitioned into submatrices*, $i, j = 1, 2$, if $\boldsymbol{A}$ can be written as

$$\boldsymbol{A} = \begin{bmatrix} \boldsymbol{A}_{11} & \boldsymbol{A}_{12} \\ \boldsymbol{A}_{21} & \boldsymbol{A}_{22} \end{bmatrix},$$

where $A_{11} = (a_{ij})$, $i = 1, \dots, m$, $j = 1, \dots, q$; $A_{12} = (a_{ij})$, $i = 1, \dots, m$, $j = q+1, \dots, p$; $A_{21} = (a_{ij})$, $i = m+1, \dots, n$, $j = 1, \dots, q$; and $A_{22} = (a_{ij})$, $i = m+1, \dots, n$, $j = q+1, \dots, p$.

If two matrices A, B of the same dimension are similarly partitioned, then

$$A + B = \begin{bmatrix} A_{11} + B_{11} & A_{12} + B_{12} \\ A_{21} + B_{21} & A_{22} + B_{22} \end{bmatrix}.$$

Also, if C is a $q \times l$ matrix partitioned as

$$C = \begin{bmatrix} C_{11} & C_{12} \\ C_{21} & C_{22} \end{bmatrix}$$

where C_{11}: $q \times r$, C_{12}: $q \times (l - r)$, C_{21}: $(p - q) \times r$ and C_{22}: $(p - q) \times (l - r)$, then

$$AC = \begin{bmatrix} A_{11}C_{11} + A_{12}C_{21} & A_{11}C_{12} + A_{12}C_{22} \\ A_{21}C_{11} + A_{22}C_{21} & A_{21}C_{12} + A_{22}C_{22} \end{bmatrix}.$$

The following results are extremely useful:

(1) Let A be a $p \times p$ nonsingular matrix, and let $B = A^{-1}$. Partition A and B in the same fashion

$$\text{(1.1.5)} \qquad A = \begin{bmatrix} A_{11} & A_{12} \\ A_{21} & A_{22} \end{bmatrix}, \qquad B = \begin{bmatrix} B_{11} & B_{12} \\ B_{21} & B_{22} \end{bmatrix},$$

where A_{11}: $q \times q$ and A_{22}: $(p - q) \times (p - q)$. Write

$$\text{(1.1.6)} \qquad A_{11.2} = A_{11} - A_{12}A_{22}^{-1}A_{21}, \qquad A_{22.1} = A_{22} - A_{21}A_{11}^{-1}A_{12}.$$

Then

$$\text{(1.1.7)} \qquad \begin{matrix} B_{11} = A_{11.2}^{-1} & B_{12} = -A_{11}^{-1}A_{12}A_{22.1}^{-1} \\ B_{21} = -A_{22}^{-1}A_{21}A_{11.2}^{-1} & B_{22} = A_{22.1}^{-1} \end{matrix},$$

or

$$\text{(1.1.8)} \qquad \begin{matrix} B_{11} = A_{11}^{-1} + A_{11}^{-1}A_{12}A_{22.1}^{-1}A_{21}A_{11}^{-1} & B_{12} = -A_{11}^{-1}A_{12}A_{22.1}^{-1} \\ B_{21} = -A_{22.1}^{-1}A_{21}A_{11}^{-1} & B_{22} = A_{22.1}^{-1} \end{matrix},$$

or

$$\text{(1.1.9)} \qquad \begin{matrix} B_{11} = A_{11.2}^{-1} & B_{12} = -A_{11.2}^{-1}A_{12}A_{22}^{-1} \\ B_{21} = -A_{22}^{-1}A_{21}A_{11.2}^{-1} & B_{22} = A_{22}^{-1}A_{21}A_{11.2}^{-1}A_{12}A_{22}^{-1} + A_{22}^{-1}. \end{matrix}$$

(2) Let A be partitioned as in (1.1.5) and let $A_{11.2}$ and $A_{22.1}$ be defined by (1.1.6).

(i) If $|A_{22}| \neq 0$, then $|A| = |A_{22}|\,|A_{11.2}|$.

(ii) If $|A_{11}| \neq 0$, then $|A| = |A_{11}|\,|A_{22.1}|$.

(iii) If $|A_{11}| \neq 0$, $|A_{22}| \neq 0$, then $|A_{11}|\,|A_{22.1}| = |A_{22}|\,|A_{11.2}|$.

(3) Let A and B be nonsingular $p \times p$ and $q \times q$ matrices, respectively, and let C be $p \times q$ and D be $q \times p$. Then

$$\text{(1.1.10)} \qquad (A + CBD)^{-1} = A^{-1} - A^{-1}CB(B + BDA^{-1}CB)^{-1}BDA^{-1}.$$

In particular, take $B = I$, $C = u$ and $D = -v$. Then

$$\text{(1.1.11)} \qquad (A - uv')^{-1} = A^{-1} + (A^{-1}uv'A^{-1})/(1 - v'A^{-1}u).$$

(4) If $\boldsymbol{X}$ is an $n \times p$ matrix denoted similarly by (1.1.2), then

$$\boldsymbol{X}'\boldsymbol{X} = \begin{bmatrix} \boldsymbol{x}_1'\boldsymbol{x}_1 & \cdots & \boldsymbol{X}_1'\boldsymbol{X}_p \\ \vdots & & \vdots \\ \boldsymbol{x}_p'\boldsymbol{x}_1 & \cdots & \boldsymbol{x}_p'\boldsymbol{x}_p \end{bmatrix} = \sum_{i=1}^{n} \boldsymbol{x}_{(i)}\boldsymbol{x}_{(i)}'. \tag{1.1.12}$$

1.1.5. Rank of a Matrix

Definition 1.1.5. If $\boldsymbol{A}$ is a nonzero $n \times p$ matrix, it is said to have *rank* r, written as $\text{rk}(\boldsymbol{A}) = r$, if at least one of its minors of order r is different from zero while every minor of order $(r+1)$ is zero.

Clearly, $\text{rk}(\boldsymbol{O}) = 0$ and $\text{rk}(\boldsymbol{A}) = p$ if $\boldsymbol{A}$ is a $p \times p$ nonsingular matrix. The following properties can be verified:

(1) $\text{rk}(\boldsymbol{A}) = \text{rk}(\boldsymbol{A}') = \text{rk}(\boldsymbol{A}'\boldsymbol{A}) = \text{rk}(\boldsymbol{A}\boldsymbol{A}')$.
(2) $\text{rk}(\boldsymbol{A}) \leqslant \min(n, p)$, where $\boldsymbol{A}: n \times p$.
(3) $\text{rk}(\boldsymbol{AB}) \leqslant \min\,(\text{rk}(\boldsymbol{A}), \text{rk}(\boldsymbol{B}))$.
(4) $\text{rk}(\boldsymbol{A} + \boldsymbol{B}) \leqslant \text{rk}(\boldsymbol{A}) + \text{rk}(\boldsymbol{B})$.
(5) $\text{rk}(\boldsymbol{ABC}) = \text{rk}(\boldsymbol{B})$ if $\boldsymbol{A}$ and $\boldsymbol{C}$ are nonsingular square matrices.
(6) If $\boldsymbol{A}$ is $p \times q$ and $\boldsymbol{B}$ is $q \times r$ such that $\boldsymbol{AB} = \boldsymbol{O}$, then $\text{rk}(\boldsymbol{B}) \leqslant q - \text{rk}(\boldsymbol{A})$.

1.1.6. Trace of a Matrix

Definition 1.1.6. The *trace* of a square matrix $\boldsymbol{A} = (a_{ij})$ of order p is defined by the sum of its diagonal elements and is denoted by $\text{tr}(\boldsymbol{A}) = \sum_1^p a_{ii}$.

The following facts are basic:

(1) $\text{tr}(\boldsymbol{A}) = \text{tr}(\boldsymbol{A}')$.
(2) $\text{tr}(\boldsymbol{A} + \boldsymbol{B}) = \text{tr}(\boldsymbol{A}) + \text{tr}(\boldsymbol{B})$.
(3) $\text{tr}(\boldsymbol{AB}) = \text{tr}(\boldsymbol{BA})$.
(4) $\text{tr}(c\boldsymbol{A}) = c\ \text{tr}(\boldsymbol{A})$.

Throughout this book, $\exp(x) = e^x$ and

$$\exp(\boldsymbol{A}) = \exp(\text{tr}(\boldsymbol{A})). \tag{1.1.13}$$

1.1.7. Characteristic Roots and Characteristic Vectors

Definition 1.1.7. The *characteristic roots* (or eigenvalues or latent roots) of a square matrix $\boldsymbol{A}$ of order p are given by the roots of the characteristic equation

$$|\boldsymbol{A} - \lambda \boldsymbol{I}| = 0. \tag{1.1.14}$$

The left side of (1.1.14) is a polynomial of degree p in λ so that this equation has exactly p roots. These roots are not necessarily distinct and may be real, or complex, or both. If λ is an eigenvalue of $\boldsymbol{A}$ then $|\boldsymbol{A} - \lambda\boldsymbol{I}| = 0$ so that $\boldsymbol{A} - \lambda\boldsymbol{I}$ is singular. Hence there is a nonzero vector $\boldsymbol{x}$ such that $(\boldsymbol{A} - \lambda \boldsymbol{I})\boldsymbol{x} = \boldsymbol{O}$, called a *characteristic vector* (or eigenvector or latent vector) of $\boldsymbol{A}$ corresponding to λ. If $\boldsymbol{A}$ has a characteristic root λ of multiplicity r there exist r orthogonal characteristic vectors corresponding to λ.

The following results are very useful:

(1) If $\boldsymbol{A}$ is a real symmetric matrix of order p, then all its characteristic roots are real. Let $\lambda_1 \geqslant \lambda_2 \geqslant \dots \geqslant \lambda_p$ be the characteristic roots of $\boldsymbol{A}$. Put $\lambda(\boldsymbol{A}) = \operatorname{diag}(\lambda_1, \dots, \lambda_p)$.

(2) The characteristic vectors corresponding to distinct characteristic roots of a symmetric matrix are orthogonal.

(3) If $\boldsymbol{B} = \boldsymbol{PAP}^{-1}$, where $\boldsymbol{A}$, $\boldsymbol{B}$ and $\boldsymbol{P}$ are square matrices and $\boldsymbol{P}$ is nonsingular, then $\boldsymbol{A}$ and $\boldsymbol{B}$ have the same eigenvalues.

(4) The characteristic roots of $\boldsymbol{A}$ and $\boldsymbol{A}'$ are the same.

(5) The nonzero characteristic roots of $\boldsymbol{AB}$ and $\boldsymbol{BA}$ are the same. In particular, the nonzero characteristic roots of $\boldsymbol{AA}'$ and $\boldsymbol{A}'\boldsymbol{A}$ are the same.

(6) If $\boldsymbol{A} = \operatorname{diag}(a_{11}, \dots, a_{pp})$ then $a_{11}, \dots, a_{pp}$ are the characteristic roots of $\boldsymbol{A}$ and the vector $\boldsymbol{e}_1' = (1, 0, \dots, 0)$, $\boldsymbol{e}_2' = (0, 1, 0, \dots, 0), \dots, \boldsymbol{e}_p' = (0, \dots, 0, 1)$ are the associated latent vectors.

(7) If the characteristic roots of $\boldsymbol{A}$ are $\lambda_1, \dots, \lambda_p$, then the characteristic roots of $\boldsymbol{A}^{-1}$ are $\lambda_1^{-1}, \dots, \lambda_p^{-1}$.

(8) If $\boldsymbol{A} \in O(p)$ then all its characteristic roots have absolute value 1.

(9) If $\boldsymbol{A} \in \mathrm{UT}(p)$ (or $\mathrm{LT}(p)$) then the characteristic roots of $\boldsymbol{A}$ are $a_{11}, \dots, a_{pp}$ (the diagonal elements).

(10) If $\boldsymbol{A}$ has characteristic roots $\lambda_1, \dots, \lambda_p$ then $\boldsymbol{A} - k\boldsymbol{I}$ has characteristic roots $\lambda_1 - k, \dots, \lambda_p - k$.

1.1.8. Positive Definite Matrices

Definition 1.1.8. A $p \times p$ symmetric matrix $\boldsymbol{A}$ is called *positive (negative) definite* if $\boldsymbol{x}'\boldsymbol{Ax} > 0 (< 0)$ for each $\boldsymbol{x} \neq \boldsymbol{0}$, written as $\boldsymbol{A} > 0$ $(\boldsymbol{A} < 0)$; and is called *positive (negative) semidefinite* if $\boldsymbol{x}'\boldsymbol{Ax} \geqslant 0$ $(\leqslant 0)$ for each $\boldsymbol{x} \neq 0$, written as $\boldsymbol{A} \geqslant 0$ $(\boldsymbol{A} \leqslant 0)$.

The determinant of a submatrix (of $\boldsymbol{A}$) of dimension $r \times r$ whose diagonal elements are also the diagonal elements of $\boldsymbol{A}$ is called a *principal minor* of order r, written as $\boldsymbol{A}_{i_1, \dots, i_r}$, where $i_1, \dots, i_r$ are the orders of the diagonal elements contained in this submatrix. When $(i_1, \dots, i_r) = (1, \dots, r)$ we write $\boldsymbol{A}_r$ instead of $\boldsymbol{A}_{1, \dots, r}$.

The following facts are needed in this book:

(1) $\boldsymbol{A} > 0$ if and only if $\boldsymbol{A}_r > 0$, $r = 1, \dots, p$.

(2) $\boldsymbol{A} > 0$ if and only if $\boldsymbol{A}^{-1} > 0$.

(3) If $\boldsymbol{A} > 0$ and $\boldsymbol{A}$ is partitioned as (1.1.5), then $\boldsymbol{A}_{11} > 0$, $\boldsymbol{A}_{22} > 0$, $\boldsymbol{A}_{11.2} > 0$ and $\boldsymbol{A}_{22.1} > 0$ where $\boldsymbol{A}_{11.2}$ and $\boldsymbol{A}_{22.1}$ are defined by (1.1.6).

(4) A symmetric matrix is positive definite (positive semidefinite) if and only if all of its characteristic roots are positive (non-negative).

(5) If $\boldsymbol{A} > 0$ is $p \times p$ and $\boldsymbol{B}$ is $q \times p$ *of rank* r then $\boldsymbol{BAB}' > 0$ if $r = q$ and $\boldsymbol{BAB}' \geqslant 0$ if $r \leqslant q$. In particular, $\boldsymbol{BB}' \geqslant 0$ for any $\boldsymbol{B}$.

(6) If $\boldsymbol{A} > 0$, $\boldsymbol{B} > 0$ and $\boldsymbol{A} - \boldsymbol{B} > 0$ then $\boldsymbol{B}^{-1} - \boldsymbol{A}^{-1} > 0$ and $|\boldsymbol{A}| > |\boldsymbol{B}|$. In this case, we write $\boldsymbol{A} > \boldsymbol{B}$.

1.1.9. Projection Matrices

Definition 1.1.9. A matrix $\boldsymbol{A}$ is called to be *idempotent* if $\boldsymbol{A}^2 = \boldsymbol{A}$ and to be *tripotent* if $\boldsymbol{A}^3 = \boldsymbol{A}$. A symmetric idempotent matrix is called a *projection matrix.*

In this book, we use projection matrices frequently. The following are some useful properties:

(1) If $\boldsymbol{A}$ is a projection matrix, then $\boldsymbol{I} - \boldsymbol{A}$ is a projection matrix, too.
(2) If $\boldsymbol{A}$ is a projection matrix, then $\operatorname{tr}(\boldsymbol{A}) = \operatorname{rk}(\boldsymbol{A})$.
(3) The characteristic roots of a projection matrix are 0 or 1.

1.2. Some Matrix Factorizations

Matrix factorizations are extremely useful in multivariate analysis. The appendices in most of the textbooks on multivariate analysis contain some results of matrix factorization. The excellent reference for matrix factorization is the appendix in Muirhead (1982). Here we only list some useful results without proofs. The reader can find the proofs of them in Murihead's book.

(1) If $\boldsymbol{A}$ is a real $p \times p$ matrix with real characteristic roots then

(1.2.1) $$\boldsymbol{A} = \boldsymbol{HTH}'$$

where $\boldsymbol{H} \in O(p)$, $\boldsymbol{T} \in \mathrm{UT}(p)$ whose diagonal elements are the characteristic roots of $\boldsymbol{A}$.

(2) If $\boldsymbol{A}$ is a symmetric $p \times p$ matrix with eigenvalues $\lambda_1, \ldots, \lambda_p$, then

(1.2.2) $$\boldsymbol{A} = \boldsymbol{H}' \Lambda \boldsymbol{H},$$

where $\boldsymbol{H} \in O(p)$ and $\Lambda = \operatorname{diag}(\lambda_1, \ldots, \lambda_p)$. If $\boldsymbol{H}' = (\boldsymbol{h}_1, \ldots, \boldsymbol{h}_p)$ then $\boldsymbol{h}_i$ is an eigenvector of $\boldsymbol{A}$ corresponding to the eigenvalue λ_i, $i = 1, \ldots, p$. Very often we assume $\lambda_1 \geqslant \lambda_2 \geqslant \cdots \geqslant \lambda_p$. Moreover, if $\lambda_1, \ldots, \lambda_p$ are distinct the representation (1.2.2) is unique up to sign changes in each element of first row of $\boldsymbol{H}$. We can rewrite (1.2.2) in the following form

$$\boldsymbol{A} = \sum_{i=1}^{p} \lambda_i \boldsymbol{h}_i \boldsymbol{h}_i',$$

which is called the spectral decomposition of $\boldsymbol{A}$. When $\lambda_1 = \ldots = \lambda_r = 1$ and $\lambda_{r+1} = \ldots = \lambda_p = 0$, the above decomposition reduces to

$$\boldsymbol{A} = \sum_{i=1}^{r} \boldsymbol{h}_i \boldsymbol{h}_i' = (\boldsymbol{h}_1, \ldots, \boldsymbol{h}_r) \begin{pmatrix} \boldsymbol{h}_1' \\ \vdots \\ \boldsymbol{h}_r' \end{pmatrix} \hat{=} \boldsymbol{H}_1' \boldsymbol{H}_1,$$

say.

(3) If $\boldsymbol{A} > 0 (\geqslant 0)$ then there exists $\boldsymbol{A}^{\frac{1}{2}} > 0 (\geqslant 0)$ such that $\boldsymbol{A} = \boldsymbol{A}^{\frac{1}{2}} \boldsymbol{A}^{\frac{1}{2}}$. Throughout this book denote $(\boldsymbol{A}^{\frac{1}{2}})^{-1}$ by $\boldsymbol{A}^{-\frac{1}{2}}$. Here $\boldsymbol{A}^{\frac{1}{2}}$ can be taken as follows

(1.2.3) $$\boldsymbol{A}^{\frac{1}{2}} = \boldsymbol{H}' \Lambda^{\frac{1}{2}} \boldsymbol{H},$$

where $\boldsymbol{H}$ and Λ are defined by (1.2.2) and $\Lambda^{\frac{1}{2}} = \operatorname{diag}(\lambda_1^{\frac{1}{2}}, \ldots, \lambda_p^{\frac{1}{2}})$. More generally, we can write

(1.2.4) $$f(\boldsymbol{A}) = \boldsymbol{H}' f(\Lambda) \boldsymbol{H},$$

where $f(\Lambda) = \operatorname{diag}(f(\lambda_1), \ldots, f(\lambda_p))$ and $f(\cdot)$ is a Borel function.

(4) If $\boldsymbol{A} \geqslant 0$ is a $p \times p$ matrix of rank $r(\leqslant p)$ then:

(i) There exists a $p \times r$ matrix $\boldsymbol{B}$ of rank r such that

$$\boldsymbol{A} = \boldsymbol{B}\boldsymbol{B}'. \tag{1.2.5}$$

(ii) There exists a $p \times p$ nonsingular matrix $\boldsymbol{C}$ such that

$$\boldsymbol{A} = \boldsymbol{C}\begin{bmatrix} \boldsymbol{I}_r & \boldsymbol{O} \\ \boldsymbol{O} & \boldsymbol{O} \end{bmatrix}\boldsymbol{C}'. \tag{1.2.6}$$

(iii) There exists $\boldsymbol{T} \in \mathrm{UT}(p)$ such that

$$\boldsymbol{A} = \boldsymbol{T}'\boldsymbol{T}. \tag{1.2.7}$$

When the diagonal elements of $\boldsymbol{T}$ are nonnegative, the decomposition (1.2.7) is called the *Cholesky decomposition*. The Cholesky decomposition is unique if $\boldsymbol{A} > 0$.

(5) If $\boldsymbol{A}$ is an $n \times p$ $(n \geqslant p)$ matrix, then $\boldsymbol{A}$ can be decomposed in the following form:

$$\boldsymbol{A} = \boldsymbol{U}\boldsymbol{B} \tag{1.2.8}$$

where $\boldsymbol{U}$ is an $n \times p$ matrix with $\boldsymbol{U}'\boldsymbol{U} = \boldsymbol{I}_p$ and $\boldsymbol{B} \geqslant 0$. If $\mathrm{rk}(\boldsymbol{A}) = p$ then $\boldsymbol{B} > 0$. Or

$$\boldsymbol{A} = \boldsymbol{U}\begin{bmatrix} \boldsymbol{I}_p \\ \boldsymbol{O} \end{bmatrix}\boldsymbol{B}, \tag{1.2.9}$$

where $\boldsymbol{U} \in \mathcal{O}(n)$ and $\boldsymbol{B}$ is defined as (1.2.8). Or

$$\boldsymbol{A} = \boldsymbol{U}\boldsymbol{T}, \tag{1.2.10}$$

where $\boldsymbol{U}$ is an $n \times p$ matrix with $\boldsymbol{U}'\boldsymbol{U} = \boldsymbol{I}_p$ and $\boldsymbol{T} \in \mathrm{UT}(p)$ with nonnegative diagonal elements. When $\mathrm{rk}(\boldsymbol{A}) = p$ all the diagonal elements are positive.

If $\mathrm{rk}(\boldsymbol{A}) = r$, there exist $n \times r$ and $r \times p$ matrices $\boldsymbol{F}$ and $\boldsymbol{G}$ such that

$$\boldsymbol{A} = \boldsymbol{F}\boldsymbol{G}. \tag{1.2.11}$$

(6) Suppose that $\boldsymbol{A}$ and $\boldsymbol{B}$ are $k \times m$ and $k \times n$ matrices with $m \leqslant n$ respectively. Then $\boldsymbol{A}\boldsymbol{A}' = \boldsymbol{B}\boldsymbol{B}'$ if and only if there exists an $m \times n$ matrix $\boldsymbol{H}$ with $\boldsymbol{H}\boldsymbol{H}' = \boldsymbol{I}_m$ such that $\boldsymbol{A}\boldsymbol{H} = \boldsymbol{B}$.

(7) If $\boldsymbol{A}$ is an $n \times p$ matrix $(n \geqslant p)$ then

$$\boldsymbol{A} = \boldsymbol{U}\Lambda\boldsymbol{V}, \tag{1.2.12}$$

where $\boldsymbol{U}$ is an $n \times p$ matrix with $\boldsymbol{U}'\boldsymbol{U} = \boldsymbol{I}_p$, $\boldsymbol{V} \in \mathcal{O}(p)$, $\Lambda = \mathrm{diag}(\lambda_1, \ldots, \lambda_p)$ and $\lambda_1^2, \ldots, \lambda_p^2$ are the characteristic roots of $\boldsymbol{A}'\boldsymbol{A}$. Or

$$\boldsymbol{A} = \boldsymbol{H}(\Lambda \ \ \boldsymbol{O})'\boldsymbol{V}, \tag{1.2.13}$$

where $\boldsymbol{H} \in \mathcal{O}(n)$, $\boldsymbol{V}$ and Λ are the same as in (1.2.12) and $\boldsymbol{O}'$ is an $n \times (n-p)$ matrix. The decomposition (1.2.12) or (1.2.13) is called the singular value decomposition.

(8) If $\boldsymbol{A}_1, \ldots, \boldsymbol{A}_k$ are symmetric matrices such that $\boldsymbol{A}_i\boldsymbol{A}_j = \boldsymbol{O}$ for $i \neq j$, $i, j = 1, \ldots, k$, then there exists an orthogonal matrix $\boldsymbol{H}$ such that $\boldsymbol{H}'\boldsymbol{A}_i\boldsymbol{H} = \Lambda_i$ with Λ_i being diagonal, $i = 1, \ldots, k$.

(9) Let $\boldsymbol{A}$ be an $n \times n$ nonsingular matrix. If the principle minor of order i is nonzero for $i = 1, \ldots, n$, then

$$\boldsymbol{A} = \boldsymbol{T}\boldsymbol{U}, \tag{1.2.14}$$

where $T \in \mathrm{LT}(n)$ and $U = (u_{ij}) \in \mathrm{UT}(n)$ with $u_{ii} = 1$ for $i = 1, \dots, n$.

(10) If A and B are $n \times n$ matrices, $A > 0$ and $B' = B$, there exists an $n \times n$ nonsingular matrix H such that

$$A = HH' \text{ and } B = H\Lambda H', \tag{1.2.15}$$

where $\Lambda = \mathrm{diag}(\lambda_1, \dots, \lambda_n)$ and $\lambda_1, \dots, \lambda_n$ are the eigenvalues of $A^{-1}B$. If $B > 0$ and $\lambda_1, \dots, \lambda_n$ are all distinct, H is unique up to sign changes in each element of first row of H.

1.3. Generalized Inverse of Matrix

Definition 1.3.1. Given an $n \times p$ matrix A if there exists a $p \times n$ matrix X such that

$$AXA = A \tag{1.3.1.}$$

then X is called a *generalized inverse matrix* of A, written as $X = A^{-}$.

First of all, we point out that there exists a generalized inverse matrix for any A. If $\mathrm{rk}(A) = r$, by (1.2.11) A can be represented as follows

$$A = P\begin{bmatrix} I_r & O \\ O & O \end{bmatrix}Q, \tag{1.3.2.}$$

where P and Q are $n \times n$ and $p \times p$ nonsingular matrices. We have

$$AXA = A \Longleftrightarrow P\begin{bmatrix} I_r & O \\ O & O \end{bmatrix}QXP\begin{bmatrix} I_r & O \\ O & O \end{bmatrix}Q = P\begin{bmatrix} I_r & O \\ O & O \end{bmatrix}Q$$

$$\Longleftrightarrow \begin{bmatrix} I_r & O \\ O & O \end{bmatrix}QXP\begin{bmatrix} I_r & O \\ O & O \end{bmatrix} = \begin{bmatrix} I_r & O \\ O & O \end{bmatrix}.$$

Denote

$$QXP = \begin{bmatrix} T_{11} & T_{12} \\ T_{21} & T_{22} \end{bmatrix}$$

with T_{11}: $r \times r$. Then $AXA = A$ if and only if $T_{11} = I_r$, and

$$A^{-} = X = Q^{-1}\begin{bmatrix} I_r & T_{12} \\ T_{21} & T_{22} \end{bmatrix}P^{-1}, \tag{1.3.3}$$

where T_{12}, T_{21} and T_{22} can be arbitrary. By (1.3.3) we obtain the following facts immediately.

(1) $\mathrm{rk}(A^{-}) \geqslant \mathrm{rk}(A)$.

(2) A^{-} is unique if A is a nonsingular square matrix. In this case $A^{-} = A^{-1}$.

(3) $\mathrm{rk}(A) = \mathrm{rk}(AA^{-}) = \mathrm{rk}(A^{-}A) = \mathrm{tr}(AA^{-}) = \mathrm{tr}(A^{-}A)$, because

$$AA^{-} = P\begin{bmatrix} I_r & T_{12} \\ O & O \end{bmatrix}P^{-1} \text{ and } A^{-}A = Q^{-1}\begin{bmatrix} I_r & O \\ T_{21} & O \end{bmatrix}Q. \tag{1.3.4}$$

Hence both AA^{-} and $A^{-}A$ are idempotent.

(4) If $\mathrm{rk}(A) = p$, then $A^{-}A = I_p$; if $\mathrm{rk}(A) = n$, then $AA^{-} = I_n$.

(5) For every A,

$$A'A(A'A)^{-}A' = A' \text{ and } A(A'A)^{-}A'A = A. \tag{1.3.5}$$

As $Ax = 0 \Rightarrow A'Ax = 0 \Rightarrow x'A'Ax = 0 \Rightarrow Ax = 0$ we have

$$AA'x = AA'y \Longleftrightarrow A'x = A'y \tag{1.3.6}$$

and

$$A'Ax = A'Ay \Longleftrightarrow Ax = Ay. \tag{1.3.7}$$

By these two relationships then (1.3.5) follows.

(6) $A(A'A)^- A'$ is a projection matrix and is independent of what is taken as $(A'A)^-$.

Let $(A'A)_1^-$ and $(A'A)_2^-$ be two of the generalized inverse matrices of $(A'A)$. By definition

$$A'A(A'A)_1^- A'A = A'A = A'A(A'A)_2^- A'A.$$

Thus $A(A'A)_1^- A' = A(A'A)_2^- A'$ by (1.3.6). This means that $A(A'A)^- A'$ is independent of what is taken as $(A'A)^-$; In particular, we can take a symmetric one as $(A'A)^-$. By using (1.3.5) we have

$$(A(A'A)^- A')^2 = A(A'A)^- A'A(A'A)^- A' = A(A'A)^- A'.$$

Hence $A(A'A)^- A'$ is a projection matrix.

As A^- is not unique, it is not convenient for many cases. One wishes to define a kind of generalized inverse matrix which is unique.

Definition 1.3.2. Let A be an $n \times p$ matrix. If there exists a $p \times n$ matrix X such that

$$\begin{cases} AXA = A & XAX = X \\ (AX)' = AX & (XA)' = XA, \end{cases} \tag{1.3.8}$$

then X is called a *Moore-Penrose inverse matrix* of A, written as $X = A^+$.

Now we show the existence and uniqueness of A^+. Let $r = \text{rk}(A)$. If $r = 0$, then $A = O$ and take $A^+ = O$. If $r > 0$, there exist $P(n \times r)$ and $Q(p \times r)$ such that $r = \text{rk}(P) = \text{rk}(Q)$ and $A = PQ'$ (cf. (1.2.11)). Therefore $(P'P)^{-1}$ and $(Q'Q)^{-1}$ exist. Let

$$X = Q(Q'Q)^{-1}(P'P)^{-1}P'. \tag{1.3.9}$$

Now we check that X satisfies the conditions of (1.3.8).

$$AXA = PQ'Q(Q'Q)^{-1}(P'P)^{-1}P'PQ' = PQ' = A,$$

$$XAX = Q(Q'Q)^{-1}(P'P)^{-1}P'PQ'Q(Q'Q)^{-1}(P'P)^{-1}P'$$

$$= Q(Q'Q)^{-1}(P'P)^{-1}P' = X$$

$$AX = P(P'P)^{-1}P' \qquad \text{is symmetric,}$$

and

$$XA = Q(Q'Q)^{-1}Q' \qquad \text{is symmetric.}$$

Thus X is a Moore-Penrose inverse matrix of A. Let A_1^+ and A_2^+ be two of the Moore-Penrose inverse matrices of A. Then

$$A_1^+ = A_1^+AA_1^+ = A_1^+(A_1^+)'A' = A_1^+(A_1^+)'A'(A_2^+)'A'$$

$$= A_1^+(AA_1^+)'(AA_2^+)' = A_1^+AA_1^+AA_2^+ = A_1^+AA_2^+,$$

and

$$A_2^+ = A_2^+ A A_2^+ = A'(A_2^+)' A_2^+ = A'(A_1^+)' A'(A_2^+)' A_2^+$$
$$= (A_1^+ A)'(A_2^+ A)' A_2^+ = A_1^+ A A_2^+ A A_2^+ = A_1^+ A A_2^+,$$

and $A_1^+ = A_2^+$.

The following are some basic properties of Moore-Penrose inverse matrices. These can be verified directly by the definition and (1.3.9).

(1) If $\mathrm{rk}(A) = n$, $A^+ = A'(A'A)^{-1}$; if $\mathrm{rk}(A) = p$, $A^+ = (A'A)^{-1}A'$; if $\mathrm{rk}(A) = n = p$, $A^+ = A^{-1}$.

(2) $(A^+)^+ = A$.

(3) $A^+ = (A'A)^+ A' = A'(AA')^+$.

(4) $(A'A)^+ = A^+(A^+)'$.

(5) Let $A = PQ'$ with $\mathrm{rk}(A) = \mathrm{rk}(P) = \mathrm{rk}(Q) = r$. Then $A^+ = (Q^+)' P^+$.

(6) If A is a projection matrix, then $A^+ = A$.

(7) If $A' = A$, then $A = H'\Lambda H$ (cf. (1.2.2)), where $H \in O(n)$ and $\Lambda = \mathrm{diag}(\lambda_1, \ldots, \lambda_n)$. Let

$$\lambda^+ = \begin{cases} \lambda^{-1} & \text{for } \lambda \neq 0, \\ 0 & \text{for } \lambda = 0. \end{cases}$$

Then $A^+ = H' \mathrm{diag}\,(\lambda_1^+, \ldots, \lambda_n^+) H$.

(8) AA^+ and A^+A are projection matrices.

1.4. "vec" Operator and Kronecker Products

Throughout this book, let $e_i(n) = (0, \ldots, 0, 1, 0, \ldots, 0)'$ be an $n \times 1$ vector with 1 at the i-th position and $E_{ij}(m, n) = e_i(m)e_j'(n)$. Sometimes write e_i or E_{ij} for simplicity. Clearly, we can obtain the following elementary results.

(1) $e_i'e_j = \delta_{ij}$, where $\delta_{ii} = 1$ and $\delta_{ij} = 0$ for $i \neq j$.

(2) $E_{ij}e_r = \delta_{jr}e_i$ and $e_r'E_{ij} = \delta_{ri}e_j'$.

(3) $E_{ij}E_{rs} = \delta_{jr}E_{is}$.

(4) $I = \sum_i E_{ii} = \sum_i e_ie_i'$.

(5) $E_{ij}'(m, n) = [E_{ij}(m, n)]' = E_{ji}(n, m)$.

1.4.1. "vec" Operator

Definition 1.4.1. Let $A = (a_1, \ldots, a_p)$ be $n \times p$ matrix. Define an np-*dimensional vector*

$$\mathrm{vec}(A) = \begin{bmatrix} a_1 \\ \vdots \\ a_p \end{bmatrix}, \tag{1.4.1}$$

where "vec" can be realized as an operator.

Clearly,

$$\mathrm{vec}(A') = \begin{bmatrix} a_{(1)} \\ \vdots \\ a_{(n)} \end{bmatrix}, \qquad \text{where } A = \begin{bmatrix} a_{(1)}' \\ \vdots \\ a_{(n)}' \end{bmatrix}. \tag{1.4.2}$$

The following results are basic:

(1) $\mathrm{vec}(c\boldsymbol{A}+d\boldsymbol{B}) = c\,\mathrm{vec}(\boldsymbol{A}) + d\,\mathrm{vec}(\boldsymbol{B})$ where c and d are real numbers.

(2) $\boldsymbol{A} = \sum_{i,j} a_{ij}\boldsymbol{E}_{ij} = \sum_{i,j} a_{ij}\boldsymbol{e}_i\boldsymbol{e}_j' = \sum_j \boldsymbol{a}_j\boldsymbol{e}_j' = \sum_i \boldsymbol{e}_i\boldsymbol{a}_{(i)}'$.

(3) $\boldsymbol{a}_j = \boldsymbol{A}\boldsymbol{e}_j = \sum_i a_{ij}\boldsymbol{e}_i$ and

$$\boldsymbol{a}_{(i)} = \boldsymbol{A}'\boldsymbol{e}_i = \sum_j a_{ij}\boldsymbol{e}_j.$$

(4) $a_{ij} = \boldsymbol{e}_i'\boldsymbol{A}\boldsymbol{e}_j$.

(5) $\mathrm{tr}(\boldsymbol{E}_{rs}'\boldsymbol{A}) = a_{rs}$.

(6) $\mathrm{tr}(\boldsymbol{A}\boldsymbol{B}) = \sum_{i,j} a_{ij}\,b_{ji} = (\mathrm{vec}(\boldsymbol{A}'))'(\mathrm{vec}(\boldsymbol{B}))$.

We only prove (6). By (2), (3) and (4) we have

$$\mathrm{tr}(\boldsymbol{A}\boldsymbol{B}) = \sum_i \boldsymbol{e}_i'\boldsymbol{A}\boldsymbol{B}\boldsymbol{e}_i = \sum_i \boldsymbol{a}_{(i)}'\,\boldsymbol{b}_i = (\mathrm{vec}(\boldsymbol{A}'))'(\mathrm{vec}(\boldsymbol{B})).$$

1.4.2. Kronecker Products

Definition 1.4.2. If $\boldsymbol{A} = (a_{ij})$ and $\boldsymbol{B}$ be $n \times p$ and $m \times q$ matrices, the *Kronecker product* of $\boldsymbol{A}$ and $\boldsymbol{B}$ is an $nm \times pq$ matrix defined as follows:

(1.4.3)
$$\boldsymbol{A} \otimes \boldsymbol{B} = (a_{ij}\boldsymbol{B}) = \begin{bmatrix} a_{11}\boldsymbol{B} & \cdots & a_{1p}\boldsymbol{B} \\ \vdots & & \vdots \\ a_{n1}\boldsymbol{B} & \cdots & a_{np}\boldsymbol{B} \end{bmatrix}.$$

The Kronecker product is a useful tool in multivariate analysis. By definition (1.4.3) we obtain immediately the following results:

(1) $(\alpha\boldsymbol{A}) \otimes \boldsymbol{B} = \boldsymbol{A} \otimes (\alpha\boldsymbol{B}) = \alpha(\boldsymbol{A} \otimes \boldsymbol{B})$ where α is real.

(2) $\boldsymbol{A} \otimes (\boldsymbol{B}+\boldsymbol{C}) = \boldsymbol{A} \otimes \boldsymbol{B} + \boldsymbol{A} \otimes \boldsymbol{C}$,
$(\boldsymbol{B}+\boldsymbol{C}) \otimes \boldsymbol{A} = \boldsymbol{B} \otimes \boldsymbol{A} + \boldsymbol{C} \otimes \boldsymbol{A}$.

(3) $(\boldsymbol{A} \otimes \boldsymbol{B}) \otimes \boldsymbol{C} = \boldsymbol{A} \otimes (\boldsymbol{B} \otimes \boldsymbol{C})$.

(4) $\boldsymbol{I}_{mn} = \boldsymbol{I}_m \otimes \boldsymbol{I}_n = \boldsymbol{I}_n \otimes \boldsymbol{I}_m$.

(5) $(\boldsymbol{A} \otimes \boldsymbol{B})' = \boldsymbol{A}' \otimes \boldsymbol{B}'$.

(6) $(\boldsymbol{A} \otimes \boldsymbol{B})(\boldsymbol{C} \otimes \boldsymbol{D}) = (\boldsymbol{A}\boldsymbol{C}) \otimes (\boldsymbol{B}\boldsymbol{D})$.

(7) If $\boldsymbol{A}$ and $\boldsymbol{B}$ are nonsingular square matrices, then

$$(\boldsymbol{A} \otimes \boldsymbol{B})^{-1} = \boldsymbol{A}^{-1} \otimes \boldsymbol{B}^{-1}.$$

This is because $(\boldsymbol{A} \otimes \boldsymbol{B})(\boldsymbol{A}^{-1} \otimes \boldsymbol{B}^{-1}) = (\boldsymbol{A}\boldsymbol{A}^{-1}) \otimes (\boldsymbol{B}\boldsymbol{B}^{-1}) = \boldsymbol{I} \otimes \boldsymbol{I} = \boldsymbol{I}$ by (6) and (4).

(8) Let $\boldsymbol{A}$, $\boldsymbol{X}$ and $\boldsymbol{B}$ be $n \times m$, $m \times p$ and $p \times q$ matrices, respectively. Then

(1.4.4)
$$\mathrm{vec}(\boldsymbol{A}\boldsymbol{X}\boldsymbol{B}) = (\boldsymbol{B}' \otimes \boldsymbol{A})\,\mathrm{vec}(\boldsymbol{X}).$$

Denote the ith column and the jth row of $\boldsymbol{A}$ by $(\boldsymbol{A})_i$ and $(\boldsymbol{A})_{(j)}$ respectively. We have

$$(\boldsymbol{A}\boldsymbol{X}\boldsymbol{B})_k = \boldsymbol{A}\boldsymbol{X}\boldsymbol{B}\boldsymbol{e}_k = \boldsymbol{A}(\sum_j (\boldsymbol{X})_j\,\boldsymbol{e}_j')\boldsymbol{B}\boldsymbol{e}_k$$

$$= \boldsymbol{I}_n \otimes \boldsymbol{I}_m = \boldsymbol{I}_{mn}.$$ □

$$= \sum_j A(X)_j(e'_j B e_k) = \sum_j b_{jk} A(X)_j$$

$$= (b_{1k}A, \ldots, b_{pk}A) \begin{bmatrix} (X)_1 \\ \vdots \\ (X)_p \end{bmatrix} = ((B)'_k \otimes A)\,\mathrm{vec}(X),$$

thus

$$\mathrm{vec}(AXB) = \begin{bmatrix} (AXB)_1 \\ \vdots \\ (AXB)_q \end{bmatrix} = \begin{bmatrix} (B)'_1 \otimes A \\ \vdots \\ (B)'_q \otimes A \end{bmatrix} \mathrm{vec}(X) = (B' \otimes A)\,\mathrm{vec}(X).$$

(9) $\mathrm{tr}(A \otimes B) = (\mathrm{tr}(A))(\mathrm{tr}(B))$.

(10) Let x and y be column vectors. Then $xy' = x \otimes y' = y' \otimes x$.

(11) Let A and B be $n \times n$ and $m \times m$ matrices with eigenvalues $\{\lambda_i, i=1, \ldots, n\}$ and $\{\mu_j, j=1, \ldots, m\}$ respectively. Let x_i and y_j be eigenvectors of A and B corresponding to λ_i and μ_j respectively. Then $\{\lambda_i\mu_j, i=1, \ldots, n; j=1, \ldots, m\}$ are the eigenvalues of $A \otimes B$ and $\{x_i \otimes y_j, i=1, \ldots, n; j=1, \ldots, m\}$ are the corresponding eigenvectors.

Now we wish to prove this assertion. By (1.2.1) there exist $H \in O(n)$ and $P \in O(m)$ such that

$$A = HTH' \qquad \text{and} \qquad B = PVP',$$

where $T \in \mathrm{UT}(n)$ with $\lambda_1, \ldots, \lambda_n$ as its diagonal elements and $V \in \mathrm{UT}(m)$ with $\mu_1, \ldots, \mu_m$ as its diagonal elements. Thus

$$A \otimes B = (HTH') \otimes (PVP') = (H \otimes P)(T \otimes V)(H \otimes P)'.$$

Clearly $H \otimes P \in O(nm)$, $T \otimes V \in \mathrm{UT}(nm)$ with $\{\lambda_i \mu_j, i=1, \ldots, n; j=1, \ldots, m\}$ as its diagonal elements, i.e., $\{\lambda_i\mu_j, i=1, \ldots, n; j=1, \ldots, m\}$ are the eigenvalues of $A \otimes B$. As

$$(A \otimes B)(x_i \otimes y_j) = (Ax_i) \otimes (By_j) = (\lambda_i x_i) \otimes (\mu_j y_j) = \lambda_i \mu_j (x_i \otimes y_j),$$

$x_i \otimes y_j$ is the eigenvector of $A \otimes B$ corresponding to $\lambda_i\mu_j$, $i=1, \ldots, n; j=1, \ldots, m$.

By (11), we find the following property directly.

(12) If A and B are $n \times n$ and $m \times m$ matrices respectively, then

$$|A \otimes B| = |A|^m |B|^n. \tag{1.4.5}$$

1.4.3. Permutation Matrix

In many cases, one wants to establish the relationship between $\mathrm{vec}(X)$ and $\mathrm{vec}(X')$. To this end, we need the permutation matrix which is defined as follows.

Definition 1.4.3. The matrix

$$K_{mn} = \sum_{i=1}^{m} \sum_{j=1}^{n} E_{ij}(m, n) \otimes E'_{ij}(m, n) \tag{1.4.6}$$

is called the *permutation matrix* of order $m \times n$.

The following are some properties of the permutation matrix.

(1) Let $A = (a_{ij})$ be an $m \times n$ matrix. Then

$$\mathrm{vec}(A') = K_{mn}\,\mathrm{vec}(A). \tag{1.4.7}$$

Proof.
$$\begin{aligned}\operatorname{vec}(\boldsymbol{A}') &= \operatorname{vec}\left(\sum_{i=1}^{m}\sum_{j=1}^{n} a_{ij}\boldsymbol{E}'_{ij}(m,n)\right)\\ &= \operatorname{vec}\left(\sum_{i=1}^{m}\sum_{j=1}^{n} \boldsymbol{e}_j(n)a_{ij}\boldsymbol{e}'_i(m)\right)\\ &= \operatorname{vec}\left(\sum_{i=1}^{m}\sum_{j=1}^{n} \boldsymbol{e}_j(n)\boldsymbol{e}'_i(m)\boldsymbol{A}\boldsymbol{e}_j(n)\boldsymbol{e}'_i(m)\right)\\ &= \operatorname{vec}\left(\sum_{i=1}^{m}\sum_{j=1}^{n} \boldsymbol{E}'_{ij}(m,n)\boldsymbol{A}\boldsymbol{E}'_{ij}(m,n)\right)\\ &= \sum_{i=1}^{m}\sum_{j=1}^{n} \operatorname{vec}\left(\boldsymbol{E}'_{ij}(m,n)\boldsymbol{A}\boldsymbol{E}'_{ij}(m,n)\right) \qquad \text{(By (1.4.4))}\\ &= \sum_{i=1}^{m}\sum_{j=1}^{n} \boldsymbol{E}_{ij}(m,n)\otimes\boldsymbol{E}'_{ij}(m,n)\operatorname{vec}(\boldsymbol{A})\\ &= \boldsymbol{K}_{mn}\operatorname{vec}(\boldsymbol{A}). \qquad \square\end{aligned}$$

(2) Let $\boldsymbol{A}$ and $\boldsymbol{B}$ be $n\times s$ and $m\times t$ matrices respectively. Then

$$\boldsymbol{K}_{mn}(\boldsymbol{A}\otimes\boldsymbol{B})\boldsymbol{K}_{st} = \boldsymbol{B}\otimes\boldsymbol{A}.$$

Proof. For any $s\times t$ matrix $\boldsymbol{X}$ we have

$$\begin{aligned}\boldsymbol{K}_{mn}(\boldsymbol{A}\otimes\boldsymbol{B})\boldsymbol{K}_{st}\operatorname{vec}(\boldsymbol{X}) &= \boldsymbol{K}_{mn}(\boldsymbol{A}\otimes\boldsymbol{B})\operatorname{vec}(\boldsymbol{X}') = \boldsymbol{K}_{mn}\operatorname{vec}(\boldsymbol{B}\boldsymbol{X}'\boldsymbol{A}')\\ &= \operatorname{vec}(\boldsymbol{B}\boldsymbol{X}'\boldsymbol{A}')' = \operatorname{vec}(\boldsymbol{A}\boldsymbol{X}\boldsymbol{B}') = (\boldsymbol{B}\otimes\boldsymbol{A})\operatorname{vec}(\boldsymbol{X})\end{aligned}$$

which completes the proof. $\square$

(3) $\boldsymbol{K}'_{mn} = \boldsymbol{K}_{nm}$, $\boldsymbol{K}_{1n} = \boldsymbol{K}_{n1} = \boldsymbol{I}_n$.

(4) $\boldsymbol{K}_{mn} = [\operatorname{vec}(\boldsymbol{E}'_{11}), \operatorname{vec}(\boldsymbol{E}'_{21}), \ldots, \operatorname{vec}(\boldsymbol{E}'_{m1}), \ldots, \operatorname{vec}(\boldsymbol{E}'_{1n}), \operatorname{vec}(\boldsymbol{E}'_{2n}), \ldots, \operatorname{vec}(\boldsymbol{E}'_{mn})]$.

Proof. (3) is obvious by definition 1.4.3. We only prove (4).

$$\begin{aligned}\boldsymbol{K}_{mn}\operatorname{vec}(\boldsymbol{A}) &= \operatorname{vec}(\boldsymbol{A}') = \operatorname{vec}\left(\sum_{i=1}^{m}\sum_{j=1}^{n} a_{ij}\boldsymbol{E}_{ij}\right)' = \sum_{i=1}^{m}\sum_{j=1}^{n} a_{ij}\operatorname{vec}(\boldsymbol{E}'_{ij})\\ &= [\operatorname{vec}(\boldsymbol{E}'_{11}), \operatorname{vec}(\boldsymbol{E}'_{21}), \ldots, \operatorname{vec}(\boldsymbol{E}'_{m1}), \ldots, \operatorname{vec}(\boldsymbol{E}'_{1n}), \operatorname{vec}(\boldsymbol{E}'_{2n}), \ldots, \operatorname{vec}(\boldsymbol{E}'_{mn})]\end{aligned}$$

$\times\operatorname{vec}(\boldsymbol{A})$. $\square$

(5) $\boldsymbol{K}_{mn} = \sum_{j=1}^{n}(\boldsymbol{e}'_j(n)\otimes\boldsymbol{I}_m\otimes\boldsymbol{e}_j(n)) = \sum_{i=1}^{m}(\boldsymbol{e}_i(m)\otimes\boldsymbol{I}_n\otimes\boldsymbol{e}'_i(m))$.

(6) $\boldsymbol{K}'_{mn}\boldsymbol{K}_{mn} = \boldsymbol{I}_{mn}$, i.e., $\boldsymbol{K}_{mn}$ is an orthogonal matrix.

We only prove (6). By the definition of $\boldsymbol{K}_{mn}$ we have

$$\begin{aligned}\boldsymbol{K}'_{mn}\boldsymbol{K}_{mn} &= \left(\sum_i\sum_j \boldsymbol{E}'_{ij}\otimes\boldsymbol{E}_{ij}\right)\left(\sum_s\sum_t \boldsymbol{E}_{st}\otimes\boldsymbol{E}'_{st}\right)\\ &= \sum_i\sum_j\sum_s\sum_t(\boldsymbol{E}'_{ij}\boldsymbol{E}_{st})\otimes(\boldsymbol{E}_{ij}\boldsymbol{E}'_{st})\\ &= \sum_i\sum_j\sum_s\sum_t(\delta_{is}\boldsymbol{E}_{jt})\otimes(\delta_{jt}\boldsymbol{E}_{is})\\ &= \sum_i\sum_j \boldsymbol{E}_{jj}\otimes\boldsymbol{E}_{ii} = \left(\sum_j \boldsymbol{E}_{jj}\right)\otimes\left(\sum_i \boldsymbol{E}_{ii}\right)\end{aligned}$$

Some further properties can be found in Exercise 1.12, 1.13 and 1.14.

1.5. Derivatives of Matrices and the Matrix Differential

1.5.1. The Derivatives of Matrices with Respect to a Scalar

Definition 1.5.1. Let $\boldsymbol{Y}=(y_{ij}(t))$ be a $p\times q$ matrix, where $y_{ij}(t)$ is a function of t. Write

$$(1.5.1)\qquad \frac{\partial \boldsymbol{Y}}{\partial t}=\begin{bmatrix} \frac{\partial y_{11}}{\partial t} & \cdots & \frac{\partial y_{1q}}{\partial t} \\ \vdots & & \vdots \\ \frac{\partial y_{p1}}{\partial t} & \cdots & \frac{\partial y_{pq}}{\partial t} \end{bmatrix}=\left(\frac{\partial y_{ij}(t)}{\partial t}\right).$$

By this definition the following results are easily found:

$$(1)\qquad \frac{\partial(\boldsymbol{X}+\boldsymbol{Y})}{\partial t} = \frac{\partial \boldsymbol{X}}{\partial t}+\frac{\partial \boldsymbol{Y}}{\partial t}.$$

$$(2)\qquad \frac{\partial(\boldsymbol{X}\boldsymbol{Y})}{\partial t} = \frac{\partial \boldsymbol{X}}{\partial t}\boldsymbol{Y}+\boldsymbol{X}\frac{\partial \boldsymbol{Y}}{\partial t}.$$

$$(3)\qquad \frac{\partial(\boldsymbol{X}\otimes\boldsymbol{Y})}{\partial t} = \frac{\partial \boldsymbol{X}}{\partial t}\otimes\boldsymbol{Y}+\boldsymbol{X}\otimes\frac{\partial \boldsymbol{Y}}{\partial t}.$$

$$(4)\qquad \left(\frac{\partial \boldsymbol{X}}{\partial t}\right)' = \frac{\partial \boldsymbol{X}'}{\partial t}.$$

$$(5)\qquad \frac{\partial \boldsymbol{X}}{\partial x_{ij}} = \boldsymbol{E}_{ij}\qquad \text{where } \boldsymbol{X}=(x_{ij}).$$

$$(6)\qquad \frac{\partial(\boldsymbol{A}\boldsymbol{X}\boldsymbol{B})}{\partial x_{ij}} = \boldsymbol{A}\boldsymbol{E}_{ij}\boldsymbol{B},\qquad \text{where } \boldsymbol{A} \text{ and } \boldsymbol{B} \text{ are constant matrices.}$$

$$(7)\qquad \frac{\partial(\boldsymbol{X}'\boldsymbol{A}\boldsymbol{X})}{\partial x_{ij}} = \boldsymbol{E}_{ij}'\boldsymbol{A}\boldsymbol{X}+\boldsymbol{X}'\boldsymbol{A}\boldsymbol{E}_{ij},\qquad \text{where } \boldsymbol{A} \text{ is a}$$

constant matrix.

(8) If $\boldsymbol{X}$ is a nonsingular matrix, and $\boldsymbol{A}$ and $\boldsymbol{B}$ are constant matrices, then

$$\frac{\partial(\boldsymbol{A}\boldsymbol{X}^{-1}\boldsymbol{B})}{\partial x_{ij}} = -\boldsymbol{A}\boldsymbol{X}^{-1}\boldsymbol{E}_{ij}\boldsymbol{X}^{-1}\boldsymbol{B}.$$

Proof. In fact we only need to prove

$$\frac{\partial \boldsymbol{X}^{-1}}{\partial x_{ij}} = -\boldsymbol{X}^{-1}\boldsymbol{E}_{ij}\boldsymbol{X}^{-1}.$$

Using (3) to differentiate $\boldsymbol{X}^{-1}\boldsymbol{X}=\boldsymbol{I}$ we obtain

$$\frac{\partial \boldsymbol{X}}{\partial x_{ij}}\boldsymbol{X}^{-1}+\boldsymbol{X}\frac{\partial \boldsymbol{X}^{-1}}{\partial x_{ij}}=\boldsymbol{O}.$$

Hence

$$\frac{\partial \boldsymbol{X}^{-1}}{\partial x_{ij}} = -\boldsymbol{X}^{-1}\frac{\partial \boldsymbol{X}}{\partial x_{ij}}\boldsymbol{X}^{-1} = -\boldsymbol{X}^{-1}\boldsymbol{E}_{ij}\boldsymbol{X}^{-1}. \qquad \square$$

(9) $$\frac{\partial \boldsymbol{X}^n}{\partial x_{ij}} = \sum_{k=0}^{n-1} \boldsymbol{X}^k \boldsymbol{E}_{ij} \boldsymbol{X}^{n-k-1}.$$

(10) $$\frac{\partial \boldsymbol{X}^{-n}}{\partial x_{ij}} = -\boldsymbol{X}^{-n}\left[\sum_{k=0}^{n-1} \boldsymbol{X}^k \boldsymbol{E}_{ij} \boldsymbol{X}^{n-k-1}\right]\boldsymbol{X}^{-n}.$$

1.5.2. The Derivative of Scalar Functions of a Matrix with Respect to the Matrix

Definition 1.5.2. Let $\boldsymbol{X}=(x_{ij})$ be an $m \times n$ matrix and let $y= f(\boldsymbol{X})$ be a scalar function of $\boldsymbol{X}$. The *derivative of y with respect to* $\boldsymbol{X}$ is defined as the following $m \times n$ matrix

(1.5.2) $$\frac{\partial y}{\partial \boldsymbol{X}} = \begin{bmatrix} \frac{\partial y}{\partial x_{11}} & \cdots & \frac{\partial y}{\partial x_{1n}} \\ \vdots & & \vdots \\ \frac{\partial y}{\partial x_{m1}} & \cdots & \frac{\partial y}{\partial x_{mn}} \end{bmatrix} = \left(\frac{\partial y}{\partial x_{ij}}\right).$$

By this definition, the following facts are obvious:

(1) $$\left(\frac{\partial f(\boldsymbol{X})}{\partial \boldsymbol{X}}\right)' = \frac{\partial f(\boldsymbol{X})}{\partial \boldsymbol{X}'}.$$

(2) $$\frac{\partial \operatorname{tr}(\boldsymbol{X})}{\partial \boldsymbol{X}} = \boldsymbol{I} \quad \text{and} \quad \frac{\partial \operatorname{tr}(\boldsymbol{AXB})}{\partial \boldsymbol{X}} = \boldsymbol{A}'\boldsymbol{B}'.$$

(3) $$\frac{\partial \operatorname{tr}(\boldsymbol{X}'\boldsymbol{AXB})}{\partial \boldsymbol{X}} = \boldsymbol{AXB} + \boldsymbol{A}'\boldsymbol{XB}'.$$

Proof. Using (7) of Section 1.5.1 we have

$$\frac{\partial \operatorname{tr}(\boldsymbol{X}'\boldsymbol{AXB})}{\partial x_{ij}} = \operatorname{tr}(\boldsymbol{E}_{ij}'\boldsymbol{AXB}) + \operatorname{tr}(\boldsymbol{X}'\boldsymbol{AE}_{ij}\boldsymbol{B}).$$

Hence

$$\begin{aligned} \frac{\partial \operatorname{tr}(\boldsymbol{X}'\boldsymbol{AXB})}{\partial \boldsymbol{X}} &= \sum_i \sum_j \frac{\partial \operatorname{tr}(\boldsymbol{X}'\boldsymbol{AXB})}{\boldsymbol{x}_{ij}} \boldsymbol{E}_{ij} \\ &= \sum_i \sum_j [\operatorname{tr}(\boldsymbol{e}_i'\boldsymbol{AXBe}_j) + \operatorname{tr}(\boldsymbol{e}_j'\boldsymbol{BX}'\boldsymbol{Ae}_i)]\, \boldsymbol{E}_{ij} \\ &= \boldsymbol{AXB} + \boldsymbol{A}'\boldsymbol{XB}'. \end{aligned}$$ □

(4) If $\dfrac{\partial f(\boldsymbol{X})}{\partial x_{ij}} = \operatorname{tr}(\boldsymbol{E}_{ij}'\boldsymbol{A})$, then $\dfrac{\partial f(\boldsymbol{X})}{\partial \boldsymbol{X}} = \boldsymbol{A}.$

Proof. As

$$\frac{\partial f(\boldsymbol{X})}{\partial x_{ij}} = \operatorname{tr}(\boldsymbol{E}_{ij}'\boldsymbol{A}) = a_{ij},$$

the assertion follows. □

By using (4) we can easily calculate some useful derivatives.

(5) If $\mathbf{X} = (x_{ij})$ is an $n \times n$ nonsingular matrix, then

$$\frac{\partial|\mathbf{X}|}{\partial \mathbf{X}} = \begin{cases} |\mathbf{X}|\,(\mathbf{X}^{-1})' \\ 2|\mathbf{X}|\,(\mathbf{X}^{-1})' - \operatorname{diag}(X_{11}, \ldots, X_{nn}) \text{ if } \quad \mathbf{X}' = \mathbf{X}, \end{cases}$$

where X_{ij} is the cofactor of x_{ij}.

Proof. Let

$$\boldsymbol{Z} = \begin{bmatrix} X_{11} & \cdots & X_{1n} \\ \vdots & & \vdots \\ X_{n1} & \cdots & X_{nn} \end{bmatrix}.$$

As

$$|\mathbf{X}| = \sum_j x_{ij} X_{ij}$$

we obtain

$$\frac{\partial|\mathbf{X}|}{\partial x_{ij}} = X_{ij} = \operatorname{tr}(\boldsymbol{E}_{ij}' \boldsymbol{Z}).$$

Using (4)

$$\frac{\partial|\mathbf{X}|}{\partial \mathbf{X}} = \boldsymbol{Z} = |\mathbf{X}|\,(\mathbf{X}^{-1})'.$$

If $\mathbf{X} = \mathbf{X}'$, then

$$\begin{aligned} \frac{\partial|\mathbf{X}|}{\partial x_{ij}} &= \operatorname{tr}\left[\left(\frac{\partial \mathbf{X}}{\partial x_{ij}}\right)' \boldsymbol{Z}\right] = \operatorname{tr}[(\boldsymbol{E}_{ij}' + \boldsymbol{E}_{ji}')\boldsymbol{Z}] \\ &= X_{ij} + X_{ji} \qquad \text{if } i \neq j, \end{aligned}$$

and

$$\frac{\partial|\mathbf{X}|}{\partial x_{ii}} = X_{ii}.$$

The assertion follows. □

Similarly we can find

$$\frac{\partial \log|\mathbf{X}|}{\partial \mathbf{X}} = \begin{cases} (\mathbf{X}^{-1})' \\ |\mathbf{X}|[2(\mathbf{X}^{-1})' - \operatorname{diag}(X_{11}, \ldots, X_{nn})] \qquad \text{if } \mathbf{X}' = \mathbf{X}. \end{cases}$$

and

$$\frac{\partial|\boldsymbol{A}\mathbf{X}\boldsymbol{B}|}{\partial \mathbf{X}} = |\boldsymbol{A}\mathbf{X}\boldsymbol{B}|\boldsymbol{A}'\,(\boldsymbol{B}'\mathbf{X}'\boldsymbol{A}')^{-1}\boldsymbol{B}'.$$

(6)

$$\frac{\partial \operatorname{tr}(\boldsymbol{A}\mathbf{X})}{\partial \mathbf{X}} = \begin{cases} \boldsymbol{A}' \\ \boldsymbol{A} + \boldsymbol{A}' - \operatorname{diag}(a_{11}, \ldots, a_{nn}) \qquad \text{if } \mathbf{X} = \mathbf{X}', \end{cases}$$

where $\boldsymbol{A} = (a_{ij})$.

The following theorem establishes the relationship between the derivatives of matrices with respect to a scalar and the derivatives of scalar functions of a matrix with respect to the matrix.

Theorem 1.5.1. *Let* $\mathbf{X}, \mathbf{Y}, \boldsymbol{A}, \mathbf{B}, \mathbf{C}$ *and* $\mathbf{D}$ *be* $m \times n$, $p \times q$, $p \times m$, $n \times q$, $p \times n$ *and* $m \times q$ *matrices, respectively, The following facts are equivalent:*

(1) $$\frac{\partial \mathbf{Y}}{\partial x_{rs}} = \boldsymbol{A}\boldsymbol{E}_{rs}(m, n)\boldsymbol{B} + \boldsymbol{C}\boldsymbol{E}_{rs}'(m, n)\mathbf{D}, \; r = 1, \ldots, m; \; s = 1, \ldots, n.$$

(2) $$\frac{\partial y_{ij}}{\partial \boldsymbol{X}} = \boldsymbol{A}'\boldsymbol{E}_{ij}(p, q)\boldsymbol{B}' + \boldsymbol{D}\boldsymbol{E}'_{ij}(p, q)\boldsymbol{C}, \; i=1, \ldots, p;\; j=1, \ldots, q.$$

Proof. The assertion follows from the fact that

$$\begin{aligned} \boldsymbol{e}'_i(\boldsymbol{A}\boldsymbol{E}_{rs}\boldsymbol{B} + \boldsymbol{C}\boldsymbol{E}'_{rs}\boldsymbol{D})\boldsymbol{e}_j &= (\boldsymbol{e}'_i\boldsymbol{A}\boldsymbol{e}_r)(\boldsymbol{e}'_s\boldsymbol{B}\boldsymbol{e}_j) + (\boldsymbol{e}'_i\boldsymbol{C}\boldsymbol{e}_s)(\boldsymbol{e}'_r\boldsymbol{D}\boldsymbol{e}_j) \\ &= (\boldsymbol{e}'_r\boldsymbol{A}'\boldsymbol{e}_i)(\boldsymbol{e}'_j\boldsymbol{B}'\boldsymbol{e}_s) + (\boldsymbol{e}'_r\boldsymbol{D}\boldsymbol{e}_j)(\boldsymbol{e}'_i\boldsymbol{C}\boldsymbol{e}_s) \\ &= \boldsymbol{e}'_r(\boldsymbol{A}'\boldsymbol{E}_{ij}\boldsymbol{B}' + \boldsymbol{D}\boldsymbol{E}'_{ij}\boldsymbol{C})\boldsymbol{e}_s \end{aligned}$$

for $r=1, \ldots, m$, $s=1, \ldots, n$, $i=1, \ldots, p$ and $j=1, \ldots, q$. □

By using Theorem 1.5.1, we can usually obtain $\partial y_{ij}/\partial \boldsymbol{X}$ from $\partial \boldsymbol{Y}/\partial x_{rs}$. For example, if $\boldsymbol{Y}=\boldsymbol{A}\boldsymbol{X}\boldsymbol{B}$, we have

$$\frac{\partial \boldsymbol{Y}}{\partial x_{rs}} = \boldsymbol{A}\boldsymbol{E}_{rs}\boldsymbol{B},$$

thus

$$\frac{\partial y_{ij}}{\partial \boldsymbol{X}} = \boldsymbol{A}'\boldsymbol{E}_{ij}\boldsymbol{B}'.$$

Similarly, we can find

$$\frac{\partial y_{ij}}{\partial \boldsymbol{X}} = -(\boldsymbol{X}^{-1})'\boldsymbol{A}'\boldsymbol{E}_{ij}\boldsymbol{B}'(\boldsymbol{X}^{-1})'$$

if $\boldsymbol{Y}=\boldsymbol{A}\boldsymbol{X}^{-1}\boldsymbol{B}$, and

$$\frac{\partial y_{ij}}{\partial \boldsymbol{X}} = \boldsymbol{A}\boldsymbol{X}\boldsymbol{E}'_{ij} + \boldsymbol{A}'\boldsymbol{X}\boldsymbol{E}_{ij}$$

if $\boldsymbol{Y}=\boldsymbol{X}'\boldsymbol{A}\boldsymbol{X}$.

1.5.3. The Derivatives of Vectors

Definition 1.5.3. Let $\boldsymbol{x}$ and $\boldsymbol{y}$ be vectors of order n and m respectively. The *derivative of the vector* $\boldsymbol{y}$ with respect to vector $\boldsymbol{x}$ is the matrix

(1.5.3) $$\frac{\partial \boldsymbol{y}'}{\partial \boldsymbol{x}} = \begin{bmatrix} \dfrac{\partial y_1}{x_1} & \cdots & \dfrac{\partial y_m}{x_1} \\ \vdots & & \vdots \\ \dfrac{\partial y_1}{\partial x_n} & \cdots & \dfrac{\partial y_m}{\partial x_n} \end{bmatrix} = \begin{bmatrix} \dfrac{\partial \boldsymbol{y}'}{\partial x_1} \\ \vdots \\ \dfrac{\partial \boldsymbol{y}'}{\partial x_n} \end{bmatrix}.$$

Theorem 1.5.2. (The chain rule for vectors). *Let* $\boldsymbol{x}' = (x_1, \ldots, x_n)$, $\boldsymbol{y}' = (y_1, \ldots, y_m)$ *and* $\boldsymbol{z}' = (z_1, \ldots, z_p)$. *Then*

(1.5.4) $$\frac{\partial \boldsymbol{z}'}{\partial \boldsymbol{x}} = \frac{\partial \boldsymbol{y}'}{\partial \boldsymbol{x}}\frac{\partial \boldsymbol{z}'}{\partial \boldsymbol{y}}.$$

Proof. As

$$\begin{aligned} \boldsymbol{e}'_i\left(\frac{\partial \boldsymbol{y}'}{\partial \boldsymbol{x}}\frac{\partial \boldsymbol{z}'}{\partial \boldsymbol{y}}\right)\boldsymbol{e}_j &= \sum_k \left(\boldsymbol{e}'_i\frac{\partial \boldsymbol{y}'}{\partial \boldsymbol{x}}\boldsymbol{e}_k\right)\left(\boldsymbol{e}'_k\frac{\partial \boldsymbol{z}'}{\partial \boldsymbol{y}}\boldsymbol{e}_j\right) \\ &= \sum_k \frac{\partial y_k}{\partial x_i}\frac{\partial z_j}{\partial y_k} = \frac{\partial z_j}{\partial x_i} = \boldsymbol{e}'_i\frac{\partial \boldsymbol{z}'}{\partial \boldsymbol{x}}\boldsymbol{e}_j \end{aligned}$$

for $i=1, \ldots, n$ and $j=1, \ldots, p$, (1.5.4) follows. ∎

Theorem 1.5.3. *Let* $\boldsymbol{X}$ *and* $\boldsymbol{Y}$ *be* $m\times n$ *and* $u\times v$ *matrices. If*

$$\frac{\partial \boldsymbol{Y}}{\partial x_{ij}} = \boldsymbol{A}\boldsymbol{E}_{ij}(m, n)\,\boldsymbol{B} + \boldsymbol{C}\boldsymbol{E}_{ij}'(m, n)\,\boldsymbol{D}, \tag{1.5.5}$$

then

$$\frac{\partial(\mathrm{vec}(\boldsymbol{Y}))'}{\partial(\mathrm{vec}(\boldsymbol{X}))} = \boldsymbol{B}\otimes\boldsymbol{A}' + \boldsymbol{K}_{mn}'(\boldsymbol{D}\otimes\boldsymbol{C}'). \tag{1.5.6}$$

Proof. It can be shown that

$$\boldsymbol{I}_{mn} = (\mathrm{vec}(\boldsymbol{E}_{11}), \mathrm{vec}(\boldsymbol{E}_{21}), \ldots, \mathrm{vec}(\boldsymbol{E}_{m1}), \ldots, \mathrm{vec}(\boldsymbol{E}_{1n}), \ldots, \mathrm{vec}(\boldsymbol{E}_{mn})). \tag{1.5.7}$$

From the assumption (1.5.5) we get

$$\mathrm{vec}\left(\frac{\partial \boldsymbol{Y}}{\partial x_{ij}}\right) = (\boldsymbol{B}'\otimes\boldsymbol{A})\,\mathrm{vec}(\boldsymbol{E}_{ij}) + (\boldsymbol{D}'\otimes\boldsymbol{C})\,\mathrm{vec}\boldsymbol{E}_{ij}'.$$

Hence

$$\frac{\partial(\mathrm{vec}(\boldsymbol{Y}))'}{\partial x_{ij}} = (\mathrm{vec}(\boldsymbol{E}_{ij}))'(\boldsymbol{B}\otimes\boldsymbol{A}') + (\mathrm{vec}(\boldsymbol{E}_{ij}'))'(\boldsymbol{D}\otimes\boldsymbol{C}'),$$

and

$$\begin{aligned}\frac{\partial(\mathrm{vec}(\boldsymbol{Y}))'}{\partial(\mathrm{vec}(\boldsymbol{X}))} &= (\mathrm{vec}(\boldsymbol{E}_{11}), \ldots, \mathrm{vec}(\boldsymbol{E}_{m1}), \ldots, \mathrm{vec}(\boldsymbol{E}_{1n}), \ldots, \mathrm{vec}(\boldsymbol{E}_{mn}))'(\boldsymbol{B}\otimes\boldsymbol{A}')\\ &\quad + (\mathrm{vec}(\boldsymbol{E}_{11}'), \ldots, \mathrm{vec}(\boldsymbol{E}_{m1}'), \ldots, \mathrm{vec}(\boldsymbol{E}_{1n}'), \ldots, \mathrm{vec}(\boldsymbol{E}_{mn}'))'(\boldsymbol{D}\otimes\boldsymbol{C}').\end{aligned}$$

Using (1.5.7) and the fourth property of $\boldsymbol{K}_{mn}$ the formula (1.5.6) follows. ∎

By the definition and the above theorems, the following useful results can be easily obtained.

(1) If $\boldsymbol{y} = \boldsymbol{A}\boldsymbol{x}$, then $\dfrac{\partial \boldsymbol{y}'}{\partial \boldsymbol{x}} = \boldsymbol{A}'$.

(2) If $y = \boldsymbol{x}'\boldsymbol{A}\boldsymbol{x}$, then $\dfrac{\partial y}{\partial \boldsymbol{x}} = (\boldsymbol{A} + \boldsymbol{A}')\boldsymbol{x}$.

(3) If $\boldsymbol{Y} = \boldsymbol{A}\boldsymbol{X}\boldsymbol{B}$ where $\boldsymbol{A}, \boldsymbol{B}, \boldsymbol{X}$ and $\boldsymbol{Y}$ are given in Theorem 1.5.1, then

$$\frac{\partial(\mathrm{vec}(\boldsymbol{Y}))'}{\partial(\mathrm{vec}(\boldsymbol{X}))} = \boldsymbol{B}\otimes\boldsymbol{A}'; \tag{1.5.8}$$

and if $\boldsymbol{Y} = \boldsymbol{A}\boldsymbol{X}'\boldsymbol{B}$, then

$$\frac{\partial(\mathrm{vec}(\boldsymbol{Y}))'}{\partial(\mathrm{vec}(\boldsymbol{X}))} = \boldsymbol{K}_{nm}(\boldsymbol{B}\otimes\boldsymbol{A}'). \tag{1.5.9}$$

(4) If $\boldsymbol{Y} = \boldsymbol{X}'\boldsymbol{A}\boldsymbol{X}$ where $\boldsymbol{Y}{:}n\times m$ and $\boldsymbol{X}{:}m\times n$, then

$$\frac{\partial(\mathrm{vec}(\boldsymbol{Y}))'}{\partial(\mathrm{vec}(\boldsymbol{X}))} = \boldsymbol{K}_{nm}(\boldsymbol{A}\boldsymbol{X}\otimes\boldsymbol{I}_n) + (\boldsymbol{I}_n\otimes\boldsymbol{A}'\boldsymbol{X}) \tag{1.5.10}$$

(5) If $\boldsymbol{Y} = \boldsymbol{A}\boldsymbol{X}^{-1}\boldsymbol{B}$, then

$$\frac{\partial(\mathrm{vec}(\boldsymbol{Y}))'}{\partial(\mathrm{vec}(\boldsymbol{X}))} = -(\boldsymbol{X}^{-1}\boldsymbol{B})\otimes(\boldsymbol{A}\boldsymbol{X}^{-1})'. \tag{1.5.11}$$

1.5.4. The Matrix Differential

For a scalar function $f(\boldsymbol{x})$ where $\boldsymbol{x} = (x_1, ..., x_n)'$, the differential df is defined as

$$df = \sum_{i=1}^{n} \frac{\partial f}{\partial x_i} dx_i = \frac{\partial f}{\partial \boldsymbol{x}'} d\boldsymbol{x} \tag{1.5.12}$$

where $d\boldsymbol{x} = (dx_1, ..., dx_n)'$. Corresponding to this definition we define the matrix differential $d\boldsymbol{X}$ for the matrix $\boldsymbol{X} = (x_{ij})$ of order $m \times n$ to be

$$d\boldsymbol{X} = \begin{bmatrix} dx_{11} & \cdots & dx_{1n} \\ \vdots & & \vdots \\ dx_{m1} & \cdots & dx_{mn} \end{bmatrix} \tag{1.5.13}$$

For a scalar function $f(\boldsymbol{X})$, now the differential df is

$$df = \sum_{i=1}^{m} \sum_{j=1}^{n} \frac{\partial f}{\partial x_{ij}} dx_{ij} = \operatorname{tr}\left[\left(\frac{\partial f}{\partial \boldsymbol{X}}\right)' d\boldsymbol{X}\right]. \tag{1.5.14}$$

The following results follow immediately:

(1) $d(\boldsymbol{X}+\boldsymbol{Y}) = d\boldsymbol{X} + d\boldsymbol{Y}$.

(2) $d(c\boldsymbol{X}) = c\, d\boldsymbol{X}$ where c is a constant.

(3) $(d\boldsymbol{X})' = d\boldsymbol{X}'$.

(4) $d(\operatorname{tr}(\boldsymbol{X})) = \operatorname{tr}(d\boldsymbol{X})$.

(5) $d(\boldsymbol{X}\boldsymbol{Y}) = (d\boldsymbol{X})\boldsymbol{Y} + \boldsymbol{X}(d\boldsymbol{Y})$.

Proof. The (i, j)–element of $d(\boldsymbol{X}\boldsymbol{Y})$ is

$$d\left(\sum_k x_{ik} y_{kj}\right) = \sum_k (dx_{ik}) y_{kj} + \sum_k x_{ik} (dy_{kj})$$

which is the (i, j)-element of $(d\boldsymbol{X})\boldsymbol{Y} + \boldsymbol{X}(d\boldsymbol{Y})$. □

(6) $d(\boldsymbol{X} \otimes \boldsymbol{Y}) = (d\boldsymbol{X}) \otimes \boldsymbol{Y} + \boldsymbol{X} \otimes (d\boldsymbol{Y})$.

(7) If $f(\boldsymbol{X})$ is a scalar function and

$$df = \operatorname{tr}(\boldsymbol{A}' d\boldsymbol{X})$$

for some $\boldsymbol{A}$, then

$$\frac{\partial f}{\partial \boldsymbol{X}} = \boldsymbol{A}.$$

The proof is easy and is left to the reader. Using this relationship between df and $\partial f / \partial \boldsymbol{X}$, we can obtain many derivatives by means of matrix defferential.

Example 1.5.1. As $d \operatorname{tr}(\boldsymbol{A}\boldsymbol{X}) = \operatorname{tr}(d(\boldsymbol{A}\boldsymbol{X})) = \operatorname{tr}(\boldsymbol{A} d\boldsymbol{X})$, we have

$$\frac{\partial \operatorname{tr}(\boldsymbol{A}\boldsymbol{X})}{\partial \boldsymbol{X}} = \boldsymbol{A}'.$$

Example 1.5.2. As

$$\begin{aligned} d \operatorname{tr}(\boldsymbol{X}'\boldsymbol{A}\boldsymbol{X}) &= \operatorname{tr}[(d\boldsymbol{X}')\boldsymbol{A}\boldsymbol{X} + \boldsymbol{X}'\boldsymbol{A} d\boldsymbol{X}] \\ &= \operatorname{tr}[(\boldsymbol{A}\boldsymbol{X})' d\boldsymbol{X} + \boldsymbol{X}'\boldsymbol{A} d\boldsymbol{X}] \\ &= \operatorname{tr}[(\boldsymbol{A}\boldsymbol{X} + \boldsymbol{A}'\boldsymbol{X})' d\boldsymbol{X}], \end{aligned}$$

we find

$$\frac{\partial \operatorname{tr}(\boldsymbol{X}'\boldsymbol{A}\boldsymbol{X})}{\partial \boldsymbol{X}} = (\boldsymbol{A} + \boldsymbol{A}')\boldsymbol{X}.$$

Example 1.5.3. Let $\boldsymbol{X}$ be a square matrix and let X_{ij} be the cofactor of x_{ij} and $\boldsymbol{Z}$

$= (X_{ij})$. Then

$$d|\mathbf{X}| = \sum_{i,j} \frac{\partial|\mathbf{X}|}{\partial x_{ij}} dx_{ij} = \sum_{i,j} X_{ij} dx_{ij} = \mathrm{tr}(\boldsymbol{Z}' d\boldsymbol{X}),$$

so that

$$\frac{\partial|\mathbf{X}|}{\partial \mathbf{X}} = \boldsymbol{Z} = |\mathbf{X}|\,(\mathbf{X}^{-1})'.$$

Example 1.5.4. If $|\boldsymbol{X}'\boldsymbol{A}\boldsymbol{X}| > 0$, then, by the above examples

$$\begin{aligned} d\log|\boldsymbol{X}'\boldsymbol{A}\boldsymbol{X}| &= \mathrm{tr}[(\boldsymbol{X}'\boldsymbol{A}\boldsymbol{X})^{-1} d(\boldsymbol{X}'\boldsymbol{A}\boldsymbol{X})] \\ &= \mathrm{tr}[(\boldsymbol{X}'\boldsymbol{A}\boldsymbol{X})^{-1} \boldsymbol{X}'(\boldsymbol{A} + \boldsymbol{A}')d\boldsymbol{X}]. \end{aligned}$$

Thus

$$\frac{\partial \log|\boldsymbol{X}'\boldsymbol{A}\boldsymbol{X}|}{\partial \boldsymbol{X}} = (\boldsymbol{A} + \boldsymbol{A}')\boldsymbol{X}\,(\boldsymbol{X}'\boldsymbol{A}'\boldsymbol{X})^{-1}.$$

1.6. Evaluation of the Jacobians of Some Transformations

In order to calculate mcdf's of many statistics, a change of variables in multiple integration will be met frequently. We consider a multiple integral over a subset R of an n-dimensional space

$$\int_R g(x_1, \ldots, x_n)\, dx_1 \ldots dx_n. \tag{1.6.1}$$

Let $x_1, \ldots, x_n$ be one-to-one transformed to new independent variables $y_1, \ldots, y_n$ through relations $y_i = f_i(x_1, \ldots, x_n)$, $i = 1, \ldots, n$ where $\{f_i\}$ are continuously differentiable. These relations will be written as $\boldsymbol{y} = f(\boldsymbol{x})$ and $\boldsymbol{x} = f^{-1}(\boldsymbol{y})$. The absolute value of the determinant of $\partial\boldsymbol{x}'/\partial\boldsymbol{y}$, denoted by

$$J(\boldsymbol{x} \to \boldsymbol{y}) = \left|\frac{\partial \boldsymbol{x}'}{\partial \boldsymbol{y}}\right|_+ \quad \text{or} \quad \left|\frac{\partial(x_1, \ldots, x_n)}{\partial(y_1, \ldots, y_n)}\right|_+$$

will be called the *Jacobian of the transformation of* $\boldsymbol{x}$ *to* $\boldsymbol{y}$. Now (1.6.1) can be expressed as

$$\int_T g(f^{-1}(\boldsymbol{y})) J(\boldsymbol{x} \to \boldsymbol{y}) d\boldsymbol{y}, \tag{1.6.2}$$

where

$$T = \{\boldsymbol{y} \mid \boldsymbol{y} = f(\boldsymbol{x}),\ \boldsymbol{x} \in R\}. \tag{1.6.3}$$

Theorem 1.6.1.

(1) $J(\boldsymbol{y} \to \boldsymbol{x}) = J(\boldsymbol{x} \to \boldsymbol{y})^{-1}$.

(2) *If* $\boldsymbol{y} = f(\boldsymbol{x})$ *and* $\boldsymbol{z} = g(\boldsymbol{y})$, *then*

$$J(\boldsymbol{x} \to \boldsymbol{z}) = J(\boldsymbol{x} \to \boldsymbol{y})\, J(\boldsymbol{y} \to \boldsymbol{z}). \tag{1.6.4}$$

(3) *If* $d\boldsymbol{x} = \boldsymbol{A}\, d\boldsymbol{y}$, *then* $J(\boldsymbol{x} \to \boldsymbol{y})$ *is the absolute value of* $|\boldsymbol{A}|$. *Or write* $J(d\boldsymbol{x} \to d\boldsymbol{y}) = J(\boldsymbol{x} \to \boldsymbol{y})$.

(4) *If* $\boldsymbol{y}_1 = \boldsymbol{f}_1(\boldsymbol{x}_1, \ldots, \boldsymbol{x}_n)$, $\boldsymbol{y}_i = \boldsymbol{f}_i(\boldsymbol{y}_1, \ldots, \boldsymbol{y}_{i-1}, \boldsymbol{x}_i, \ldots, \boldsymbol{x}_n)$, $i = 2, \ldots, n-1$, $\boldsymbol{y}_n = \boldsymbol{f}_n(\boldsymbol{y}_1, \ldots, \boldsymbol{y}_{n-1}, \boldsymbol{x}_n)$ *where* $\boldsymbol{y}_i$ *and* $\boldsymbol{x}_i$ *are* $q_i \times 1$ *vectors*, $i = 1, \ldots, n$, *then*

(1.6.5) $$J[(y_1, \ldots, y_n) \to (x_1, \ldots, x_n)] = \prod_{i=1}^{n} \left| \frac{\partial f_i'}{\partial x_i} \right|_+ .$$

Proof. By Theorem 1.5.2 we have

$$\frac{\partial z'}{\partial x} = \frac{\partial z'}{\partial y}\frac{\partial y'}{\partial x},$$

and (2) follows. Setting $z = x$ in (2) we get (1). From (1.5.11)

$$dy = \frac{\partial y}{\partial x'} dx, \text{ hence } J(dx \to dy) = \left| \frac{\partial x'}{\partial y} \right|_+ = J(x \to y) = |A|_+ .$$

Finally, we prove (4). Let

$$T_{ij} = \begin{cases} \dfrac{\partial f_i}{\partial x_j'} & i \leq j, \\ O & i > j, \end{cases}$$

$$G_{ij} = \begin{cases} -\dfrac{\partial f_i}{\partial y_j'} & j < i \\ I_{qi} & j = i, \\ 0 & j > i \end{cases}$$

$T = (T_{ij})$, $G = (G_{ij})$, $dx = (dx_1', \ldots, dx_n')'$ and $dy = (dy_1', \ldots, dy_n')'$. As $y_i = f_i(y_1, \ldots, y_{i-1}, x_i, \ldots, x_n)$ the differential of y_i is

$$dy_i = \sum_{j=1}^{i-1} \frac{\partial f_i}{\partial y_j'} dy_j + \sum_{j=i}^{n} \frac{\partial f_i}{\partial x_j'} dx_j = -\sum_{j=i}^{i-1} G_{ij} dy_j + \sum_{j=i}^{n} T_{ij} dx_j,$$

i.e.,

$$\sum_{j=1}^{n} G_{ij} dy_j = \sum_{j=1}^{n} T_{ij} dx_j$$

or

$$Gdy = Tdx \quad \text{and} \quad dy = G^{-1}Tdx. \text{ Therefore}$$

$$J(y \to x) = J(dy \to dx) = |G^{-1}T|_+ = |T|_+ = \prod_{i=1}^{n} |T_{ii}|_+ = \prod_{i=1}^{n} \left| \frac{\partial f_i'}{\partial x_i} \right|_+ . \qquad \square$$

By Theorem 1.6.1. we can evaluate the following useful Jacobians.

Example 1.6.1 If $Y = AXB$ where $Y{:}n \times p$, $X{:}n \times p$, $A{:}n \times n$ and $B{:}p \times p$, then $J(Y \to X) = |A|_+^p |B|_+^n$.

Proof. First of all, if $y = Ax$ where y and x are vectors, we can show $J(y \to x) = |A|_+$ by (3) of Theorem 1.6.1. Using (1.4.4)

$\mathrm{vec}(Y) = (B' \otimes A)\,\mathrm{vec}(X)$ and $J(Y \to X) = J(\mathrm{vec}(Y) \to \mathrm{vec}(X)) = |B' \otimes A|_+ = |A|_+^p\, |B|_+^n$.

Example 1.6.2. If $Y = B'XB$ where $|B| \neq 0$ and X, Y and B are all $n \times n$ matrices.

(1) If $X' = X$, then $J(Y \to X) = |B|_+^{n+1} = |B'B|^{\frac{1}{2}(n+1)}$,

(2) If $X' = -X$, then $J(Y \to X) = |B|_+^{n-1} = |B'B|^{\frac{1}{2}(n-1)}$.

Proof. To prove (1) let B be an upper triangular matrix. Then we can express the transformation as

$$\begin{bmatrix} \boldsymbol{Y}_{11} & \boldsymbol{y} \\ \boldsymbol{y}' & y_{nn} \end{bmatrix} = \begin{bmatrix} \boldsymbol{B}'_{11} & \boldsymbol{0} \\ \boldsymbol{b}' & \boldsymbol{b}_{nn} \end{bmatrix} \begin{bmatrix} \boldsymbol{X}_{11} & \boldsymbol{x} \\ \boldsymbol{x}' & x_{nn} \end{bmatrix} \begin{bmatrix} \boldsymbol{B}_{11} & \boldsymbol{b} \\ \boldsymbol{0} & b_{nn} \end{bmatrix}$$

which gives

$$\begin{aligned} \boldsymbol{Y}_{11} &= \boldsymbol{B}'_{11}\boldsymbol{X}_{11}\boldsymbol{B}_{11} \\ \boldsymbol{y} &= \boldsymbol{B}'_{11}\boldsymbol{X}_{11}\boldsymbol{b} + \boldsymbol{B}'_{11}\boldsymbol{x}b_{nn} \\ y_{nn} &= b_{nn}^2 x_{nn} + 2b_{nn}\boldsymbol{x}'\boldsymbol{b} + \boldsymbol{b}'\boldsymbol{X}_{11}\boldsymbol{b}. \end{aligned}$$

From (4) of Theorem 1.6.1

$$\begin{aligned} J(\boldsymbol{Y} \to \boldsymbol{X}) &= J(\boldsymbol{Y}_{11} \to \boldsymbol{X}_{11}) J(\boldsymbol{y} \to \boldsymbol{x}) J(y_{nn} \to x_{nn}) \\ &= J(\boldsymbol{Y}_{11} \to \boldsymbol{X}_{11}) |\boldsymbol{B}'_{11}|_+ |b_{nn}^{n-1}|_+ b_{nn}^2 \\ &= J(\boldsymbol{Y}_{11} \to \boldsymbol{X}_{11}) |\boldsymbol{B}_{11}|_+ |b_{nn}^{n+1}|_+, \end{aligned}$$

because $J(\boldsymbol{y} \to \boldsymbol{x}) = |\boldsymbol{B}'_{11}|_+ |b_{nn}^{n-1}|_+$ and $J(y_{nn} \to x_{nn}) = b_{nn}^2$. Hence by induction $J(\boldsymbol{Y}_{11} \to \boldsymbol{X}_{11}) = |\boldsymbol{B}_{11}|_+^n$ and $J(\boldsymbol{Y} \to \boldsymbol{X}) = |\boldsymbol{B}|_+^{n+1}$.

Similarly, if $\boldsymbol{B}$ is a lower triangular matrix, then $J(\boldsymbol{Y} \to \boldsymbol{X}) = |\boldsymbol{B}|_+^{n+1}$.

When $\boldsymbol{B}$ is a nonsingular matrix, its rows and columns can be interchanged without affecting the Jacobian of the transformation, so that the leading principale minors of order i $(i=1, \ldots, n)$ are nonzero. From (1.2.14) we have the factorization $\boldsymbol{B} = \boldsymbol{T}\boldsymbol{U}$, where $\boldsymbol{T} \in \mathrm{LT}(n)$ and $\boldsymbol{U} \in \mathrm{UT}(n)$. Hence

$$\boldsymbol{Y} = \boldsymbol{U}'(\boldsymbol{T}'\boldsymbol{X}\boldsymbol{T})\boldsymbol{U} = \boldsymbol{U}'\boldsymbol{X}^*\boldsymbol{U},$$

where $\boldsymbol{X}^* = \boldsymbol{T}'\boldsymbol{X}\boldsymbol{T}$. By (2) of Theorem 1.6.1

$$J(\boldsymbol{Y} \to \boldsymbol{X}) = J(\boldsymbol{Y} \to \boldsymbol{X}^*) J(\boldsymbol{X}^* \to \boldsymbol{X}) = |\boldsymbol{U}|_+^{n+1} |\boldsymbol{T}|_+^{n+1} = |\boldsymbol{B}|_+^{n+1}.$$

Similarly, we can prove (2). □

Example 1.6.3. Let $\boldsymbol{Y} = \boldsymbol{A}\boldsymbol{X}$ where $\boldsymbol{A}$, $\boldsymbol{X}$ and $\boldsymbol{Y}$ are $n \times n$ matrices.

(1) If $\boldsymbol{X}$, $\boldsymbol{Y}$ and $\boldsymbol{A}$ all are lower triangular matrices, then

$$J(\boldsymbol{Y} \to \boldsymbol{X}) = \prod_{i=1}^{n} |a_{ii}|^i.$$

(2) If $\boldsymbol{X}$, $\boldsymbol{Y}$ and $\boldsymbol{A}$ all are upper triangular matrices, then

$$J(\boldsymbol{Y} \to \boldsymbol{X}) = \prod_{i=1}^{n} |a_{ii}|^{n+1-i}.$$

Proof. We only prove (1). As $y_{ij} = \sum_{k=1}^{i} a_{ik}x_{kj}$ for $1 \leqslant j \leqslant i \leqslant n$,

$$\frac{\partial(y_{11}, y_{21}, y_{22}, y_{31}, y_{32}, y_{33}, \ldots, y_{n1}, \ldots, y_{nn})}{\partial(x_{11}, x_{21}, x_{22}, x_{31}, x_{32}, x_{33}, \ldots, x_{n1}, \ldots, x_{nn})}$$

is a lower triangular matrix with $a_{11}, a_{22}, a_{22}, a_{33}, a_{33}, a_{33}, \ldots, a_{pp}, \ldots, a_{pp}$ as its diagonal elements and (1) follows. □

Example 1.6.4. If $\boldsymbol{Y} = \boldsymbol{X} + \boldsymbol{X}'$ where $\boldsymbol{X}$ is an $n \times n$ lower triangular matrix, then $J(\boldsymbol{Y} \to \boldsymbol{X}) = 2^n$.

Proof. As

$$y_{ij} = \begin{cases} 2x_{ii} & \text{if } i=j, \\ x_{ij} & \text{if } i \neq j, \end{cases}$$

the matrix $\dfrac{\partial(y_{11}, y_{21}, y_{22}, \ldots, y_{n1}, \ldots, y_{nn})}{\partial(x_{11}, x_{21}, x_{22}, \ldots, x_{n1}, \ldots, x_{nn})}$ is a lower triangular matrix with n 2's and

$\frac{1}{2}n(n-1)$ 1's as its diagonal elements that means $J(Y \to X) = 2^n$. □

Example 1.6.5 Let $Y = XA' + AX'$ where A, X and Y are $n \times n$ matrices

(1) If X and A are lower triangular matrices, then

$$J(Y \to X) = 2^n \prod_1^n |a_{ii}|^{n+i-1}.$$

(2) If X and A are upper triangular matrices, then

$$J(Y \to X) = 2^n \prod_1^n |a_{ii}|^i.$$

Proof. (1) Let $Z = A^{-1}YA'^{-1} = A^{-1}X + X'A'^{-1} = W + W'$, where $W = A^{-1}X$ is a lower triangular matrix. Thus, by (2) of Theorem 1.6.1, Example 1.6.2, Example 1.6.4 and Example 1.6.3, we have

$$J(Y \to X) = J(Y \to Z)\, J(Z \to W)\, J(W \to X)$$

$$= |A|_+^{n+1}\, 2^n \prod_1^n |a_{ii}|^{-i} = 2^n \prod_1^n |a_{ii}|^{n+1-i}.$$

Similarly, we can prove (2). □

Example 1.6.6. Let $Y = XX' > 0$ be the Cholesky decomposition of Y where $X \in \mathrm{LT}(n)$ with positive diagonal elements (cf. (1.2.7)). Then

$$J(Y \to X) = 2^n \prod_1^n x_{ii}^{n-i+1}.$$

Proof. As $dY = (dX)X' + X(dX)'$, by Theorem 1.6.1 and Example 1.6.5, we have

$$J(Y \to X) = J(dY \to dX) = 2^n \prod_1^n x_{ii}^{n+1-i}.$$ □

Example 1.6.7. Let S be an $n \times n$ symmetric matrix such that all the eigenvalues are distinct and nonzero. Let $S = HD_\lambda H'$, where $D_\lambda = \mathrm{diag}(\lambda_1, \lambda_2, \ldots, \lambda_n)$, $\lambda_1 > \lambda_2 > \ldots > \lambda_n$ ($\lambda_i \neq 0$), and $H \in O(n)$ whose elements in the first row are positive or whose diagonal elements are all positive. Then

$$J(S \to H, D_\lambda) = f_n(H) \prod_{i=1}^{n-1} \prod_{j=i+1}^{n} (\lambda_i - \lambda_j),$$

where

$$f_n(H) = \frac{1}{\prod_{i=1}^n |H_{(i)}|_+},$$

and $H_{(i)}$ is the first i–dimensional principal minor of H.

Proof. Since $H'H = I$ and $(dH)'H + H'(dH) = O$, we get $R + R' = O$ where $R = H'(dH)$. Hence R is a skew–symmetric matrix. Let $J(R \to dH) = f_n(H)$. Suppose the independent variables in H are $\{h_{12}, \ldots, h_{1n}, h_{23}, \ldots, h_{2n}, \ldots, h_{n-1,n}\}$ and $dH = HR$. Taking $r_{ij} = -r_{ji}$ for $i < j$, we get

$$dh_{ij} = \sum_{k=1}^{j-1} h_{ik} r_{kj} - \sum_{k=j}^{n} h_{ik} r_{jk},\ i < j = 1, \ldots, n.$$

Noticing that the transformations are conditional and by Theorem 1.6.1 (5) we have

$$J(dH \to R) = \prod_{j=2}^n J(dh_{1j}, \ldots, dh_{j-1,j} \to r_{1j}, r_{2j}, \ldots, r_{j-1,j})$$

$$= \prod_{j=2}^{n} |H_{(j-1)}|_+ = \prod_{j=1}^{n-1} |H_{(j)}| = \prod_{j=1}^{n} |H_{(j)}|_+$$

Taking the differentials of $\boldsymbol{S} = \boldsymbol{H}\boldsymbol{D}_\lambda \boldsymbol{H}'$, we have

$$d\boldsymbol{S} = (d\boldsymbol{H})\boldsymbol{D}_\lambda \boldsymbol{H}' + \boldsymbol{H}(d\boldsymbol{D}_\lambda)\boldsymbol{H}' + \boldsymbol{H}\boldsymbol{D}_\lambda(d\boldsymbol{H})',$$

and, say,

$$d\boldsymbol{W} \equiv \boldsymbol{H}'(d\boldsymbol{S})\boldsymbol{H} = \boldsymbol{R}\boldsymbol{D}_\lambda - \boldsymbol{D}_\lambda \boldsymbol{R} + d\boldsymbol{D}_\lambda.$$

Comparing elements on the two sides, we find that $dw_{ii} = d\lambda_i$ and $dw_{ij} = (\lambda_i - \lambda_j) r_{ij}$ for $i<j$. Hence

$$\begin{aligned} J(\boldsymbol{S} \to \boldsymbol{H}, \boldsymbol{D}_\lambda) &= J(d\boldsymbol{S} \to (d\boldsymbol{H}, d\boldsymbol{D}_\lambda)) = J(d\boldsymbol{S} \to d\boldsymbol{W}) J(d\boldsymbol{W} \to (d\boldsymbol{H}, d\boldsymbol{D}_\lambda)) \\ &= J(d\boldsymbol{W} \to (\boldsymbol{R}, d\boldsymbol{D}_\lambda)) J(\boldsymbol{R} \to d\boldsymbol{H}) = \prod_{i<j} (\lambda_i - \lambda_j) / J(d\boldsymbol{H} \to \dot{\boldsymbol{R}}) \end{aligned}$$

which is the required result. □

Example 1.6.8. The generalized spherical coordinate transformation is

$$\begin{cases} x_j = r\left(\prod_{k=1}^{j-1} \sin\varphi_k\right)\cos\varphi_j, & 1 \leqslant j \leqslant n-2 \\ x_{n-1} = r\left(\prod_{k=1}^{n-2} \sin\varphi_k\right)\cos\theta, & 0 \leqslant \varphi_k \leqslant \pi,\ 1 \leqslant k \leqslant n-2 \\ x_n = r\left(\prod_{k=1}^{n-2} \sin\varphi_k\right)\sin\theta, & 0 \leqslant \theta \leqslant 2\pi,\ 0 \leqslant r < \infty. \end{cases}$$

Then

$$J(x_1, \ldots, x_n \to r, \varphi_1, \ldots, \varphi_{n-2}, \theta) = r^{n-1}\left(\prod_{k=1}^{n-2} \sin^{n-k-1}\varphi_k\right).$$

Proof. Clearly, we have

$$x_n^2 = r^2\left(\prod_{k=1}^{n-2} \sin^2\varphi_k\right)\sin^2\theta,$$

$$x_{n-1}^2 + x_n^2 = r^2 \prod_{k=1}^{n-2} \sin^2\varphi_k,$$

....

$$x_2^2 + \ldots + x_n^2 = r^2 \sin^2\varphi_1,$$

$$x_1^2 + \ldots + x_n^2 = r^2.$$

By using (1.6.5), it can be shown that

$$\begin{aligned} &J(x_1, \ldots, x_n \to r, \varphi_1, \ldots, \varphi_{n-2}, \theta) \\ &= J(x_n \to \theta) J(x_{n-1} \to \varphi_{n-2}) \ldots J(x_2 \to \varphi_1)\, J(x_1 \to r) \\ &= (1/x_n) r^2\left(\prod_{k=1}^{n-2} \sin^2\varphi_k\right)\sin\theta\cos\theta \cdot (1/x_{n-1}) r^2 \\ &\cdot\left(\prod_{k=1}^{n-3} \sin^2\varphi_k\right)\sin\varphi_{n-2}\cos\varphi_{n-2} \ldots (1/x_2) r^2 \sin\varphi_1 \cos\varphi_1 \cdot (1/x_1) r \\ &= \left[r\left(\prod_{k=1}^{n-2} \sin\varphi_k\right)\cos\theta\right] \cdot \left[(r/\cos\theta)\left(\prod_{k=1}^{n-3} \sin\varphi_k\right)\cos\varphi_{n-2}\right] \ldots \end{aligned}$$

$$[(r/\cos\varphi_2)\cos\varphi_1]\cdot\left[\frac{1}{\cos\varphi_1}\right]=r^{n-1}\left(\prod_{k=1}^{n-2}\sin^{n-k-1}\varphi_k\right). \quad \square$$

1.7. Groups and Invariance

In multivariate analysis the class of invariant tests based on statistics that are invariant under some group of transformations is important. Some likelihood ratio tests under general conditions are such invariant tests.

Let G denote a *group* of transformations from a space $\mathfrak{X}$ into itself. This means that

(1) if $g_1 \in G$ and $g_2 \in G$, then $g_1 g_2 \in G$ where $g_1 g_2$ is defined as $(g_1 g_2)x = g_1(g_2 x)$;

(2) if $g \in G$, then $g^{-1} \in G$ where g^{-1} satisfies $gg^{-1}=g^{-1}g=e$, with e being the identity transformation in G.

Definition 1.7.1. Two points x_1 and x_2 in $\mathfrak{X}$ are said to be *equivalent under* G, if there exists a $g \in G$ such that $x_2 = gx_1$. We write $x_1 \sim x_2$ (mod G).

Obviously, the equivalent relation has the following properties:

(1) $x \sim x$ (mod G);

(2) $x \sim y$ (mod G) implies $y \sim x$ (mod G); and

(3) $x \sim y$ (mod G) and $y \sim z$ (mod G) imply $x \sim z$ (mod G).

The set $\{gx \mid g \in G\}$ is called the *orbit of* x *under* G. Clearly, two orbits are either identical or disjoint, and the orbits form a partition of $\mathfrak{X}$. If $x_1 \sim x_2$ (mod$G$) for all x_1, x_2 in $\mathfrak{X}$ then the group G is said to *act transitively on* $\mathfrak{X}$, and $\mathfrak{X}$ is said to be *homogeneous with respect to* G. Hence, G acts transitively on $\mathfrak{X}$ if there is only one orbit, namely, $\mathfrak{X}$ itself.

Definition 1.7.2. A function $f(x)$ on $\mathfrak{X}$ is said to be *invariant under* G if

$$f(gx) = f(x) \qquad \text{for each } x \in \mathfrak{X} \text{ and each } g \in G.$$

Hence f is invariant under G if and only if it is constant on each orbit under G.

Definition 1.7.3. A function $f(x)$ on $\mathfrak{X}$ is said to be a *maximal invariant under* G if it is invariant under G and if

$$f(x_1) = f(x_2) \qquad \text{implies } x_1 \sim x_2 \quad (\text{mod } G).$$

Clearly, f is a maximal invariant if and only if it is constant on each orbit and takes a different value for each orbit. The following theorem shows that any invariant function is a function of a maximal invariant.

Theorem 1.7.1. *Assume* $f(x)$ *on* $\mathfrak{X}$ *is a maximal invariant under* G. *Then a function* $h(x)$ *on* $\mathfrak{X}$ *is invariant under* G *if and only if* h *is a function of* $f(x)$.

Proof. If h is a function of $f(x)$, then there exists a function q such that $h(x) = q(f(x))$ for all $x \in \mathfrak{X}$. Hence for all $g \in G$, $x \in \mathfrak{X}$ we get $h(gx) = q(f(gx)) = q(f(x)) = h(x)$, i.e., h is invariant.

Now suppose h is invariant. As f is a maximal invariant, $f(x_1) = f(x_2)$ implies $x_1 \sim x_2$ (mod G), i.e., $x_2 = gx_1$ for some $g \in G$. Hence $h(x_2) = h(gx_1) = h(x_1)$ which establishes that $h(x)$ depends on x only through $f(x)$. $\square$

The following examples are useful in other chapters.

Example 1.7.1. Let $\mathscr{X} = R^n$ and $G=O(n)$, the group of $n \times n$ orthogonal matrices. The action of $\boldsymbol{H} \in O(n)$ on $\boldsymbol{x} \in R^n$ is

$$\boldsymbol{x} \to \boldsymbol{H}\boldsymbol{x}$$

and the group operation is matrix multiplication.

We want to show that a maximal invariant under G is $f(\boldsymbol{x})=\boldsymbol{x}'\boldsymbol{x}$. First, f is invariant because $f(\boldsymbol{H}\boldsymbol{x})=\boldsymbol{x}'\boldsymbol{H}'\boldsymbol{H}\boldsymbol{x}=\boldsymbol{x}'\boldsymbol{x}$ for all $\boldsymbol{x} \in R^n$ and $\boldsymbol{H} \in O(n)$. If $\boldsymbol{x}_1'\boldsymbol{x}_1 = f(\boldsymbol{x}_1) = f(\boldsymbol{x}_2) = \boldsymbol{x}_2'\boldsymbol{x}_2$, there exists $\boldsymbol{H} \in O(n)$ such that $\boldsymbol{x}_2 = \boldsymbol{H}\boldsymbol{x}_1$ (cf. Section 1.2, (6)), i.e., $\boldsymbol{x}_1 \sim \boldsymbol{x}_2$. This means a maximal invariant under G is $\boldsymbol{x}'\boldsymbol{x}$ and any invariant function is a function of $\boldsymbol{x}'\boldsymbol{x}$.

Similarly, let $\mathscr{X}$ be the set of all $n \times m$ matrices and $G=O(n)$. The action of $\boldsymbol{H} \in O(n)$ on $\boldsymbol{X} \in \mathscr{X}$ is $\boldsymbol{X} \to \boldsymbol{H}\boldsymbol{X}$, and the group operation is matrix multiplication. A maximal invariant under G is $\boldsymbol{X}'\boldsymbol{X}$. □

Example 1.7.2. Let $\mathscr{X} = \{(\mu, \Sigma) \mid \mu \in \mathrm{R}^n, \Sigma > 0 \text{ and } \Sigma{:}n \times n\}$ and $G=\{(\boldsymbol{H}, \boldsymbol{c}) \; \boldsymbol{H} \in O(n) \text{ and } \boldsymbol{c} \in R^n\}$. The action of G on $\mathscr{X}$ is

$$\mu \to \boldsymbol{H}\mu + \boldsymbol{c} \qquad \text{and} \qquad \Sigma \to \boldsymbol{H}\Sigma\boldsymbol{H}',$$

and the group operation is matrix multiplication and sum. Then a maximal invariant under G is $\lambda(\Sigma)=\mathrm{diag}(\lambda_1, \ldots, \lambda_n)$, the eigenvalues of Σ, $\lambda_1 \geqslant \lambda_2 \geqslant \ldots \geqslant \lambda_n > 0$.

Proof. Let $f(\mu, \Sigma)=\lambda(\Sigma)$ and note that $f(\mu, \Sigma)$ is invariant, because $\boldsymbol{H}\Sigma\boldsymbol{H}'$ and Σ have the same eigenvalues for each $\boldsymbol{H} \in O(n)$. To prove it is the maximal invariant, let $f(\mu, \Sigma)=f(v, \boldsymbol{V})$, i.e., Σ and $\boldsymbol{V}$ have the same eigenvalues $\lambda_1, \ldots, \lambda_n$. Take $\boldsymbol{H}_1 \in O(n)$ and $\boldsymbol{H}_2 \in O(n)$ such that

$$\boldsymbol{H}_1\Sigma\boldsymbol{H}_1' = \mathrm{diag}(\lambda_1, \ldots, \lambda_n) = \boldsymbol{H}_2\boldsymbol{V}\boldsymbol{H}_2'.$$

Write $\boldsymbol{H} = \boldsymbol{H}_2'\boldsymbol{H}_1 \in O(n)$. Hence $\boldsymbol{V} = \boldsymbol{H}_2'\boldsymbol{H}_1\Sigma\boldsymbol{H}_1'\boldsymbol{H}_2 = \boldsymbol{H}\Sigma\boldsymbol{H}'$. Setting $\boldsymbol{c} = -\boldsymbol{H}\mu + v$, we get

$$\boldsymbol{H}\mu + \boldsymbol{c} = v \qquad \text{and} \qquad \boldsymbol{H}\Sigma\boldsymbol{H}' = \boldsymbol{V}.$$

Hence $(\mu, \Sigma) \sim (v, \boldsymbol{V}) \pmod G$. □

Example 1.7.3. Let $\mathscr{X} = \{\boldsymbol{X} \mid \boldsymbol{X} \text{ is an } n \times p \text{ matrix}\}$ and $G=\{(\boldsymbol{H}, \boldsymbol{P}) \mid \boldsymbol{H} \in O(n), \boldsymbol{P} \in O(p)\}$. The group of transformations on $\mathscr{X}$ is given by $\boldsymbol{X} \to \boldsymbol{H}\boldsymbol{X}\boldsymbol{P}$ with matrix multiplication. Then a maximal invariant under G is $\lambda(\boldsymbol{X}'\boldsymbol{X})$, the eigenvalues of $\boldsymbol{X}'\boldsymbol{X}$.

Proof. Let $f(\boldsymbol{X})=\lambda(\boldsymbol{X}'\boldsymbol{X})$. Clearly $f(\boldsymbol{X})$ is invariant, because $\boldsymbol{X}'\boldsymbol{X}$ and $(\boldsymbol{H}\boldsymbol{X}\boldsymbol{P})'(\boldsymbol{H}\boldsymbol{X}\boldsymbol{P})$ have the same eigenvalues. Assume $f(\boldsymbol{X})=f(\boldsymbol{Y})$, i.e., $\boldsymbol{X}'\boldsymbol{X}$ and $\boldsymbol{Y}'\boldsymbol{Y}$ have the same eigenvalues. We want to show $\boldsymbol{X} \sim \boldsymbol{Y} \pmod G$. In fact, by the above two examples, there exist $\boldsymbol{P} \in O(p)$ and $\boldsymbol{H} \in O(n)$ such that $\boldsymbol{Y}'\boldsymbol{Y} = \boldsymbol{P}\boldsymbol{X}'\boldsymbol{X}\boldsymbol{P}' = (\boldsymbol{X}\boldsymbol{P}')'\boldsymbol{X}\boldsymbol{P}'$ and $\boldsymbol{Y} = \boldsymbol{H}\boldsymbol{X}\boldsymbol{P}'$, i.e., $\boldsymbol{X} \sim \boldsymbol{Y}$. □

Example 1.7.4. Let $\mathscr{X} = \{(\mu, \Sigma) \mid \mu \in R^n, \Sigma > 0 \text{ and } \Sigma{:}n \times n\}$ and $G=O(n)$. The action of G on $\mathscr{X}$ is

$$\mu \to \boldsymbol{H}\mu \qquad \text{and} \qquad \Sigma \to \boldsymbol{H}\Sigma\boldsymbol{H}'.$$

Then a maximal invariant under G is $(\lambda_1, \ldots, \lambda_n, \boldsymbol{P}\mu)$, where $\lambda_1, \ldots, \lambda_n$ are the eigenvalues of Σ and $\boldsymbol{P} \in O(n)$ is such that $\boldsymbol{P}\Sigma\boldsymbol{P}' = \lambda(\Sigma)=\mathrm{diag}(\lambda_1, \ldots, \lambda_n)$.

Proof. Let $f(\mu, \Sigma)=(\lambda(\Sigma), \boldsymbol{P}\mu)$ where $\boldsymbol{P}\Sigma\boldsymbol{P}'=\lambda(\Sigma)$. Clearly, $f(\mu, \Sigma)$ is invariant, because Σ and $\boldsymbol{H}\Sigma\boldsymbol{H}'$ have the same eigenvalues, and if $\boldsymbol{Q}(\boldsymbol{H}\Sigma\boldsymbol{H}')\boldsymbol{Q}' = \lambda(\Sigma)$, then $\boldsymbol{P}\Sigma\boldsymbol{P}' = \lambda(\Sigma)$, and $\boldsymbol{Q}(\boldsymbol{H}\mu)=\boldsymbol{P}\mu$. To show it is a maximal invariant, assume $f(\mu, \Sigma)=f(v, \boldsymbol{V})$. By Example 1.7.2, there exists $\boldsymbol{H} \in O(n)$ such that $\boldsymbol{V} = \boldsymbol{H}\Sigma\boldsymbol{H}'$. Let $\boldsymbol{P} \in O(n)$ be such that $\boldsymbol{P}\Sigma\boldsymbol{P}'=\lambda(\Sigma)$ and let $\boldsymbol{Q}=\boldsymbol{P}\boldsymbol{H}'$. Then $\boldsymbol{Q}\boldsymbol{V}\boldsymbol{Q}' = \boldsymbol{P}\boldsymbol{H}'\boldsymbol{H}\Sigma\boldsymbol{H}'\boldsymbol{H}\boldsymbol{P}'=\boldsymbol{P}\Sigma\boldsymbol{P}'=\lambda(\Sigma)$, $\boldsymbol{P}\boldsymbol{H}'v$

$= \boldsymbol{Q}\nu = \boldsymbol{P}\mu$ and $\nu = \boldsymbol{H}\mu$, i.e., $(\mu, \Sigma) \sim (\nu, \boldsymbol{V}) \pmod{G}$. □

Remark. In this example, it is easy to see that $\mu'\mu$, $\mu'\Sigma^{-1}\mu$ and $\lambda(\Sigma)$ are invariants, but they are not the maximal invariants.

Similarly, we can get the following example whose proof is left to the reader.

Example 1.7.5. Let $\mathfrak{X} = \{(\mu, \Sigma) | \mu \in R^n, \Sigma > 0 \text{ and } \Sigma: n \times n\}$ and $G = \{(a, \boldsymbol{b}, \boldsymbol{H}) | a \neq 0, \boldsymbol{b} \in R^n \text{ and } \boldsymbol{H} \in O(n)\}$. The group of transformations on $\mathfrak{X}$ is given by

$$\mu \to a\boldsymbol{H}\mu + \boldsymbol{b} \quad \text{and} \quad \Sigma \to a^2 \boldsymbol{H}\Sigma\boldsymbol{H}'.$$

Under this group of transformations, a maximal invariant is

$$(\lambda_1/\lambda_m, \ldots, \lambda_{m-1}/\lambda_m),$$

where $\lambda_1 \geqslant \lambda_2 \geqslant \ldots \geqslant \lambda_m > 0$ are the eigenvalues of Σ.

Example 1.7.6. Let $\mathfrak{X}$ be the same as the last one and $G = \{\boldsymbol{B} | \boldsymbol{B}$ is an $n \times n$ nonsingular matrix$\}$. The action of G on $\mathfrak{X}$ is

$$\mu \to \boldsymbol{B}\mu \quad \text{and} \quad \Sigma \to \boldsymbol{B}\Sigma\boldsymbol{B}'$$

Then a maximal invariant is $\mu'\Sigma^{-1}\mu$.

Proof. Let $f(\mu, \Sigma) = \mu'\Sigma^{-1}\mu$. Clearly, it is invariant under G, because

$$f(\boldsymbol{B}\mu, \boldsymbol{B}\Sigma\boldsymbol{B}') = \mu'\boldsymbol{B}'(\boldsymbol{B}\Sigma\boldsymbol{B}')^{-1}\boldsymbol{B}\mu = \mu'\Sigma^{-1}\mu = f(\mu, \Sigma).$$

Assume

$$f(\mu, \Sigma) = f(\nu, \boldsymbol{V}),$$

that is $\mu'\Sigma^{-1}\mu = \nu'\boldsymbol{V}^{-1}\nu$. Then

$$(\mu'\Sigma^{-\frac{1}{2}})(\mu'\Sigma^{-\frac{1}{2}})' = (\nu'\boldsymbol{V}^{-\frac{1}{2}})(\nu'\boldsymbol{V}^{-\frac{1}{2}})'.$$

By (6) of Section 1.2, there exists $\boldsymbol{H} \in O(n)$ such that

$$\boldsymbol{H}\Sigma^{-\frac{1}{2}}\mu = \boldsymbol{V}^{-\frac{1}{2}}\nu.$$

Taking $\boldsymbol{B} = \boldsymbol{V}^{\frac{1}{2}}\boldsymbol{H}\Sigma^{-\frac{1}{2}}$, we get $\boldsymbol{B}\mu = \nu$ and $\boldsymbol{B}\Sigma\boldsymbol{B}' = \boldsymbol{V}$, i.e., $(\mu, \Sigma) \sim (\nu, \boldsymbol{V}) \pmod{G}$. Hence f is a maximal invariant. □

Example 1.7.7. Suppose that $\mathfrak{X}$ is the same as the last one and $G = \{(\boldsymbol{B}, \boldsymbol{c}) | \boldsymbol{B} = \text{diag}(b_1, \ldots, b_n) > 0, \boldsymbol{c} \in \mathrm{R}^n\}$. The group operation is defined by

$$(\boldsymbol{B}_1, \boldsymbol{c}_1)(\boldsymbol{B}_2, \boldsymbol{c}_2) = (\boldsymbol{B}_1\boldsymbol{B}_2, \boldsymbol{B}_1\boldsymbol{c}_2 + \boldsymbol{c}_1),$$

and the action of G on $\mathfrak{X}$ is given by

$$\mu \to \boldsymbol{B}\mu + \boldsymbol{c} \quad \text{and} \quad \Sigma \to \boldsymbol{B}\Sigma\boldsymbol{B}'.$$

Then a maximal invariant under G is $f(\mu, \Sigma) = \boldsymbol{R} = (r_{ij})$, where

$$r_{ij} = \frac{\sigma_{ij}}{(\sigma_{ii}\sigma_{jj})^{\frac{1}{2}}}, \; i, j = 1, \ldots, n,$$

and $\Sigma = (\sigma_{ij})$.

Proof. First, $\boldsymbol{R}$ is invariant because

$$f(\boldsymbol{B}\mu + \boldsymbol{c}, \boldsymbol{B}\Sigma\boldsymbol{B}') = \left(\frac{b_i b_j \sigma_{ij}}{(b_i^2\sigma_{ii}b_j^2\sigma_{jj})^{\frac{1}{2}}}\right) = (r_{ij}) = \boldsymbol{R} = f(\mu, \Sigma).$$

To show it is the maximal invariant, suppose $f(\mu, \Sigma) = f(\nu, \boldsymbol{V})$ and $\boldsymbol{V} = (v_{ij})$, i.e.,

$$\frac{\sigma_{ij}}{(\sigma_{ii}\sigma_{jj})^{\frac{1}{2}}} = \frac{v_{ij}}{(v_{ii}v_{jj})^{\frac{1}{2}}}, \; i, j = 1, \ldots, n.$$

Let $b_i = (v_{ii}/\sigma_{ii})^{\frac{1}{2}}$, $i=1, ..., n$ and $\boldsymbol{c} = -\boldsymbol{B}\mu + v$. We have

$$\boldsymbol{V} = \boldsymbol{B}\Sigma\boldsymbol{B}' \qquad \text{and} \qquad v = \boldsymbol{B}\mu + \boldsymbol{c}$$

so that $(\mu, \Sigma) \sim (v, V)$ (mod G). Hence $\boldsymbol{R}$ is a maximal invariant. □

More generally, we get the following example.

Example 1.7.8. Let $\mathfrak{X} = \{(\mu_1, \mu_2, \Sigma_1, \Sigma_2) | \mu_1 \in \mathrm{R}^n, \mu_2 \in \mathrm{R}^n, \Sigma_1 > 0, \Sigma_2 > 0: n \times n\}$ and $G = \{(\boldsymbol{c}, \boldsymbol{d}, \boldsymbol{B}) | \boldsymbol{c} \in \mathrm{R}^n, \boldsymbol{d} \in \mathrm{R}^n, \boldsymbol{B}$ is an $n \times n$ nonsingular matrix$\}$ where the group operation is

$$(\boldsymbol{B}_1, \boldsymbol{c}_1, \boldsymbol{d}_1)(\boldsymbol{B}_2, \boldsymbol{c}_2, \boldsymbol{d}_2) = (\boldsymbol{B}_1\boldsymbol{B}_2, \boldsymbol{B}_1\boldsymbol{c}_2 + \boldsymbol{c}_1, \boldsymbol{B}_1\boldsymbol{d}_2 + \boldsymbol{d}_1).$$

The action of G on $\mathfrak{X}$ is

$$(\mu_1, \mu_2) \to (\boldsymbol{B}\mu_1 + \boldsymbol{c}, \boldsymbol{B}\mu_2 + \boldsymbol{d}) \quad \text{and} \quad (\Sigma_1, \Sigma_2) \to (\boldsymbol{B}\Sigma_1\boldsymbol{B}', \boldsymbol{B}\Sigma_2\boldsymbol{B}').$$

Under the group of transformations G a maximal invariant is $(\lambda_1, ..., \lambda_n)$, where $\lambda_1 \geqslant \lambda_2 \geqslant ... \geqslant \lambda_n > 0$ are the eigenvalues of $\Sigma_1\Sigma_2^{-1}$.

Proof. Let $f(\mu_1, \mu_2, \Sigma_1, \Sigma_2) = (\lambda_1, ..., \lambda_n)$. It is easy to see that f is invariant. Suppose $f(\mu_1, \mu_2, \Sigma_1, \Sigma_2) = f(v_1, v_2, \boldsymbol{V}_1, \boldsymbol{V}_2)$, i.e., $\Sigma_1\Sigma_2^{-1}$ and $\boldsymbol{V}_1\boldsymbol{V}_2^{-1}$ have the same eigenvalues $\Lambda = \mathrm{diag}(\lambda_1, ..., \lambda_n)$. By (10) of Section 1.2 there exist nonsingular matrices $\boldsymbol{B}_1$ and $\boldsymbol{B}_2$ such that

$$\boldsymbol{B}_1\Sigma_1\boldsymbol{B}_1' = \Lambda \qquad \boldsymbol{B}_1\Sigma_2\boldsymbol{B}_1' = \boldsymbol{I}_n,$$
$$\boldsymbol{B}_2\boldsymbol{V}_1\boldsymbol{B}_2' = \Lambda \qquad \boldsymbol{B}_2\boldsymbol{V}_2\boldsymbol{B}_2' = \boldsymbol{I}_n.$$

Then

$$\boldsymbol{V}_1 = \boldsymbol{B}_2^{-1}\Lambda\boldsymbol{B}_2' = \boldsymbol{B}_2^{-1}\boldsymbol{B}_1\Sigma_1\boldsymbol{B}_1'\boldsymbol{B}_2'^{-1} = \boldsymbol{B}\Sigma_1\boldsymbol{B}',$$

and

$$\boldsymbol{V}_2 = \boldsymbol{B}_2^{-1}\boldsymbol{B}_2'^{-1} = \boldsymbol{B}_2^{-1}\boldsymbol{B}_1\Sigma_2\boldsymbol{B}_1'\boldsymbol{B}_2'^{-1} = \boldsymbol{B}\Sigma_2\boldsymbol{B}',$$

where $\boldsymbol{B} = \boldsymbol{B}_2^{-1}\boldsymbol{B}_1$. Putting $\boldsymbol{c} = -\boldsymbol{B}\mu_1 + v_1$ and $\boldsymbol{d} = -\boldsymbol{B}\mu_2 + v_2$ we get $(\mu_1, \mu_2, \Sigma_1, \Sigma_2) \sim (v_1, v_2, \boldsymbol{V}_1, \boldsymbol{V}_2)$, i.e., Λ is a maximal invariant. □

References

Bellman (1970), Graham (1981), Muirhead (1982), Rao (1973), Srivastava and Khatri (1979), Zhang and Fang (1982).

Exercises 1

1.1. Show that $|\boldsymbol{I}_p + \boldsymbol{AB}| = |\boldsymbol{I}_q + \boldsymbol{BA}|$ where $\boldsymbol{A}: p \times q$ and $\boldsymbol{B}: q \times p$. (Hint: Taking

$$\boldsymbol{V} = \begin{bmatrix} \boldsymbol{I}_p & \boldsymbol{A} \\ \boldsymbol{B} & \boldsymbol{I}_q \end{bmatrix}$$

and using (1.1.7) and (1.1.9))

1.2. Prove that if $\boldsymbol{A} = (a_{ij}) \in LT(p)$ then $\boldsymbol{A}^{-1} = (a^{ij}) \in LT(p)$. Find the formula expressing a^{ij} in terms of a_{ij}.

1.3. Prove the formula (1.1.10).

1.4. Show that $\mathrm{rk}(\boldsymbol{A}) = \mathrm{rk}(\boldsymbol{A}'\boldsymbol{A}) = \mathrm{rk}(\boldsymbol{AA}')$.

1.5. Let $\boldsymbol{A}$ be an $n \times n$ square matrix such that $\boldsymbol{A}^m = \boldsymbol{A}$ and $\boldsymbol{A}^i \neq \boldsymbol{A}$ for $1 \leqslant i < m$. Find the eigenvalues of $\boldsymbol{A}$. $\boldsymbol{A}$ is called idempotent, tripotent or nilpotent according to $i = 2$, $i = 3$ or $i > 3$.

1.6. If $\boldsymbol{A} > 0$, then all the principals square submatrices of $\boldsymbol{A}$ are positive definite matrices.

1.7. Show that if $\boldsymbol{A} > \boldsymbol{B}$, then $\lambda_i > \mu_i$, $i = 1, \ldots, n$, where $\lambda_1 \geqslant \lambda_2 \geqslant ... \geqslant \lambda_n > 0$ and $\mu_1 \geqslant \mu_2 \geqslant ... \geqslant \mu_n$ are the eigenvalues of $\boldsymbol{A}$ and $\boldsymbol{B}$, respectively.

1.8. Prove that the nonzero eigenvalues of $\boldsymbol{AB}$ and $\boldsymbol{BA}$ are the same. Give the relationship between eigenvectors of $\boldsymbol{AB}$ and $\boldsymbol{BA}$.

1.9. Suppose $A > 0$ with eigenvalues $\lambda_1 \geqslant \lambda_2 \geqslant \dots \geqslant \lambda_n$. Prove that

$$\sup_{x \neq 0} \frac{x'Ax}{x'x} = \lambda_1 \text{ and } \inf_{x \neq 0} \frac{x'Ax}{x'x} = \lambda_n.$$

1.10. Suppose that A and B are $n \times n$ matrices, $A' = A$ and $B > 0$. Show that

$$\sup_{x \neq 0} \frac{x'Ax}{x'Bx} = \lambda_1 \text{ and } \inf_{x \neq 0} \frac{x'Ax}{x'Bx} = \lambda_n,$$

where $\lambda_1 \geqslant \lambda_2 \dots \geqslant \lambda_n$ are the eigenvalues of AB^{-1}.

1.11. Let A be an $n \times m$ matrix of rank k, $\lambda_1, \dots, \lambda_k$ be the nonzero eigenvalues and $h_1, \dots, h_k$ be the corresponding eigenvectors of $A'A$. Prove that $\lambda_1^{-1}h_1h_1' + \dots + \lambda_k^{-1}h_kh_k'$ is a generalized inverse of $A'A$.

1.12. Let A and B be two $m \times n$ matrices. Then

$$\operatorname{tr}[K_{mn}(A' \otimes B)] = \operatorname{tr}(A'B) = (\operatorname{vec}(A))'(\operatorname{vec}(B)).$$

1.13. Prove that $\operatorname{tr}(K_{mn}) = 1 + d(m-1, n-1)$ where $d(m, n)$ denotes the maximum common factor of m and n and $d(0, n) = d(n, 0) = n$.

1.14. Prove that K_{mn} has $\frac{1}{2}n(n+1)$ 1's and $\frac{1}{2}n(n-1)$ -1's as its eigenvalues. Thus $|K_{mn}| = (-1)^{\frac{1}{2}n(n-1)}$.

1.15. Find a generalized inverse of $\begin{bmatrix} A'A & B \\ B' & O \end{bmatrix}$.

1.16. If $A = \begin{bmatrix} A_{11} & A_{12} \\ A_{21} & A_{22} \end{bmatrix} > 0$, then $A_{22}A_{22}^{-}A_{21} = A_{21}$ for any choice of the generalized inverse.

1.17. Let A and B be $m \times m$ and $n \times n$ matrices respectively. Define $A \oplus B = A \otimes I_n + I_m \otimes B$. Let $\{\lambda_1, \dots, \lambda_m\}$ and $\{\mu_1, \dots, \mu_n\}$ be the eigenvalues of A and B respectively, and $\{x_1, \dots, x_m\}$ and $\{y_1, \dots, y_n\}$ be the corresponding eigenvectors. Show that $\{\lambda_i + \mu_j,\ i = 1, \dots, m;\ j = 1, \dots, n\}$ are all the eigenvalues of $A \oplus B$ and $\{x_i \otimes y_j,\ i = 1, \dots, m;\ j = 1, \dots, n\}$ are the associated eigenvectors. Thus

$$|A \oplus B| = \prod_{i=1}^{m} \prod_{j=1}^{n} (\lambda_i + \mu_j).$$

1.18. Calculate that

$$\frac{\partial \operatorname{tr}(X'AXB)}{\partial X} = AXB + A'XB'.$$

1.19. By using Theorem 1.5.1. and

$$\frac{\partial X^n}{\partial x_{ij}} = \sum_{k=0}^{n-1} X^k E_{ij} X^{n-k-1},$$

show that

$$\frac{\partial y_{ij}}{\partial X} = \sum_{k=0}^{n-1} (X')^k E_{ij} (X')^{n-k-1},$$

where $Y = X^n$.

1.20. Let $X = (x_{ij})$ and $Y = (y_{ij})$ be $m \times n$ and $p \times q$ matrices. Define the derivative $\partial Y / \partial X$ of Y with respect to X to be an $(m+p) \times (n+q)$ matrix as follows:

$$\frac{\partial \boldsymbol{Y}}{\partial \boldsymbol{X}} = \boldsymbol{K}_{pm} \begin{bmatrix} \frac{\partial \boldsymbol{Y}}{\partial x_{11}} & \cdots & \frac{\partial \boldsymbol{Y}}{x_{1n}} \\ \vdots & & \vdots \\ \frac{\partial \boldsymbol{Y}}{\partial x_{m1}} & \cdots & \frac{\partial \boldsymbol{Y}}{\partial x_{mn}} \end{bmatrix} = \boldsymbol{K}_{pm} \sum_{i=1}^{m} \sum_{j=1}^{n} (\boldsymbol{E}_{ij}(m,n) \otimes \frac{\partial \boldsymbol{Y}}{\partial x_{ij}}).$$

Show that

(1) $\dfrac{\partial(\boldsymbol{Y}+\boldsymbol{Z})}{\partial \boldsymbol{X}} = \dfrac{\partial \boldsymbol{Y}}{\partial \boldsymbol{X}} + \dfrac{\partial \boldsymbol{Z}}{\partial \boldsymbol{X}}$.

(2) $\dfrac{\partial(\boldsymbol{X}\boldsymbol{Y})}{\partial \boldsymbol{Z}} = \dfrac{\partial \boldsymbol{X}}{\partial \boldsymbol{Z}}(\boldsymbol{I} \otimes \boldsymbol{Y}) + (\boldsymbol{X} \otimes \boldsymbol{I})\dfrac{\partial \boldsymbol{Y}}{\partial \boldsymbol{Z}}$.

(3) If $\boldsymbol{X}$, $\boldsymbol{Y}$ and $\boldsymbol{Z}$ are $m \times n$, $u \times v$ and $p \times q$ matrices, then

$$\frac{\partial(\boldsymbol{X} \otimes \boldsymbol{Y})}{\partial \boldsymbol{Z}} = (\boldsymbol{I}_m \otimes \boldsymbol{K}_{up})(\frac{\partial \boldsymbol{X}}{\partial \boldsymbol{Z}} \otimes \boldsymbol{Y}) + (\boldsymbol{X} \otimes \frac{\partial \boldsymbol{Y}}{\partial \boldsymbol{Z}})(\boldsymbol{K}_{nq} \otimes \boldsymbol{I}_v).$$

(4) $\dfrac{\partial \boldsymbol{X}'}{\partial \boldsymbol{X}} = \boldsymbol{I}_{mn}$.

(5) $\dfrac{\partial(\boldsymbol{X}^{-1})'}{\partial \boldsymbol{X}} = -(\boldsymbol{X}^{-1} \otimes \boldsymbol{X}^{-1})'$.

(6) $\dfrac{\partial(\boldsymbol{A}\boldsymbol{X}'\boldsymbol{B})}{\partial \boldsymbol{X}} = \boldsymbol{A} \otimes \boldsymbol{B}$.

1.21. Let $\boldsymbol{Y}$ and $\boldsymbol{X}$ be upper triangular matrices and $\boldsymbol{Y} = \boldsymbol{X}^{-1}$. Find $J(\boldsymbol{X} \to \boldsymbol{Y})$.

1.22. Let $\boldsymbol{X}$ be an $n \times p$ matrix and $\boldsymbol{X} = \boldsymbol{U}\boldsymbol{A}$, where $\boldsymbol{A} \in UT(p)$ with positive diagonal elements, $\boldsymbol{U}: n \times p$ and $\boldsymbol{U}'\boldsymbol{U} = \boldsymbol{I}_p$. Then show that

$$J(\boldsymbol{X} \to (\boldsymbol{U}, \boldsymbol{A})) = (\prod_{i=1}^{p} a_{ii}^{n-i}) g_{n,p}(\boldsymbol{U})$$

where $g_{n,p}(\boldsymbol{U})$ is a function of the elements of $\boldsymbol{U}$.

1.23. Let $\boldsymbol{Y} = \boldsymbol{X}^{-1}$ be an $n \times n$ matrix. Show that

$$J(\boldsymbol{Y} \to \boldsymbol{X}) = |\boldsymbol{X}|^{-2n},$$

and

$$J(\boldsymbol{Y} \to \boldsymbol{X}) = |\boldsymbol{X}|^{-(n+1)} \qquad \text{if } \boldsymbol{X}' = \boldsymbol{X}.$$

1.24. Suppose that $\mathfrak{X} = \{\boldsymbol{U} \mid \boldsymbol{U}: n \times p,\ \boldsymbol{U}'\boldsymbol{U} = \boldsymbol{I}_p\}$, the Stiefel manifold of $n \times p$ matrices with orthonormal columns, and $G = O(n)$. The action of $\boldsymbol{H} \in O(n)$ on $\boldsymbol{U} \in \mathfrak{X}$ is given by

$$\boldsymbol{U} \to \boldsymbol{H}\boldsymbol{U},$$

with the group operation being matrix multiplication. Find what a maximal invariant under G is.

1.25. Show that $\boldsymbol{A}^- = \boldsymbol{Q}(\boldsymbol{P}\boldsymbol{A}\boldsymbol{Q})^-\boldsymbol{P}$, where $\boldsymbol{P}$ and $\boldsymbol{Q}$ are nonsingular.

1.26. Show that $\begin{bmatrix} \boldsymbol{A} & \boldsymbol{O} \\ \boldsymbol{O} & \boldsymbol{B} \end{bmatrix}^- = \begin{bmatrix} \boldsymbol{A}^- & \boldsymbol{X} \\ \boldsymbol{Y} & \boldsymbol{B}^- \end{bmatrix}$ where $\boldsymbol{X}$ and $\boldsymbol{Y}$ can be arbitrary up to $\boldsymbol{A}\boldsymbol{X}\boldsymbol{B} = \boldsymbol{O}$ and $\boldsymbol{B}\boldsymbol{Y}\boldsymbol{A} = \boldsymbol{O}$.

(Remark. The equation $\begin{bmatrix} \boldsymbol{A} & \boldsymbol{O} \\ \boldsymbol{A} & \boldsymbol{B} \end{bmatrix}^- = \begin{bmatrix} \boldsymbol{A}^- & \boldsymbol{X} \\ \boldsymbol{Y} & \boldsymbol{B}^- \end{bmatrix}$ holds in the sense that for any given $\begin{bmatrix} \boldsymbol{A} & \boldsymbol{O} \\ \boldsymbol{O} & \boldsymbol{B} \end{bmatrix}_0^-$ there exist $\boldsymbol{A}_0^-$, $\boldsymbol{B}_0^-$, $\boldsymbol{X}$, and $\boldsymbol{Y}$ such that the equation holds.)

1.27. If A_{11} is nonsingular, show that

$$\begin{bmatrix} A_{11} & A_{12} \\ A_{21} & A_{22} \end{bmatrix}^{-} = \begin{bmatrix} A_{11}^{-1} - A_{11}^{-1}A_{12}Y - XA_{21}A_{11}^{-1} & X \\ Y & O \end{bmatrix} + \begin{bmatrix} A_{11}^{-1}A_{12} \\ -I \end{bmatrix} A_{22.1}^{-}(A_{21}A_{11}^{-1}, -I),$$

where X and Y can be arbitrary up to $XA_{22.1} = O$ and $A_{22.1}Y = O$.

1.28. If $A \geqslant 0$, show that the formula in Exercise 1.27 holds, but X and Y satisfy $A_{11}XA_{22.1} = O$ and $A_{22.1}YA_{11} = O$.

CHAPTER II

ELLIPTICALLY CONTOURED DISTRIBUTIONS

In this chapter elliptically contoured distributions (ECD) which form the base of generalized multivariate analysis are defined and their properties discussed. As specific elements of ECD, multivariate normal distributions and spherical distributions are, in advance, treated in Sections 3 and 5. In Section 7 some characterizations of normality will be given. There is a close relationship between spherical distributions and Dirichlet distributions. The latter are discussed in Section 4. Sections 8 and 9 show how to calculate distributions of quadratic forms, and some generalized non–central distributions (Chi squared distributions, generalized non–central t–distributions and generalized non–central F–distributions) respectively. Section 1 and Section 2 introduce some elementary concepts and tools for the whole book.

2.1. Multivariate Distributions

2.1.1. Multivariate Cumulative Distribution Function (mcdf)

Let $\boldsymbol{x} = (X_1, \dots, X_p)'$ be a $p \times 1$ random vector. Its joint distribution is defined by

$$F(\boldsymbol{x}) \equiv F(x_1, \dots, x_p) \equiv P\{X_1 \leqslant x_1, \dots, X_p \leqslant x_p\}.$$

Clearly, every multivariate cumulative distribution function (mcdf) has the following properties:

(1) $F(\boldsymbol{x})$ is monotone nondecreasing and right–continuous in each component of $\boldsymbol{x}$;

(2) $0 \leqslant F(\boldsymbol{x}) \leqslant 1$;

(3) $F(-\infty, x_2, \dots, x_p) = F(x_1, -\infty, x_3, \dots, x_p) = \dots = F(x_1, \dots, x_{p-1}, -\infty) = 0$;

(4) $F(+\infty, \dots, +\infty) = 1$.

2.1.2. Density

Given a mcdf $F(\boldsymbol{x})$ if there exists a nonnegative function $f(\boldsymbol{x})$ such that

$$F(\boldsymbol{x}) = \int_{-\infty}^{x_1} \dots \int_{-\infty}^{x_p} f(\boldsymbol{t})\, d\boldsymbol{t} \tag{2.1.1}$$

for every $\boldsymbol{x} = (x_1, \dots, x_p)' \in R^p$, we say that the mcdf $F(\boldsymbol{x})$ has a *density* $f(\boldsymbol{x})$. Every density $f(\boldsymbol{x})$ has the following properties:

(1) $f(\boldsymbol{x}) \geqslant 0$;

(2) $\int_{R^p} f(\boldsymbol{x})\, d\boldsymbol{x} = 1$;

(3) For every Borel set B in R^p, we have

$$P\{\boldsymbol{x}\in B\} = \int_B f(\boldsymbol{x})\,d\boldsymbol{x}. \tag{2.1.2}$$

In particular,

$$P\{a_i \leqslant X_i \leqslant b_i,\ i=1,\dots,p\} = \int_{a_1}^{b_1}\cdots\int_{a_p}^{b_p} f(\boldsymbol{x})\,d\boldsymbol{x}.$$

(4) For every continuous point $\boldsymbol{x}$ of $f(\boldsymbol{x})$, we have

$$f(x_1,\dots,x_p) = \frac{\partial^p F(x_1,\dots,x_p)}{\partial x_1 \partial x_2 \dots \partial x_p}, \tag{2.1.3}$$

where $F(\boldsymbol{x})$ is the mcdf of $f(\boldsymbol{x})$.

2.1.3. Marginal Distributions

Let $\boldsymbol{x}$ be a p-dimensional random vector and $\boldsymbol{x}^{(1)}$ be its subvector with q components of $\boldsymbol{x}$. The distribution of $\boldsymbol{x}^{(1)}$ is called the *marginal distribution of* $\boldsymbol{x}$.

Without loss of generality we can assume $\boldsymbol{x} = \begin{bmatrix}\boldsymbol{x}^{(1)}\\ \boldsymbol{x}^{(2)}\end{bmatrix}$, i.e., the components of $\boldsymbol{x}^{(1)}$ are the first q components of $\boldsymbol{x}$. Then the marginal distribution of $\boldsymbol{x}^{(1)}$ is $F(x_1,\dots,x_q, +\infty,\dots,+\infty)$.

If $\boldsymbol{x}$ has a density $f(\boldsymbol{x})$, then $\boldsymbol{x}^{(1)}$ has a density, too. As

$$F(x_1,\dots,x_q,+\infty,\dots,+\infty) = \int_{-\infty}^{x_1}\cdots\int_{-\infty}^{x_q}\int_{-\infty}^{+\infty}\cdots\int_{-\infty}^{+\infty} f(y_1,\dots,y_p)\,dy_1\dots dy_p,$$

the marginal density of $\boldsymbol{x}^{(1)}$ is

$$f^{(1)}(x_1,\dots,x_q) = \int_{-\infty}^{\infty}\cdots\int_{-\infty}^{\infty} f(x_1,\dots,x_p)\,dx_{q+1}\dots dx_p. \tag{2.1.4}$$

2.1.4. Conditional Distributions

In probability theory it is a well-known fact that if A and B are two events, the *conditional probability* of B, given that A has occurred, is given by

$$P\{B\,|\,A\} = \frac{P\{AB\}}{P\{A\}} \tag{2.1.5}$$

provided $P\{A\} \neq 0$. Now let A and B *be the events* $A = \{a \leqslant X \leqslant b\}$ and $B = \{c \leqslant Y \leqslant d\}$, where X and Y are two random variables. We have

$$P\{a \leqslant X \leqslant b\,|\,c \leqslant Y \leqslant d\} = \frac{P\{a \leqslant X \leqslant b,\ c \leqslant Y \leqslant d\}}{P\{c \leqslant Y \leqslant d\}},$$

and

(2.1.6)
$$P\{a \leqslant X \leqslant b \mid c \leqslant Y \leqslant d\} = \frac{\int_a^b \int_c^d f(x,y)\,dx\,dy}{\int_c^d g(y)dy},$$

where $f(x,y)$ is the joint density of X and Y and $g(y)$ is the marginal density of Y (if they exist), provided $P\{c \leqslant Y \leqslant d\} \neq 0$. In the expression, the conditional density of X, given $Y=y$, is defined as

(2.1.7)
$$f(x \mid y) = \frac{f(x,y)}{g(y)}.$$

In general, if $\boldsymbol{x} = \begin{bmatrix} \boldsymbol{x}^{(1)} \\ \boldsymbol{x}^{(2)} \end{bmatrix}$ and $\boldsymbol{x}^{(2)}$ has densities $f(\boldsymbol{x}^{(1)}, \boldsymbol{x}^{(2)})$ and $g(\boldsymbol{x}^{(2)})$ respectively, then the conditional density of $\boldsymbol{x}^{(1)}$, given $\boldsymbol{x}^{(2)}$, is

(2.1.8)
$$f(\boldsymbol{x}^{(1)} \mid \boldsymbol{x}^{(2)}) = \frac{f(\boldsymbol{x}^{(1)}, \boldsymbol{x}^{(2)})}{g(\boldsymbol{x}^{(2)})}.$$

When $\boldsymbol{x}$ does not have a density, we can still define the conditional distribution of $\{\boldsymbol{x}^{(1)} \mid \boldsymbol{x}^{(2)}\}$. Its definition and properties can be found in many textbooks, for example in Loève's book (1960).

2.1.5. Independence

Definition 2.1.1. Two random vectors $\boldsymbol{x}: p \times 1, \boldsymbol{y}: q \times 1$ are said to be *independent* if $F(\boldsymbol{x},\boldsymbol{y}) = G(\boldsymbol{x})H(\boldsymbol{y})$ where $F(\boldsymbol{x},\boldsymbol{y})$, $G(\boldsymbol{x})$ and $H(\boldsymbol{y})$ are the mcdf's of $(\boldsymbol{x},\boldsymbol{y})$, $\boldsymbol{x}$ and $\boldsymbol{y}$, respectively. If they have the densities $f(\boldsymbol{x},\boldsymbol{y})$, $g(\boldsymbol{x})$ and $h(\boldsymbol{y})$ respectively, then $\boldsymbol{x}$ and $\boldsymbol{y}$ are independent iff

(2.1.9)
$$f(\boldsymbol{x},\boldsymbol{y}) = g(\boldsymbol{x})h(\boldsymbol{y}),$$

or

(2.1.10)
$$f(\boldsymbol{x} \mid \boldsymbol{y}) = g(\boldsymbol{x}).$$

Definition 2.1.2. Suppose that $F(x_1, \dots, x_p)$ is the distribution of a random vector $\boldsymbol{x} = (X_1, \dots, X_p)'$ and $F_j(x_j)$, $j = 1, \dots, p$, are the marginal distributions. The random variables $X_1, \dots, X_p$ are said to be *independent* if

(2.1.11)
$$F(x_1, \dots, x_p) = \prod_{j=1}^{p} F_j(x_j).$$

It is a well-known fact that independence of $X_1, \dots, X_p$ implies independence between X_i and X_j for any $i \neq j$, but the converse is not always true. There are some examples that X_i and X_j are independent for any $i \neq j$, but $X_1, \dots, X_p$ are not independent (cf. Exercise 2.1).

2.1.6. Characteristic Functions

Definition 2.1.3. Let $\boldsymbol{x}$ be a $p \times 1$ random vector with a mcdf $F(\boldsymbol{x})$ and $\boldsymbol{t}$ be a $p \times 1$ constant vector. The *characteristic function* (c.f.) of $\boldsymbol{x}$ is defined as

(2.1.12) $$\phi_{\boldsymbol{x}}(\boldsymbol{t}) = \int_{-\infty}^{\infty} \cdots \int_{-\infty}^{\infty} e^{i\boldsymbol{t}'\boldsymbol{x}}\, dF(\boldsymbol{x}),$$

where $i = \sqrt{-1}$.

Let us recall the univariate analysis. Then it shows that the following properties are useful:

(a) $\phi_{\boldsymbol{x}}(\boldsymbol{0}) = 1$;

(b) $|\phi_{\boldsymbol{x}}(\boldsymbol{t})| \leqslant 1$;

(c) $\phi_{\boldsymbol{x}}(\boldsymbol{t})$ is uniformly continuous on R^p;

(d) If $\boldsymbol{x}$ and $\boldsymbol{y}$ are independent random vectors with c.f.'s $\phi_{\boldsymbol{x}}(\boldsymbol{t})$ and $\phi_{\boldsymbol{y}}(\boldsymbol{t})$, then the c.f. of $(\boldsymbol{x} + \boldsymbol{y})$ is

(2.1.13) $$\phi_{\boldsymbol{x}+\boldsymbol{y}}(\boldsymbol{t}) = \phi_{\boldsymbol{x}}(\boldsymbol{t})\, \phi_{\boldsymbol{y}}(\boldsymbol{t}).$$

For example, if the components $\{X_j\}$ of $\boldsymbol{x}$ are independent, then

(2.1.14) $$\phi_{\boldsymbol{x}}(t_1, \ldots, t_p) = \phi_{\boldsymbol{x}}(\boldsymbol{t}) = \prod_{j=1}^{p} \phi_{X_j}(t_j).$$

(e) Let $\boldsymbol{x}$ and $\boldsymbol{y}$ be two random vectors with respective mcdf's $F_1(\boldsymbol{x})$ and $F_2(\boldsymbol{y})$, and respective c.f.'s $\phi_{\boldsymbol{x}}(\boldsymbol{t})$ and $\phi_{\boldsymbol{y}}(\boldsymbol{t})$. Then $F_1(\boldsymbol{x})$ and $F_2(\boldsymbol{y})$ are identical if and only if $\phi_{\boldsymbol{x}}(\boldsymbol{t})$ and $\phi_{\boldsymbol{y}}(\boldsymbol{t})$ are identical. This means that a mcdf in R^p is uniquely determined by its c.f.

(f) Assume that $\phi_{\boldsymbol{x}}(\boldsymbol{t})$ is the c.f. of $\boldsymbol{x}$ and $\boldsymbol{y} = \boldsymbol{A}\boldsymbol{x} + \boldsymbol{a}$. Then the c.f. of $\boldsymbol{y}$ is $\exp(i\boldsymbol{a}'\boldsymbol{t})\phi_{\boldsymbol{x}}(\boldsymbol{A}'\boldsymbol{t})$.

(g) If $\boldsymbol{x}$ is a $p \times 1$ random vector, then its distribution is uniquely determined by the distributions of $\boldsymbol{a}'\boldsymbol{x}$ for every $\boldsymbol{a} \in \mathrm{R}^p$.

Proof. Since the c.f. of $\boldsymbol{a}'\boldsymbol{x}$ is

$$\phi_{\boldsymbol{a}}(t) = E(e^{ita'\boldsymbol{x}}),$$

so

$$\phi_{\boldsymbol{a}}(1) = E(e^{ia'\boldsymbol{x}}),$$

is the c.f. of $\boldsymbol{x}$ as a function of $\boldsymbol{a}$. The assertion then follows from (e). □

Definition 2.1.4. We say that a c.f. belongs to the class $(\bar{U})$ if it can equal another c.f. in the neighborhood of zero without equalling it identically. A c.f. is said to belong to (U) if it does not belong to $(\bar{U})$.

Hsu (1954, 1983) reviewed the known examples of c.f.'s belonging to $(\bar{U})$ and gave further examples and theorems. Marcinkiewicz (1938) showed that the c.f. of a nonnegative (or alternatively a nonpositive) random variable belongs to (U) (cf. Theorem 3). Zygmund (1951) gave, for a c.f. to belong to (U), the sufficient condition that the corresponding distribution function $F(x)$ should satisfy the condition

(2.1.15) $$\int_{-\infty}^{0} e^{-rx}\, dF(x) < \infty \qquad \left(\text{or} \int_{0}^{\infty} e^{rx}\, dF(x) < \infty\right),$$

where r is a certain positive constant. Essen (1945) proved that if the distribution has finite moments $\mu_r'(r = 1, 2, \ldots)$ and is determined by these moments $\{\mu_r'\}$ uniquely, then the corresponding c.f. belongs to (U).

2.1.7. The Operation "$\overset{d}{=}$"

Definition 2.1.5. If random vectors $\boldsymbol{x}$ and $\boldsymbol{y}$ have the same distribution, we denote this fact by $\boldsymbol{x} \overset{d}{=} \boldsymbol{y}$.

The operation "$\overset{d}{=}$" will play an important role in this book. The following are some basic facts:

(a) Assume that $\boldsymbol{x} \overset{d}{=} \boldsymbol{y}$ and $f_j(\cdot)$, $j = 1, \dots, m$, are Borel functions. Then

$$\begin{bmatrix} f_1(\boldsymbol{x}) \\ \vdots \\ f_m(\boldsymbol{x}) \end{bmatrix} \overset{d}{=} \begin{bmatrix} f_1(\boldsymbol{y}) \\ \vdots \\ f_m(\boldsymbol{y}) \end{bmatrix}, \tag{2.1.16}$$

because

$$\phi_{f_1(\boldsymbol{x}), \dots, f_m(\boldsymbol{x})}(t_1, \dots, t_m) = \int e^{i(t_1 f_1(\boldsymbol{x}) + \cdots + t_m f_m(\boldsymbol{x}))} dF_1(\boldsymbol{x})$$

$$= \int e^{i(t_1 f_1(\boldsymbol{y}) + \cdots + t_m f_m(\boldsymbol{y}))} dF_2(\boldsymbol{y}) = \phi_{f_1(\boldsymbol{y}), \dots, f_m(\boldsymbol{y})}(t_1, \dots, t_m)$$

where $F_1(\boldsymbol{x})$ and $F_2(\boldsymbol{y})$ are mcdf's of $\boldsymbol{x}$ and $\boldsymbol{y}$ respectively. For instance, we have

$$\begin{bmatrix} \boldsymbol{x}'\boldsymbol{A}_1\boldsymbol{x} \\ \vdots \\ \boldsymbol{x}'\boldsymbol{A}_m\boldsymbol{x} \end{bmatrix} \overset{d}{=} \begin{bmatrix} \boldsymbol{y}'\boldsymbol{A}_1\boldsymbol{y} \\ \vdots \\ \boldsymbol{y}'\boldsymbol{A}_m\boldsymbol{y} \end{bmatrix}$$

where $\boldsymbol{A}_1, \dots, \boldsymbol{A}_m$ are constant matrices.

(b) Assume that X, Y, Z and W are random variables, X and Z are independent, and Y and W are also independent. Then

(1) $X \overset{d}{=} Y$ and $Z \overset{d}{=} W$ implies $X + Z \overset{d}{=} Y + W$.

(2) If $Z \overset{d}{=} W$ and $\phi_Z(t) \neq 0$ for almost all t or if $\phi_X(t) \in (U)$, then $X + Z \overset{d}{=} Y + W$ implies $X \overset{d}{=} Y$.

Proof. By (2.1.13) the assertion (1) is trivial. Assuming $Z \overset{d}{=} W$ and $X + Z \overset{d}{=} Y + W$, we wish to prove $X \overset{d}{=} Y$. As

$$\phi_X(t)\,\phi_Z(t) = \phi_{X+Z}(t) = \phi_{Y+W}(t) = \phi_Y(t)\phi_W(t), \tag{2.1.17}$$

if $\phi_Z(t)(= \phi_W(t)) \neq 0$ for almost all t, then $\phi_X(t) = \phi_Y(t)$ for almost all t, i.e., $X \overset{d}{=} Y$. If $\phi_X(t) \in (U)$, by (2.1.17) and the continuity of c.f. at $t = 0$, there exists a $\delta > 0$ such that

$$\phi_X(t) = \phi_Y(t) \qquad \text{for } 0 \leqslant t \leqslant \delta.$$

The assertion (2) follows from $\phi_X(t) \in (U)$. □

(c) Assume that X, Y and Z are random variables and Z is independent of X and of

Y, respectively. Then

(1) $X \stackrel{d}{=} Y$ implies $ZX \stackrel{d}{=} ZY$.

(2) If

$$P(X>0)=P(Y>0)=P(Z>0)=1 \tag{2.1.18}$$

and $\phi_{\log X}(t) \in (U)$ or $\phi_{\log Z}(t) \neq 0$ for almost all t, then $ZX \stackrel{d}{=} ZY$ implies $X \stackrel{d}{=} Y$.

(3) If $P(Z>0)=1$ and $\phi_{\log Z}(t) \neq 0$ for almost all t, then $ZX \stackrel{d}{=} ZY$ implies $X \stackrel{d}{=} Y$.

(4) If $P(Z>0)=1$, the c.f. of $\{\log X \mid X>0\}$ and the c.f. of $\{\log(-X) \mid X<0\}$ belong to (U), then $ZX \stackrel{d}{=} ZY$ implies $X \stackrel{d}{=} Y$.

Proof. Assertions (1) and (2) are trivial. We only prove (3). If X and Y satisfy (2.1.18), from assumption we have $\log Z + \log X \stackrel{d}{=} \log Z + \log Y$ and $\log X \stackrel{d}{=} \log Y$, i.e., $X \stackrel{d}{=} Y$.

Now we consider X and Y to be arbitrary random variables. Let

$$f_+(x)=\begin{cases} x & \text{if } x>0 \\ 0 & \text{if } x \leqslant 0 \end{cases}, \qquad f_-(x)=\begin{cases} 0 & \text{if } x \geqslant 0 \\ -x & \text{if } x<0 \end{cases}$$

and $X^+ = f_+(x)$, $X^- = f_-(x)$; $Y^+ = f_+(y)$ and $Y^- = f_-(y)$. We will show that

$$X \stackrel{d}{=} Y \Leftrightarrow X^+ \stackrel{d}{=} Y^+ \text{ and } X^- \stackrel{d}{=} Y^-. \tag{2.1.19}$$

As $f_+(.)$ and $f_-(.)$ are Borel functions, the "only if" part is trivial. If $X^+ \stackrel{d}{=} Y^+$ and $X^- \stackrel{d}{=} Y^-$, we have

$$\begin{aligned} E(e^{itX}) &= \int_{\{x>0\}} e^{itX}\,dF(x) + \int_{\{x<0\}} e^{itX}\,dF(x) + P(X=0) \\ &= E(e^{itX^+}) - P(X \leqslant 0) + E(e^{-itX^-}) - P(X \geqslant 0) + P(X=0) \\ &= E(e^{itX^+}) + E(e^{-itX^-}) - 1 \\ &= E(e^{itY^+}) + E(e^{-itY^-}) - 1 = E(e^{itY}), \end{aligned}$$

where $F(x)$ is the cdf of X, i.e., $X \stackrel{d}{=} Y$. If we can prove that $ZX^+ \stackrel{d}{=} ZY^+ \Leftrightarrow X^+ \stackrel{d}{=} Y^+$ and $ZX^- \stackrel{d}{=} ZY^- \Leftrightarrow X^- \stackrel{d}{=} Y^-$, then the assertion follows from

$$ZX \stackrel{d}{=} ZY \Leftrightarrow (ZX)^+ \stackrel{d}{=} (ZY)^+ \text{ and } (ZX)^- \stackrel{d}{=} (ZY)^- \Leftrightarrow ZX^+ \stackrel{d}{=} ZY^+ \text{ and } ZX^- \stackrel{d}{=} ZY^- \Leftrightarrow X^+ \stackrel{d}{=} Y^+ \text{ and } X^- \stackrel{d}{=} Y^- \Leftrightarrow X \stackrel{d}{=} Y.$$

Now we prove that $ZX^+ \stackrel{d}{=} ZY^+ \Leftrightarrow X^+ \stackrel{d}{=} Y^+$. Assume $ZX^+ \stackrel{d}{=} ZY^+$. Then

$$P(X>0)=P(X^+>0)=P(ZX^+>0)=P(ZY^+>0)=P(Y^+>0)=P(Y>0)$$

and

$$P(X>0)E(e^{itZX}\mid X>0)=E(e^{itZX^+})-P(X\leqslant 0)=E(e^{itZY^+})-P(Y\leqslant 0)$$
$$=E(e^{itZY}\mid Y>0)P(Y>0).$$

From the first part of the proof we have $E(e^{itX}\mid X>0)=E(e^{itY}\mid Y>0)$, i.e., $X^+\overset{d}{=}Y^+$.

The "if" part follows by the same technique. Similarly we have $ZX^-\overset{d}{=}ZY^-$ $\Longleftrightarrow X^-\overset{d}{=}Y^-$. By the same technique we can prove (4). □

Remark 1. Assertions (1) and (2) still hold when $\boldsymbol{x}$ and $\boldsymbol{y}$ are random vectors. In (2) the conditions apply to the components of $\boldsymbol{x}$ and $\boldsymbol{y}$.

Remark 2. In assertion (3) the condition $P(Z>0)=1$ is necessary. For example, if

$$P(Z=1,\ X=-1)=P(Z=1,X=-2)=P(Z=-1,X=-1)$$
$$=P(Z=-1,X=-2)=\frac{1}{4},$$
$$P(Z=1,Y=1)=P(Z=1,Y=2)=P(Z=-1,Y=1)$$
$$=P(Z=-1,Y=2)=\frac{1}{4},$$

it is easy to see that Z is independent of X and of Y respectively, $ZX\overset{d}{=}ZY$ and $\phi_{\log Z}(t)$ $=\frac{1}{2}(e^{it}+e^{-it})=\cosh t\neq 0$ for almost all t, but $X\overset{d}{\neq}Y$.

The above results are due to Anderson and Fang (1982a).

2.2. Moments of Multivariate Distributions

The *expectation* of a random variable X is defined by

$$E(X)=\int_{-\infty}^{\infty}x\,dF(x),$$

where $F(x)$ is the cdf of X.

Definition 2.2.1. Let $\boldsymbol{X}=(X_{ij})$ be an $n\times p$ random matrix. If every element X_{ij} of $\boldsymbol{X}$ has finite expectation, we write

(2.2.1) $$\mathcal{E}\boldsymbol{X}=(EX_{ij}).$$

When $n=p=1$, $\mathcal{E}\boldsymbol{X}=E\boldsymbol{X}$.

From this definition, there are the following properties:

(1)Let $\boldsymbol{A}=(a_{ij})$, $\boldsymbol{B}=(b_{ij})$ and $\boldsymbol{C}=(c_{ij})$ be $m\times n$, $p\times q$ and $m\times q$ constant matrices. Then

(2.2.2) $$\mathcal{E}(\boldsymbol{A}\boldsymbol{X}\boldsymbol{B}+\boldsymbol{C})=\boldsymbol{A}\mathcal{E}(\boldsymbol{X})\boldsymbol{B}+\boldsymbol{C}.$$

In particular, if $\boldsymbol{x}$ is an $n\times 1$ random vector, we have

(2.2.3) $$\mathcal{E}(\boldsymbol{A}\boldsymbol{x}) = \boldsymbol{A}\mathcal{E}(\boldsymbol{x}).$$

Here $\mathcal{E}(\boldsymbol{x})$ is called the *mean vector* of $\boldsymbol{x}$.

(2) Let $\boldsymbol{A} = (a_{ij})$ and $\boldsymbol{B} = (b_{ij})$ be $m \times n$ constant matrices, and $\boldsymbol{x}$ and $\boldsymbol{y}$ be $n \times 1$ random vectors, Then

(2.2.4) $$\mathcal{E}(\boldsymbol{A}\boldsymbol{x} + \boldsymbol{B}\boldsymbol{y}) = \boldsymbol{A}\mathcal{E}(\boldsymbol{x}) + \boldsymbol{B}\mathcal{E}(\boldsymbol{y}).$$

(3) Let $\boldsymbol{X}$ and $\boldsymbol{Y}$ be $n \times p$ random matrices, and a and b be constant scalars, Then

(2.2.5) $$\mathcal{E}(a\boldsymbol{X} + b\boldsymbol{Y}) = a\mathcal{E}(\boldsymbol{X}) + b\mathcal{E}(\boldsymbol{Y}).$$

(4) Assume that $\boldsymbol{X}$ is an $n \times p$ random matrix and $\boldsymbol{A}$ is a $p \times n$ constant matrix. Then

(2.2.6) $$E(\mathrm{tr}\boldsymbol{A}\boldsymbol{X}) = \mathrm{tr}(\mathcal{E}\boldsymbol{A}\boldsymbol{X}).$$

It is known that the *covariance* between two scalar random variables X and Y is defined as

(2.2.7) $$\mathrm{cov}(X, Y) = E(X - EX)(Y - EY).$$

Now we can generalize the concept of covariance to the case of random vectors.

Definition 2.2.2. Let $\boldsymbol{x}$ and $\boldsymbol{y}$ be $n \times 1$ and $p \times 1$ random vectors. The *covariance matrix* $\mathrm{cov}(\boldsymbol{x}, \boldsymbol{y})$ is defined by

(2.2.8) $$\mathrm{cov}(\boldsymbol{x}, \boldsymbol{y}) = (\mathrm{cov}(X_i, Y_j)),$$

i.e., $\mathrm{cov}(\boldsymbol{x}, \boldsymbol{y})$ is an $n \times p$ matrix with elements $\mathrm{cov}(X_i, Y_j)$. If $\mathrm{cov}(\boldsymbol{x}, \boldsymbol{y}) = \boldsymbol{O}$, we say that $\boldsymbol{x}$ and $\boldsymbol{y}$ are *uncorrelative*. If $\boldsymbol{x}$ and $\boldsymbol{y}$ are independent, we have $\mathrm{cov}(\boldsymbol{x}, \boldsymbol{y}) = \boldsymbol{O}$ if it exists, but the converse is not true in general. Many examples will be given in Section 2.7.

It can be proved that

(2.2.9) $$\mathrm{cov}(\boldsymbol{x}, \boldsymbol{y}) = \mathcal{E}(\boldsymbol{x} - \mathcal{E}(\boldsymbol{x}))(\boldsymbol{y} - \mathcal{E}(\boldsymbol{y}))'.$$

When $n = p$ and $\boldsymbol{x} = \boldsymbol{y}$, we write $\mathcal{D}(\boldsymbol{x})$ instead of $\mathrm{cov}(\boldsymbol{x}, \boldsymbol{y})$, and call it the *covariance matrix of vector* $\boldsymbol{x}$.

Here are some basic properties:

(1) We have

(2.2.10) $$\mathcal{D}(\boldsymbol{x}) = \mathcal{E}(\boldsymbol{x}\boldsymbol{x}') - \mathcal{E}(\boldsymbol{x})\mathcal{E}(\boldsymbol{x})'.$$

(2) If $\boldsymbol{a}$ is an $n \times 1$ constant vector, then

(2.2.11) $$\mathcal{D}(\boldsymbol{x} - \boldsymbol{a}) = \mathcal{D}(\boldsymbol{x})$$

and

(2.2.12) $$\mathcal{E}(\boldsymbol{x} - \boldsymbol{a})(\boldsymbol{x} - \boldsymbol{a}) = \mathcal{D}(\boldsymbol{x}) + (\mathcal{E}(\boldsymbol{x}) - \boldsymbol{a})\,(\mathcal{E}(\boldsymbol{x}) - \boldsymbol{a})'.$$

(3) Assume that $\boldsymbol{x}$ and $\boldsymbol{y}$ are $n \times 1$ and $m \times 1$ random vectors, and $\boldsymbol{A}$ and $\boldsymbol{B}$ are $p \times n$ and $q \times m$ constant matrices. Then

(2.2.13) $$\mathrm{cov}(\boldsymbol{A}\boldsymbol{x}, \boldsymbol{B}\boldsymbol{y}) = \boldsymbol{A}\mathrm{cov}(\boldsymbol{x}, \boldsymbol{y})\boldsymbol{B}'.$$

In particular, we have

(2.2.14) $$\mathcal{D}(\boldsymbol{A}\boldsymbol{x}) = \boldsymbol{A}\mathcal{D}(\boldsymbol{x})\boldsymbol{A}'.$$

Definition 2.2.3. The *correlation coefficient* between two scalar random variables X and Y is defined as

$$\rho = \text{corr}(X, Y) = \frac{\text{cov}(X, Y)}{[\text{var}(X)\,\text{var}(Y)]^{\frac{1}{2}}}. \tag{2.2.15}$$

The *corrlation matrix* of a random vector $\boldsymbol{x}$ is defined as $\boldsymbol{R} = (\rho_{ij})$, where $\rho_{ij} = \text{corr}(X_i, X_j)$.

Denote $\boldsymbol{D} = \text{diag}(\sigma_{11}, \dots, \sigma_{nn})$, where $\sigma_{11}, \dots, \sigma_{nn}$ are the diagonal elements of $\Sigma = \mathcal{D}(\boldsymbol{x})$ and $\sigma_j^2 = \sigma_{jj}$, $j = 1, 2, \dots, n$. The relationship between Σ and $\boldsymbol{R}$ is given by

$$\Sigma = \boldsymbol{D}^{\frac{1}{2}}\boldsymbol{R}\boldsymbol{D}^{\frac{1}{2}}, \tag{2.2.16}$$

where $\boldsymbol{D}^{\frac{1}{2}} = \text{diag}(\sigma_1, \dots, \sigma_n)$.

If all the moments of a random variable X exist, then the *semi–invariants* (or cumulants) are the coefficients κ in

$$\log\phi_X(t) = \sum_{j=0}^{\infty} \kappa_j \frac{(it)^j}{j!},$$

where $\phi_X(t)$ is the c.f. of X. The first three semi-invariants in terms of the moments $\mu_j' = E(X^j)$ of X are

$$\kappa_1 = \mu_1' = E(X),$$
$$\kappa_2 = \mu_2' - (\mu_1')^2 = \text{var}(X),$$
$$\kappa_3 = \mu_3' - 3\mu_1'\mu_2' + 2(\mu_1')^3.$$

Definition 2.2.4. If all the moments of a random vector $\boldsymbol{x}$ exist, then the *semi–invariants* are the coefficients κ in

$$\log\phi_{\boldsymbol{x}}(\boldsymbol{t}) = \sum_{s_1, \dots, s_n} \frac{(\text{i}t_1)^{s_1} \cdots (\text{i}t_n)^{s_n}}{s_1! \cdots s_n!} \kappa_{s_1, \dots, s_n}. \tag{2.2.17}$$

In general, we can define the *k–th moment of* $\boldsymbol{x}$ about the origin as follows:

$$\Gamma_k = \begin{cases} \mathcal{E}(\boldsymbol{x} \otimes \boldsymbol{x}' \otimes \boldsymbol{x} \cdots \otimes \boldsymbol{x}') & \text{if } k \text{ is even,} \\ \underbrace{\mathcal{E}(\boldsymbol{x} \otimes \boldsymbol{x}' \otimes \boldsymbol{x} \cdots \otimes \boldsymbol{x})}_{k \text{ factors}} & \text{if } k \text{ is odd,} \end{cases} \quad \text{k} = 1, 2, \dots \tag{2.2.18}$$

if it exists. For example $\Gamma_1 = (\boldsymbol{x}), \Gamma_2 = \mathcal{E}(\boldsymbol{x}\boldsymbol{x}'), \Gamma_3 = \mathcal{E}(\boldsymbol{x} \otimes \boldsymbol{x}' \otimes \boldsymbol{x})$. By induction, we can find

$$E(X_{i_1} X_{i_2} \cdots X_{i_k}) = r_{st}^{(k)}. \tag{2.2.19}$$

Here $\Gamma_k = (r_{ij}^{(k)})$;

$$s = \sum_{\alpha=1}^{\left[\frac{k+1}{2}\right]} (i_{2\alpha-1} - 1)n^{\left[\frac{k+1}{2}\right] - 1} + 1,$$

$$t = \sum_{\alpha=1}^{\left[\frac{k}{2}\right]} (i_{2\alpha-1} - 1)n^{\left[\frac{k}{2}\right] - 1} + 1,$$

with $[x]$ denoting the greatest integer less than or equal to x. The proof is left to the reader.

We know that there is a relationship between moments and c.f. of a random variable. Let $\phi(t)$ be the c.f. of a random variable X. Then $E(X^k) = 1/i^k\phi^{(k)}(0)$, $k = 1, 2, \dots$, where $i = (-1)^{\frac{1}{2}}$, if they exist. Similarly, if a random vector $\boldsymbol{x}$ with c.f. $\phi(\boldsymbol{t})$ has the k–th moment, then

$$\Gamma_k = \begin{cases} \dfrac{1}{i^k} \dfrac{\partial^k \phi(t)}{\partial t \partial t' \partial t \cdots \partial t \partial t'}\Big|_{t=0} & \text{if } k \text{ is even,} \\ \dfrac{1}{i^k} \dfrac{\partial^k \phi(t)}{\partial t \partial t' \partial t \cdots \partial t' \partial t}\Big|_{t=0} & \text{if } k \text{ is odd.} \end{cases} \tag{2.2.20}$$

Using the notation Γ_k, we can calculate moments of a quadratic form of $\boldsymbol{x}$.

Theorem 2.2.1. *Assume that $\boldsymbol{x}$ has the 2k–th moment and $\boldsymbol{A}$ is an $n \times n$ constant matrix. Then*

$$E[(\boldsymbol{x}'\boldsymbol{A}\boldsymbol{x})^k] = \mathrm{tr}\,[(\underbrace{\boldsymbol{A} \otimes \cdots \otimes \boldsymbol{A}}_{k}\Gamma_{2k}]. \tag{2.2.21}$$

Proof. By Kronecker product's properties, we have

$$\begin{aligned} E[(\boldsymbol{x}'\boldsymbol{A}\boldsymbol{x})^k] &= E[(\boldsymbol{x}'\boldsymbol{A}\boldsymbol{x}) \otimes (\boldsymbol{x}'\boldsymbol{A}\boldsymbol{x}) \otimes \cdots \otimes (\boldsymbol{x}'\boldsymbol{A}\boldsymbol{x})] \\ &= E[(\boldsymbol{x}' \otimes \cdots \otimes \boldsymbol{x}')(\boldsymbol{A} \otimes \cdots \otimes \boldsymbol{A})(\boldsymbol{x} \otimes \cdots \otimes \boldsymbol{x}) \\ &= \mathcal{E}\{\mathrm{tr}[(\boldsymbol{A} \otimes \cdots \otimes \boldsymbol{A})(\boldsymbol{x} \otimes \boldsymbol{x}' \otimes \boldsymbol{x} \otimes \cdots \otimes \boldsymbol{x}')]\} \\ &= \mathrm{tr}[(\boldsymbol{A} \otimes \cdots \otimes \boldsymbol{A})\Gamma_{2k}]. \end{aligned}$$

□

Corollary 1. *Assume that $\boldsymbol{x}$ has a finite mean vector μ and a finite covariance matrix Σ, and $\boldsymbol{A}$ is an $n \times n$ constant matrix. Then*

$$E(\boldsymbol{x}'\boldsymbol{A}\boldsymbol{x}) = \mathrm{tr}(\boldsymbol{A}\Sigma) + \mu'\boldsymbol{A}\mu. \tag{2.2.22}$$

Proof. As

$$\Gamma_2 = \mathcal{E}(\boldsymbol{x}\boldsymbol{x}') = \mathcal{D}(\boldsymbol{x}) + \mu\mu' = \Sigma + \mu\mu'$$

the formula (2.2.22) follows from (2.2.21) with $k = 1$. □

2.3. Multivariate Normal Distribution

We write $X \sim N(0, 1)$, if the density of X is $(2\pi)^{-\frac{1}{2}}e^{-x^2/2}$, and say that X is distributed according to the standard normal distribution. In this section $\boldsymbol{x}$ denotes a random vector with i.i.d. components $\{X_j\}$ and $X_j \sim N(0, 1)$. The dimension of $\boldsymbol{x}$ depends on the current situation.

Definition 2.3.1. If a $m \times 1$ random vector $\boldsymbol{y}$ can be expressed as

$$\boldsymbol{y} \stackrel{d}{=} \mu + \boldsymbol{A}'\boldsymbol{x}, \tag{2.3.1}$$

where μ is an $m \times 1$ constant vector, $\boldsymbol{A}$ is an $n \times m$ constant matrix and $\boldsymbol{x}$ is n–dimensional, we write $y \sim N_m(\mu, \boldsymbol{A}'\boldsymbol{A})$ and say that $\boldsymbol{y}$ is *distributed according to the multivariate normal distribution.*

By this definition we can write $\boldsymbol{x} \sim N_n(\boldsymbol{0}, \boldsymbol{I}_n)$ where $\boldsymbol{x}$ is in (2.3.1) and find the following facts:

(1) Assume that $\boldsymbol{y} \sim N_m(\mu', \boldsymbol{A}'\boldsymbol{A})$ and $\boldsymbol{z} = \boldsymbol{B}'\boldsymbol{y} + \boldsymbol{d}$ with $\boldsymbol{B}$: $m \times 1$ and d: $l \times 1$. Then $\boldsymbol{z} \sim N_l(\boldsymbol{B}'\mu + \boldsymbol{d}, \boldsymbol{B}'\boldsymbol{A}'\boldsymbol{A}\boldsymbol{B})$.

Proof. From assumption we have $\boldsymbol{y} \stackrel{d}{=} \mu + \boldsymbol{A}'\boldsymbol{x}$. Then

$$z \stackrel{d}{=} (B'\mu + d) + (AB)'x,$$

i.e., $z \sim N_1(B'\mu + d, B'A'AB)$. □

(2) Assume $y \sim N_m(\mu, A'A)$ and let

$$y = \begin{bmatrix} y^{(1)} \\ y^{(2)} \end{bmatrix}, \quad \mu = \begin{bmatrix} \mu^{(1)} \\ \mu^{(2)} \end{bmatrix} \quad \text{and} \quad A'A = \begin{bmatrix} \Sigma_{11} & \Sigma_{12} \\ \Sigma_{21} & \Sigma_{22} \end{bmatrix}, \tag{2.3.2}$$

where $y^{(1)}$: $r \times 1$, $\mu^{(1)}$: $r \times 1$, Σ_{11}: $r \times r$. Then

$$y^{(1)} \sim N_r(\mu^{(1)}, \Sigma_{11}) \text{ and } y^{(2)} \sim N_{m-r}(\mu^{(2)}, \Sigma_{22}). \tag{2.3.3}$$

Proof. Taking $B' = (I_r\colon O)$ with $O\colon r \times (m-r)$ and $B' = (O\colon I_{m-r})$ with O: $r \times r$, the assertion follows by (1). □

This property tells us that any marginal distributions of a multivariate normal distribution are still normal distributions. But the converse is not true in general; that is, the fact that each component of a random vector is (marginally) normal does not imply that the vector has a multivariate normal distribution. As a counterexample, assume the density of $x = (X_1, X_2)'$ is

$$f(x_1, x_2) = \frac{1}{2\pi} e^{-\frac{1}{2}(x_1^2 + x_2^2)} [1 + x_1 x_2 e^{-\frac{1}{2}(x_1^2 + x_2^2)}].$$

It is easily seen that $X_1 \sim N(0, 1)$ and $X_2 \sim N(0, 1)$, but the joint distribution of X_1 and X_2 is not a binormal distribution.

Theorem 2.3.1. *Assume that $y \sim N_m(\mu, A'A)$. Then the c.f. of y is*

$$\phi_y(t) = \exp(it'\mu - \tfrac{1}{2}t'A'At). \tag{2.3.4}$$

Proof. As $y \stackrel{d}{=} \mu + A'x$ and $x = (X_1, \cdots X_n)' \sim N_n(0, I_n)$, thus by the c.f.'s properties (d) and (f) we have

$$\phi_x(t) = \prod_1^n \phi_{X_i}(t_i) = \prod_1^n e^{-t_i^2/2} = e^{-\frac{1}{2}(t_1^2 + \cdots + t_n^2)} = e^{-\frac{1}{2}t't}$$

and

$$\begin{aligned} \phi_y(t) &= e^{it'\mu} \phi_x(At) = e^{it'\mu} e^{-\frac{1}{2}t'A'At} \\ &= \exp(it'\mu - \tfrac{1}{2}t'A'At). \end{aligned}$$ □

This theorem shows that the multivariate normal distribution $N_m(\mu, A'A)$ is uniquely determined by μ and $A'A$. From now on, denote $\Sigma = A'A$. Clearly $\Sigma \geqslant 0$. For each $\Sigma \geqslant 0$, generally the factorization of $\Sigma = A'A$ is not unique. In particular, we can take an $m \times m$ matrix as A with the corresponding x: $m \times 1$.

Theorem 2.3.2. *Assume $y \sim N_m(\mu, \Sigma)$* and $|\Sigma| \neq 0$. *Then the density of y is*

$$f_m(y) = (2\pi)^{-m/2} |\Sigma|^{-\frac{1}{2}} \exp\left(-\frac{1}{2}(y - \mu)'\Sigma^{-1}(y - \mu)\right). \tag{2.3.5.}$$

Proof. Take an $m \times m$ matrix A in (2.3.1) with $x \sim N_m(0, I_m)$. Since $A'A = \Sigma$ is nonsingular, A^{-1} exists. The density of x is

$$f_x(x) = \prod_{i=1}^m (2\pi)^{-\frac{1}{2}} \exp(\tfrac{1}{2}x_i^2) = (2\pi)^{-\frac{1}{2}m} \exp\left(-\frac{1}{2}x'x\right).$$

Consider the transformation $x = A'^{-1}(y - \mu)$ and the corresponding Jacobian is (cf. Example 1.6.1)

$$|\frac{\partial x}{\partial y}|_+ = |A'|_+^{-1} = |A'A|^{-\frac{1}{2}} = |\Sigma|^{-\frac{1}{2}},$$

Hence the density function of y is

$$f_y(y) = (2\pi)^{-\frac{1}{2}m}|\Sigma|-\frac{1}{2}\exp(-\frac{1}{2}(y-\mu)'A^{-1}(y-\mu)).$$

Then the assertion follows fram $\Sigma = A'A$. □

Example 2.3.1. The bivariate distribution. In this case

$$y = \begin{bmatrix} Y_1 \\ Y_2 \end{bmatrix}, \quad \mu = \begin{bmatrix} \mu_1 \\ \mu_2 \end{bmatrix} \text{ and } \Sigma = \begin{pmatrix} \sigma_{11} & \sigma_{12} \\ \sigma_{21} & \sigma_{22} \end{pmatrix} = \begin{pmatrix} \sigma_1^2 & \sigma_1\sigma_2\rho \\ \sigma_1\sigma_2\rho & \sigma_2^2 \end{pmatrix}.$$

By (2.3.4), its c.f. is

$$\phi_y(t_1, t_2) = \exp i(t_1\mu_1 + t_2\mu_2) - \tfrac{1}{2}(t_1^2\sigma_1^2 + 2t_1t_2\sigma_1\sigma_2 + t_2^2\sigma_2^2).$$

As $|\Sigma| = \sigma_1^2\sigma_2^2(1-\rho^2)$, if $\sigma_1 > 0$, $\sigma_2 > 0$ and $|\rho| < 1$, then $\Sigma > 0$. In this case

$$\Sigma^{-1} = \frac{1}{\sigma_1^2\sigma_2^2(1-\rho^2)}\begin{pmatrix} \sigma_2^2 & -\sigma_1\sigma_2\rho \\ -\sigma_1\sigma_2\rho & \sigma_1^2 \end{pmatrix}.$$

Thus the density of y is

$$f(y_1, y_2) = (2\pi\sigma_1\sigma_2\sqrt{1-\rho^2})^{-1}\exp\left\{-\frac{1}{2(1-\rho^2)}\left[\left(\frac{y_1-\mu_1}{\sigma_1}\right)^2 - 2\left(\frac{y_1-\mu_1}{\sigma_1}\right)\left(\frac{y_2-\mu_2}{\sigma_2}\right) + \left(\frac{y_2-\mu_2}{\sigma_2}\right)^2\right]\right\}.$$

Theorem 2.3.3. *Assume $y \sim N_m(\mu, \Sigma)$. Then*

$$\mathcal{E}(y) = \mu \text{ and } \mathcal{D}(y) = \Sigma. \tag{2.3.6}$$

Proof. By (2.3.1) and $\mathcal{E}(x) = \mathbf{0}$, $\mathcal{D}(x) = I_m$, the theorem follows as

$$\mathcal{E}(y) = \mu + \mathcal{E}(A'x) = \mu$$

and

$$\mathcal{D}(y) = A'\mathcal{D}(x)A = A'A = \Sigma. \quad □$$

Theorem 2.3.4. *Let y be a $m \times 1$ random vector. Then y is distributed according to a multivariate normal distribution if and only if $a'y$ is distributed according to a univariate normal distribution for every $a \in R^m$.*

Proof. Obviously, we need only to prove the "if" part. Assume, for every $a \in R^m$, $a'y$ is distributed according to the univariate normal distribution. It can be seen that there exist $\mathcal{E}(y)$ and $\mathcal{D}(y)$. Denote $\mu = \mathcal{E}(y)$ and $\Sigma = \mathcal{D}(y)$. We have $a'y \sim N(a'\mu, a'\Sigma a)$ by assumption, and the c.f. of $a'y$ is

$$\phi_{a'y}(t) = \exp(ita'\mu - \tfrac{1}{2}t^2a'\Sigma a).$$

Taking $t = 1$, we obtain the c.f. of y considered as a function in a which is

$$\phi_y(a) = \exp(ia'\mu - \tfrac{1}{2}a'\Sigma a).$$

The theorem follows from property (g) of the c.f. (cf. Section 2.1.6.) and Theorem 2.3.1. □

Assume $\boldsymbol{y} \sim N_m(\mu, \Sigma)$. Partition $\boldsymbol{y}$ in the fashion of (2.3.2).

Theorem 2.3.5. *Assume $\boldsymbol{y} \sim N_m(\mu, \Sigma)$ with $\Sigma > 0$. Then*

(1) *The conditional distribution of $\boldsymbol{y}^{(1)}$ given $\boldsymbol{y}^{(2)}$ is*

$$N_r(\mu_{1\cdot 2}, \Sigma_{11\cdot 2}), \tag{2.3.7}$$

where

$$\mu_{1\cdot 2} = \mu^{(1)} + \Sigma_{12}\Sigma_{22}^{-1}(\boldsymbol{y}^{(2)} - \mu^{(2)}), \tag{2.3.8}$$

and

$$\Sigma_{11\cdot 2} = \Sigma_{11} - \Sigma_{12}\Sigma_{22}^{-1}\Sigma_{21}. \tag{2.3.9}$$

(2) $\boldsymbol{y}^{(1)} - \Sigma_{12}\Sigma_{22}^{-1}\boldsymbol{y}^{(2)}$ *is independent of* $\boldsymbol{y}^{(2)}$.

(3) $\boldsymbol{y}^{(2)} - \Sigma_{21}\Sigma_{11}^{-1}\boldsymbol{y}^{(1)}$ *is independent of* $\boldsymbol{y}^{(1)}$.

Proof. Clearly, $\Sigma > 0$ implies $\Sigma_{11} > 0$ and $\Sigma_{22} > 0$. Let

$$\boldsymbol{z} = \begin{bmatrix} \boldsymbol{z}^{(1)} \\ \boldsymbol{z}^{(2)} \end{bmatrix} = \begin{bmatrix} \boldsymbol{I}_p & -\Sigma_{12}\Sigma_{22}^{-1} \\ \boldsymbol{O} & \boldsymbol{I}_q \end{bmatrix} \begin{bmatrix} \boldsymbol{y}^{(1)} \\ \boldsymbol{y}^{(2)} \end{bmatrix} \equiv \boldsymbol{B}\boldsymbol{y} = \begin{bmatrix} \boldsymbol{y}^{(1)} - \Sigma_{12}\Sigma_{22}^{-1}\boldsymbol{y}^{(2)} \\ \boldsymbol{y}^{(2)} \end{bmatrix}.$$

The mcdf of $\boldsymbol{z}$ is normal and the corresponding mean vector and covariance matrix are

$$\mathcal{E}(\boldsymbol{z}) = \boldsymbol{B}\mathcal{E}(\boldsymbol{y}) = \begin{bmatrix} \mu^{(1)} - \Sigma_{12}\Sigma_{22}^{-1}\mu^{(2)} \\ \mu^{(2)} \end{bmatrix}$$

and

$$\mathcal{D}(\boldsymbol{z}) = \boldsymbol{B}\mathcal{D}(\boldsymbol{y})\boldsymbol{B}' = \begin{pmatrix} \Sigma_{11} - \Sigma_{12}\Sigma_{22}^{-1}\Sigma_{21} & \boldsymbol{O} \\ \boldsymbol{O} & \Sigma_{22} \end{pmatrix} = \begin{pmatrix} \Sigma_{11\cdot 2} & \boldsymbol{O} \\ \boldsymbol{O} & \Sigma_{22} \end{pmatrix}.$$

Clearly, $\Sigma > 0$ implies $\Sigma_{11\cdot 2} > 0$ and $\Sigma_{22} > 0$. We see that the density of $\boldsymbol{z}$ is a product of two normal densities, i.e.,

$$N_r(\mu_{1\cdot 2}, \Sigma_{11\cdot 2})N_{m-r}(\mu^{(2)}, \Sigma_{22}).$$

Noting that the Jacobian $\left|\dfrac{\partial \boldsymbol{y}'}{\partial \boldsymbol{z}}\right|_+ = |\boldsymbol{B}|_+ = 1$, the density of $\boldsymbol{y}$ can be expressed as

$$\begin{aligned} f(\boldsymbol{y}) = f(\boldsymbol{y}^{(1)}, \boldsymbol{y}^{(2)}) &= (2\pi)^{-r/2}|\Sigma_{11\cdot 2}|^{-\frac{1}{2}}\exp\{-\tfrac{1}{2}(\boldsymbol{y}^{(1)} - \mu_{1\cdot 2})'\Sigma_{11\cdot 2}^{-1}(\boldsymbol{y}^{(1)} - \mu_{1\cdot 2})\} \\ &\quad \times (2\pi)^{-(m-r)/2}|\Sigma_{22}|^{-\frac{1}{2}}\exp\{-\frac{1}{2}(\boldsymbol{y}^{(2)} - \mu^{(2)})'\Sigma_{22}^{-1}(\boldsymbol{y}^{(2)} - \mu^{(2)})\}. \end{aligned}$$

Then the theorem is a consequence of (2.1.8) and (2.3.3). □

Now we turn to consider the independence in the case of the multivariate normal distribution.

Lemma 2.3.1. *Assume $\boldsymbol{x} \sim N_n(\boldsymbol{0}, \boldsymbol{I}_n)$, $\boldsymbol{y} = \mu + \boldsymbol{A}'\boldsymbol{x}$ and $\boldsymbol{z} = \gamma + \boldsymbol{B}'\boldsymbol{x}$, where $\boldsymbol{A}$: $n \times p$, $\boldsymbol{B}$: $n \times q$. Then $\boldsymbol{y}$ and $\boldsymbol{z}$ are independent if and only if $\boldsymbol{A}'\boldsymbol{B} = \boldsymbol{O}$.*

Proof. Write

$$\boldsymbol{w} = \begin{bmatrix} \boldsymbol{y} \\ \boldsymbol{z} \end{bmatrix} = \begin{bmatrix} \mu \\ \gamma \end{bmatrix} + \begin{bmatrix} \boldsymbol{A}' \\ \boldsymbol{B}' \end{bmatrix}\boldsymbol{x}.$$

Then

$$\boldsymbol{w} \sim N_{p+q}\left(\begin{pmatrix}\mu\\ \gamma\end{pmatrix}, \begin{pmatrix}\boldsymbol{A}'\boldsymbol{A} & \boldsymbol{A}'\boldsymbol{B}\\ \boldsymbol{B}'\boldsymbol{A} & \boldsymbol{B}'\boldsymbol{B}\end{pmatrix}\right).$$

Assume $\boldsymbol{y}$ and $\boldsymbol{z}$ are independent, We have

$$\boldsymbol{O} = \mathrm{cov}(\boldsymbol{y}, \boldsymbol{z}) = \mathrm{cov}(\mu + \boldsymbol{A}'\boldsymbol{x}, \gamma + \boldsymbol{B}'\boldsymbol{x}) = \mathrm{cov}(\boldsymbol{A}'\boldsymbol{x}, \boldsymbol{B}'\boldsymbol{x}) = \boldsymbol{A}'\mathcal{D}(\boldsymbol{x})\boldsymbol{B} = \boldsymbol{A}'\boldsymbol{B}.$$

On the other hand, if $\boldsymbol{A}'\boldsymbol{B} = \boldsymbol{O}$, then

$$\boldsymbol{w} \sim N_{p+q}\left(\begin{pmatrix}\mu\\ \gamma\end{pmatrix}, \begin{pmatrix}\boldsymbol{A}'\boldsymbol{A} & \boldsymbol{O}\\ \boldsymbol{O} & \boldsymbol{B}'\boldsymbol{B}\end{pmatrix}\right),$$

and the c.f. of $\boldsymbol{w}$ is a product of the c.f.'s of $N_p(\mu, \boldsymbol{A}'\boldsymbol{A})$ and $N_q(\gamma, \boldsymbol{B}'\boldsymbol{B})$, i.e., $\boldsymbol{y}$ and $\boldsymbol{z}$ are independent. □

Corollary 1. *Assume* $\boldsymbol{y} \sim N_m(\mu, \Sigma)$, $\boldsymbol{z} \overset{d}{=} \alpha + \boldsymbol{B}'\boldsymbol{y}$ *and* $\boldsymbol{w} \overset{d}{=} \beta + \boldsymbol{C}'\boldsymbol{y}$, *where* $\boldsymbol{B}$: $m \times p$ *and* $\boldsymbol{C}$: $m \times q$. *Then* $\boldsymbol{z}$ *and* $\boldsymbol{w}$ *are independent if and only if* $\boldsymbol{B}'\Sigma\boldsymbol{C} = \boldsymbol{O}$.

Proof. As $\boldsymbol{z} \overset{d}{=} (\boldsymbol{B}'\mu + \alpha) + (\boldsymbol{A}\boldsymbol{B})'\boldsymbol{x}$, $\boldsymbol{w} \overset{d}{=} (\boldsymbol{C}'\mu + \beta) + (\boldsymbol{A}\boldsymbol{C})'\boldsymbol{x}$, hence $\boldsymbol{z}$ and $\boldsymbol{w}$ are independent if and only if

$$\boldsymbol{O} = (\boldsymbol{A}\boldsymbol{B})'\boldsymbol{A}\boldsymbol{C} = \boldsymbol{B}'\boldsymbol{A}'\boldsymbol{A}\boldsymbol{C} = \boldsymbol{B}'\Sigma\boldsymbol{C}$$

by using Lemma 2.3.1. □

Similarly, we can find the following Corollary.

Corollary 2. *Assume* $\boldsymbol{y} \sim N_m(\mu, \Sigma)$. *Partition* $\boldsymbol{y}$ *and* Σ *into*

$$\boldsymbol{y} = \begin{bmatrix}\boldsymbol{y}^{(1)}\\ \vdots\\ \boldsymbol{y}^{(k)}\end{bmatrix} \qquad \Sigma = \begin{bmatrix}\Sigma_{11} & \cdots & \Sigma_{1k}\\ \vdots & & \vdots\\ \Sigma_{k1} & \cdots & \Sigma_{kk}\end{bmatrix}.$$

Then $\boldsymbol{y}^{(1)}, \ldots, \boldsymbol{y}^{(k)}$ are independent if and only if $\Sigma_{ij} = \boldsymbol{O}$, for $i \neq j$. Here independence is equivalent to uncorrelation. But it is not always true in general (cf. Section 2.7).

2.4. Dirichlet Distribution

Definition 2.4.1. If $\boldsymbol{x} \sim N_n(\mu, \boldsymbol{I}_n)$ we say that $Y = \boldsymbol{x}'\boldsymbol{x}$ is *distributed according to a non-central chi-squared distribution* with n degrees of freedom and non-central parameter $\delta^2 = \mu'\mu$, and write $Y \sim x_n^2(\delta^2)$. When $\delta = 0$ (i.e., $\mu = \boldsymbol{O}$), the distribution of Y is called the *chi-squared distribution with n* degrees of freedom and write $Y \sim x_n^2$.

We shall show, in Example 2.4.1., that the density of $Y \sim x_n^2$ is

$$(\Gamma(n))^{-1}\, y^{\frac{1}{2}n-1}\, e^{-\frac{1}{2}y}. \tag{2.4.1}$$

If a random variable B has the density

$$\frac{\Gamma(\alpha_1 + \alpha_2)}{\Gamma(\alpha_1)\,\Gamma(\alpha_2)} b^{\alpha_1 - 1}(1 - b)^{\alpha_2 - 1},\ 0 \leqslant b < 1 \tag{2.4.2}$$

we say that B is *distributed according to a beta-distribution* and write $B \sim B(\alpha_1, \alpha_2)$.

It is well-known fact that if $Y \sim x_n^2$ and $Z \sim x_m^2$ are independent, then $Y/(Y + Z) \sim B(\frac{1}{2}n, \frac{1}{2}m)$. An extension of beta-distribution is the so-called Dirichlet distribution.

Definition 2.4.2. If $y = (Y_1, \dots, Y_m)'$ is a random vector such that $\sum_1^m Y_i = 1$ and $Y_1, \dots, Y_{m-1}$ have joint density with $\alpha_i > 0$, $i = 1, \dots, m$,

$$(2.4.3)\quad p_m(y_1, \dots, y_{m-1}) = \frac{\Gamma(\sum_1^m \alpha_i)}{\prod_1^m \Gamma(\alpha_i)} \prod_{i=1}^{m-1} y_i^{\alpha_i - 1}\Big(1 - \sum_1^{m-1} y_j\Big)^{\alpha_m - 1}, \text{ if } y_i \geqslant 0, i = 1, \dots, m - 1 \text{ and } \sum_1^{m-1} y_i < 1;$$

$$= 0 \qquad \text{otherwise,}$$

we say that y is *distributed according to the Dirichlet distribution* and denote it as $(Y_1, \dots, Y_m) \sim D_m(\alpha_1, \dots, \alpha_{m-1}, \alpha_m)$ or $(Y_1, \dots, Y_{m-1}) \sim D_m(\alpha_1, \dots, \alpha_{m-1}; \alpha_m)$ depending on which is concerned. When $m = 2$, $D_2(\alpha_1; \alpha_2)$ reduces to Beta–distribution $B(\alpha_1, \alpha_2)$.

First, we need to show that $p_m(y_1, \dots, y_{m-1})$ defined in (2.4.3) is a density. Then it amounts to proving the following lemma.

Lemma 2.4.1. *Denote*

(2.4.4.)

$$B_m(\alpha_1, \dots, \alpha_m) = \int_{D(x_1, \dots, x_{m-1})} x_1^{\alpha_1 - 1} \dots x_{m-1}^{\alpha_{m-1} - 1}\Big(1 - \sum_1^{m-1} x_i\Big)^{\alpha_m - 1} dx_1 \dots dx_{m-1},$$

where

$$(2.4.5)\qquad D(x_1, \dots, x_{m-1}) = \{(x_1, \dots, x_{m-1}) \mid x_i \geqslant 0,\ i = 1, \dots, m - 1;\ \sum_1^{m-1} x_i < 1\}.$$

Then

$$(2.4.6)\qquad B_m(\alpha_1, \dots, \alpha_m) = \prod_{i=1}^m \Gamma(\alpha_i) / \Gamma\Big(\sum_{i=1}^m \alpha_i\Big).$$

Proof. We will prove (2.4.6) by induction. It is known that the formula (2.4.6) is true when $m = 2$. Assume that (2.4.6) is true for $m \leqslant k$. By (2.4.4),

$$B_{k+1}(\alpha_1, \dots, \alpha_{k+1}) = \int_{D(x_1, \dots, x_k)} \Big(1 - \sum_1^k x_i\Big)^{\alpha_{k+1} - 1}\Big(\prod_{i=1}^k x_i^{\alpha_i - 1} dx_i\Big)$$

$$= \int_{D(x_1, \dots, x_{k-1})} \Big(\prod_1^{k-1} x_i^{\alpha_i - 1} dx_i\Big) \int^{1 - \sum_1^{k-1} x_i} x_k^{\alpha_k - 1}\Big(1 - \sum_{i=1}^{k-1} x_i\Big)^{\alpha_{k+1} - 1} dx_k.$$

Let $u_i = x_i$, $i = 1, \dots, k - 1$, and $u_k = x_k / (1 - \sum_{i=1}^{k-1} x_i)$. Then the Jacobian of this transformation is $\left|\frac{\partial(x_1, \dots, x_k)}{\partial(u_1, \dots, u_k)}\right|_+ = 1 - \sum_{i=1}^{k-1} u_i$. By the assumption of induction we have

$$B_{k+1}(\alpha_1, \dots, \alpha_{k+1}) = \int\limits_{D(x_1,\dots,x_{k-1})} \Big(\prod_{i=1}^{k-1} u_i^{\alpha_i - 1} du_i\Big)\Big(1 - \sum_{i=1}^{k-1} u_i\Big)^{\alpha_k + \alpha_{k+1} - 1}$$

$$\cdot \int_0^1 u_k^{\alpha_k - 1}(1 - u_k)^{\alpha_{k+1} - 1} du_k = \frac{\Gamma(\alpha_k)\Gamma(\alpha_{k+1})}{\Gamma(\alpha_k + \alpha_{k+1})} B_k(\alpha_1, \dots, \alpha_{k-1}, \alpha_k + \alpha_{k+1})$$

$$= \frac{\Gamma(\alpha_k)\Gamma(\alpha_{k+1})}{\Gamma(\alpha_k + \alpha_{k+1})} \frac{\Big(\prod_1^{k-1} \Gamma(\alpha_i)\Big)\Gamma(\alpha_k + \alpha_{k+1})}{\Gamma\Big(\sum_1^{k+1} \alpha_i\Big)} = \prod_1^{k+1} \Gamma(\alpha_i) \Big/ \Gamma\Big(\sum_1^{k+1} \alpha_i\Big)$$

which completes the proof. □

Corollary 1. *The function $p_m(y_1, \dots, y_{m-1})$ defined by (2.4.1) is a density.*

Theorem 2.4.1. *Assume that $Z_i \sim x_{n_i}^2$, $i = 1, \dots, m$, are independent and $Z = Z_1 + \dots + Z_m$. Then $(Z_1/Z, \dots, Z_m/Z) \sim D_m(n_1/2, \dots, n_m/2)$.*

Proof. The joint density of $Z_1, \dots, Z_m$ is

$$\Big[\prod_1^m \Gamma\big(\tfrac{1}{2} n_i\big)\Big]^{-1} 2^{-\frac{1}{2}(n_1 + \dots + n_m)} \Big[\prod_1^m z_i^{\frac{1}{2}n_i - 1}\Big] e\Big(-\frac{1}{2}\sum_1^m z_i\Big).$$

Let

$$\begin{cases} Y_i = Z_i / Z, & i = 1, \dots, m-1 \\ Z = Z_1 + \dots + Z_m. \end{cases}$$

The Jacobian of this transformation is

$$J = \left|\frac{\partial(z_1, z_2, \dots, z_m)}{\partial(z, y_1, \dots, y_{m-1})}\right| = \operatorname{mod} \begin{vmatrix} y_1 & z & 0 & \dots & 0 \\ y_2 & 0 & z & \dots & 0 \\ \vdots & \vdots & \vdots & \dots & \vdots \\ 1 - \sum_1^{m-1} y_i & -z & -z & \dots & -z \end{vmatrix}$$

$$= \operatorname{mod}|\operatorname{diag}(z, \dots, z)| \left\{\Big(1 - \sum_1^{m-1} y_i\Big) - (y_1, \dots, y_{m-1}) \begin{bmatrix} z & & & \\ & \ddots & & \\ & & \ddots & \\ & & & z \end{bmatrix}^{-1} \begin{bmatrix} -z \\ \vdots \\ \vdots \\ \vdots \\ -z \end{bmatrix}\right\}$$

$$= z^{m-1}\Big(1 - \sum_1^{m-1} y_i + \sum_1^{m-1} y_i\Big) = z^{m-1}.$$

Hence the joint density of $Z, Y_1, \dots, Y_{m-1}$ is

$$\left[2^{\frac{1}{2}n}\prod_{1}^{m}\Gamma\left(\tfrac{1}{2}n_i\right)\right]^{-1}\left[\prod_{i=1}^{m-1}y_i^{\frac{1}{2}n_i-1}\right]\left(1-\sum_{1}^{m-1}y_i\right)^{\frac{1}{2}n_m-1}z^{\frac{1}{2}n-1}e^{-\frac{1}{2}z}$$

$$=\left\{\left[2^{\frac{1}{2}n}\Gamma\left(\tfrac{1}{2}n\right)\right]^{-1}z^{\frac{1}{2}n-1}e^{-\frac{1}{2}z}\right\}\left\{\frac{\Gamma\left(\frac{1}{2}\sum_{1}^{m}n_i\right)}{\prod_{1}^{m}\Gamma\left(\frac{1}{2}n_i\right)}\prod_{i=1}^{m-1}y_i^{\frac{1}{2}n_i-1}\left(1-\sum_{1}^{m-1}y_i\right)^{\frac{1}{2}n_m-1}\right\},$$

where $n = n_1 + \cdots + n_m$. The first factor is the density of the chi-square distribution with n degrees of freedom and the second factor is the density of $D_m(\frac{1}{2}n_1, \ldots, \frac{1}{2}n_{m-1}; \frac{1}{2}n_m)$, which completes the proof. □

Corollary 1. *$Z = Z_1 + \cdots + Z_m \sim x_n^2$ is independent of $(Z_1/Z, \ldots, Z_{m-1}/Z)$.*

Corollary 2. *Assume $\boldsymbol{x} \sim N_n(\boldsymbol{0}, \boldsymbol{I}_n)$ and it is partitioned into m parts $\boldsymbol{x}^{(1)}, \ldots, \boldsymbol{x}^{(m)}$ with $n_1, \ldots, n_m$ components of $\boldsymbol{x}$ respectively, i.e.,*

$$\boldsymbol{x} = \begin{bmatrix}\boldsymbol{x}^{(1)}\\ \vdots \\ \boldsymbol{x}^{(m)}\end{bmatrix}. \tag{2.4.7}$$

Then $\left(\frac{\|\boldsymbol{x}^{(1)}\|^2}{\|\boldsymbol{x}\|^2}, \ldots, \frac{\|\boldsymbol{x}^{(m)}\|^2}{\|\boldsymbol{x}\|^2}\right) \sim D_m(\frac{1}{2}n_1, \ldots, \frac{1}{2}n_m)$ and is independent of $\|\boldsymbol{x}\|^2$.

The mixed moments of the Dirichlet distribution can be easily evaluated by using (2.4.6.)

Theorem 2.4.2. *Assume $\boldsymbol{y} = (Y_1, \ldots, Y_{m+1})' \sim D_{m+1}(\alpha_1, \ldots, \alpha_m, \alpha_{m+1})$. Then for every $r_1, \ldots, r_m \geqslant 0$, we have*

$$\mu_{r_1,\ldots r_m} = E\left(\prod_{j=1}^{m} Y_j^{r_j}\right) = \left(\prod_{j=1}^{m}\alpha_j^{[r_j]}\right)\Big/\left(\sum_{1}^{m+1}\alpha_j\right)^{[r_1+\cdots+r_m]}, \tag{2.4.8}$$

where $x^{[r]} = x(x+1)\ldots(x+r-1)$.

Proof. As $\Gamma(x+1) = x\Gamma(x)$, we have

$$\mu_{r_1,\cdots r_m} = \frac{\Gamma\left(\sum_{1}^{m+1}\alpha_j\right)}{\prod_{j=1}^{m+1}\Gamma(\alpha_j)}\int_{E(y_1,\ldots,y_m)}\left(1-\sum_{j=1}^{m}y_j\right)^{\alpha_{m+1}-1}\left(\prod_{j=1}^{m}y_j^{\alpha_j+r_j-1}dy_j\right)$$

$$=\frac{\Gamma\left(\sum_{1}^{m+1}\alpha_j\right)\Gamma(\alpha_{m+1})\prod_{1}^{m}\Gamma(\alpha_j+r_j)}{\prod_{j=1}^{m+1}\Gamma(\alpha_j)\Gamma\left(\sum_{1}^{m}(\alpha_j+r_j)+\alpha_{m+1}\right)}$$

$$=\prod_{j=1}^{m}\alpha_j^{[r_j]}\Big/\left(\sum_{j=1}^{m+1}\alpha_j\right)^{[r_1+\cdots+r_m]}.$$ □

In particular,

$$E(Y_j) = \alpha_j/\alpha \quad \text{and} \quad \operatorname{var}(Y_j) = \frac{\alpha_j(\alpha-\alpha_j)}{\alpha^2(\alpha+1)},$$

where $\alpha = \alpha_1 + \ldots + \alpha_{m+1}$, and the covariance between Y_i and Y_j is

$$\frac{-\alpha_i\alpha_j}{\alpha^2(\alpha+1)}.$$

Theorem 2.4.1 points out that there is a close relationship between the chi-squared distribution and the Dirichlet distribution. The following is a further result.

Lemma 2.4.2. *Assume that* $(Z_1,\dots,Z_m) \sim D_{m+1}(\frac{1}{2}n_1,\dots,\frac{1}{2}n_m;\frac{1}{2}n_{m+1})$ *where* $n_1,\dots,n_{m+1}$ *are integers with* $n_1+\dots+n_{m+1}=n$, *and* $Y_0, Y_1,\dots,Y_m$ *are independently distributed according to the chi-squared distributions with* $n, n_1,\dots,n_m$ *degrees of freedom respectively. Then*

$$\phi_{\log z_1,\dots,\log z_m}(t_1,\dots,t_m) = \frac{\prod_1^m \phi_{\log} Y_j(t_j)}{\phi_{\log} Y_0(t_1+\dots+t_m)} \tag{2.4.9}$$

$$= \frac{\Gamma(\frac{1}{2}n)}{\Gamma(\frac{1}{2}n+i(t_1+\dots+t_m))}\prod_{j=1}^m \frac{\Gamma(\frac{1}{2}n_j+it_j)}{\Gamma(\frac{1}{2}n_j)}.$$

Proof. Let $\boldsymbol{x} \sim N_n(\boldsymbol{0}, \boldsymbol{I}_n)$ and $\boldsymbol{x}^{(1)},\dots,\boldsymbol{x}^{(m+1)}$ have the same meaning as in (1.5.5). From the following Example 2.4.1. we have $\|\boldsymbol{x}^{(j)}\|^2 \overset{d}{=} Y_j$, $j=1,\dots,m$, and $\|\boldsymbol{x}\|^2 \overset{d}{=} Y_0$, and

$$\begin{bmatrix} \|\boldsymbol{x}^{(1)}\|^2 \\ \vdots \\ \|\boldsymbol{x}^{(m)}\|^2 \end{bmatrix} = \begin{bmatrix} \|\boldsymbol{x}^{(1)}\|^2/\|\boldsymbol{x}\|^2 \\ \vdots \\ \|\boldsymbol{x}^{(m)}\|^2/\|\boldsymbol{x}\|^2 \end{bmatrix} \cdot \|\boldsymbol{x}\|^2.$$

$$\begin{bmatrix} \log\|\boldsymbol{x}^{(1)}\|^2 \\ \vdots \\ \log\|\boldsymbol{x}^{(m)}\|^2 \end{bmatrix} \overset{d}{=} \log\|\boldsymbol{x}\|^2 \, \mathbb{1}_m + \begin{bmatrix} \log\|\boldsymbol{x}^{(1)}\|^2/\|\boldsymbol{x}\|^2 \\ \vdots \\ \log\|\boldsymbol{x}^{(m)}\|^2/\|\boldsymbol{x}\|^2 \end{bmatrix}, \text{ where } \mathbb{1}_m = (1,\dots,1)'.$$

The first part of (2.4.9) follows from the fact that $\|\boldsymbol{x}\|^2$ and $(\|\boldsymbol{x}^{(1)}\|^2/\|\boldsymbol{x}\|^2,\dots,\|\boldsymbol{x}^{(m)}\|^2/\|\boldsymbol{x}\|^2) \overset{d}{=} (Z_1,\dots,Z_m)$ are independent. Noting that

$$E(e^{it\log Y_j}) = (2^{\frac{1}{2}n_j}\Gamma(\tfrac{1}{2}n_j))^{-1}\int_0^\infty e^{it\log y - \frac{1}{2}y}y^{\frac{1}{2}n_j-1}\,dy$$

$$= 2^{it}\Gamma(\tfrac{1}{2}n_j+it)/\Gamma(\tfrac{1}{2}n_j),$$

the second part of (2.4.9) then follows. □

By means of the density of the Dirichlet distribution we can obtain all the marginal densities of $\boldsymbol{x}/\|\boldsymbol{x}\|$. By taking $m=k+1$ and $n_1=\dots=n_k=1$ in Corollary 2 of Theorem 2.4.1., then $(X_1^2/\|\boldsymbol{x}\|^2,\dots,X_k^2/\|\boldsymbol{x}\|^2) \sim D_{k+1}(\frac{1}{2},\dots,\frac{1}{2};\frac{1}{2}(n-k))$ for $0<k$

$< n$. Form this fact, it can be verified that the density of $(|X_1|/\|\boldsymbol{x}\|, \dots, |X_k|/\|\boldsymbol{x}\|)$ is

$$\text{(2.4.10)} \qquad \frac{\Gamma(\frac{1}{2}n)2^k}{\Gamma(\frac{1}{2}(n-k))\pi^{\frac{1}{2}k}}\Big(1 - \sum_1^k x_i^2\Big)^{\frac{1}{2}(n-k)-1}, \quad \text{if } x_i \geqslant 0,\ i = 1, \dots, k$$

$$x_1^2 + \dots + x_k^2 < 1$$

and the density of $(X_1/\|\boldsymbol{x}\|, \dots, X_k/\|\boldsymbol{x}\|)$ is

$$\text{(2.4.11)} \qquad \frac{\Gamma(\frac{1}{2}n)}{\Gamma(\frac{1}{2}(n-k))\pi^{\frac{1}{2}k}}\Big(1 - \sum_1^k x_i^2\Big)^{\frac{1}{2}(n-k)-1}, \qquad \text{if } \sum_1^k x_i^2 < 1.$$

We can generalize (2.4.4) to a more general case. For any nonnegative measurable function $f(\cdot)$, let

$$\text{(2.4.12)} \qquad I_m(f \mid \alpha_1, \dots, \alpha_m) = \int\limits_{E(x_1, \dots x_m)} f\Big(\sum_1^m x_i\Big)\Big(\prod_1^m x_i^{\alpha_i - 1}\, dx_i\Big),$$

where $E(x_1, \dots, x_m) = \{(x_1, \dots, x_m) : x_i \geqslant 0,\ i = 1, \dots, m\}$.

Lemma 2.4.3. *For $\alpha_i > 0$, $i = 1, \dots, m$, we have*

$$\text{(2.4.13)} \qquad \int\limits_{R^m} f\Big(\sum_1^m x_i^2\Big)\Big(\prod_1^m |x_i|^{2\alpha_i - 1}\, \mathrm{d}x_i\Big) = I_m(f \mid \alpha_1, \dots, \alpha_m).$$

Proof. By the symmetry of $x_1, \dots, x_m$ about the origin, we have

$$I = \int\limits_{R^m} f\Big(\sum_{i=1}^m x_i^2\Big)\Big(\prod_{i=1}^m |x_i|^{2\alpha_i - 1}\, dx_i\Big) = 2^m \int\limits_{E(x_1, \dots x_m)} f\Big(\sum_{i=1}^m x_i^2\Big)\prod_{i=1}^m x_i^{2\alpha_i - 1}\, dx_i.$$

Let $u_i = x_i^2$, $i = 1, \dots, m$. The Jacobian is $\Big(2^m \prod_{i=1}^m u_i^{\frac{1}{2}}\Big)^{-1}$. Hence we have

$$I = \int\limits_{E(u_1, \dots, u_m)} f\Big(\sum_{i=1}^m u_i\Big)\Big(\prod_{i=1}^m u_i^{\alpha_i - 1}\, du_i\Big) = I(f \mid \alpha_1, \dots, \alpha_m). \qquad \square$$

Lemma 2.4.4.

$$\text{(2.4.14)} \qquad I_m(f \mid \alpha_1, \dots, \alpha_m) = B_m(\alpha_1, \dots, \alpha_m) I_1\Big(f \mid \sum_{i=1}^m \alpha_i\Big).$$

Proof. Making the transformation

$$\begin{cases} y_i = x_i, & i = 1, \dots, m-1, \\ y_m = x_1 + \dots + x_m \end{cases}$$

in (2.4.12) and noting that its Jacobian is 1, we have

$$I_m(f \mid \alpha_1, \dots, \alpha_m) = \int_D f(y_m)\left(y_m - \sum_{i=1}^{m-1} y_i\right)^{\alpha_m - 1}\left(\prod_{i=1}^{m} y_i^{\alpha_i - 1} dy_i\right),$$

where

$$D = \left\{(y_1, \dots, y_m) \mid y_i \geqslant 0,\ i = 1, \dots, m;\ \sum_{i=1}^{m-1} y_i \leqslant y_m\right\}.$$

Consider another transformation $u_i = y_i / y_m$, $i = 1, \dots, m-1$; $y = y_m$. The Jacobian is y^{m-1}. Thus

$$I_m(f \mid \alpha_1, \dots, \alpha_m) = \int_{D(u_1, \dots, u_{m-1})} \left(1 - \sum_{1}^{m-1} u_i\right)^{\alpha_m - 1}\left(\prod_{1}^{m-1} u_i^{\alpha_i - 1} du_i\right) \cdot \int_0^\infty f(y) y^{\sum_1^m \alpha_i - 1} dy$$

which completes the proof. □

Here are some applications of the above lemmas.

Example 2.4.1. Taking $\alpha_i = \frac{1}{2}$, $i = 1, \dots, m$, in (2.4.13), we obtain

$$\int f\left(\sum_1^m x_i^2\right) dx_1, \dots, dx_m = \frac{(\pi)^{\frac{1}{2}m}}{\Gamma(\frac{1}{2}m)} \int_0^\infty y^{\frac{1}{2}m - 1} f(y) dy = (\pi)^{\frac{1}{2}m}\, \Gamma(\tfrac{1}{2}m)^{-1}\, I_1(f \mid \tfrac{1}{2}m). \tag{2.4.15}$$

This formula is very useful. For example, putting

$$f(y) = \begin{cases} (2\pi)^{-\frac{1}{2}m} e^{-y/2} & \text{if } y \leqslant x, \\ 0 & \text{if } y > x, \end{cases}$$

in (2.4.15), we have

$$\int_{\sum_1^m x_i^2 \leqslant x} (2\pi)^{-\frac{1}{2}m} e^{-\frac{1}{2}(x_1^2 + \dots + x_m^2)} dx_1 \dots dx_m$$

$$= (\pi)^{\frac{1}{2}m} \Gamma(\tfrac{1}{2}m)^{-1} \int_0^x (2\pi)^{-\frac{1}{2}m} e^{-\frac{1}{2}y} y^{\frac{1}{2}m - 1} dy$$

$$= 2^{-\frac{1}{2}m}\Gamma(\tfrac{1}{2}m)^{-1}\int_0^x y^{\frac{1}{2}m-1}e^{-\frac{1}{2}y}\,dy.$$

It means that if $\boldsymbol{x} \sim N_m(\boldsymbol{0}, \boldsymbol{I}_m)$, then the distribution of $\boldsymbol{x}'\boldsymbol{x}$ is given by the above formula, i.e., the chi-squared distribution with m degrees of freedom.

Example 2.4.2. Putting

$$f(y) = \begin{cases} 1 & \text{if } y \leqslant a^2 \\ 0 & \text{if } y > a^2 \end{cases}$$

in (2.4.15), we obtain the volume of an m-dimensional sphere of radius a, which is

(2.4.16)
$$\int\limits_{x_1^2+\ldots+x_m^2\leqslant a^2} dx_1 \ldots dx_m = (\pi)^{\frac{1}{2}m}\Gamma(\tfrac{1}{2}m)^{-1}\int_0^{a^2} y^{\frac{1}{2}m-1}\,dy$$

$$= (\pi)^{\frac{1}{2}m}\Gamma(\tfrac{1}{2}(m+2))^{-1}a^m = \frac{2(\pi)^{m/2}}{m\Gamma(m/2)}a^m.$$

We can extend (2.4.15) to the multivariate case.

Lemma 2.4.5. *For every nonnegative measurable function f, we have*

(2.4.17)
$$\underset{R^{n_1}\times\cdots\times R^{n_m}}{\int\cdots\int} f\left(\sum_{i=1}^{n_1} x_{1i}^2, \ldots, \sum_{i=1}^{n_m} x_{mi}^2\right)\prod_{i=1}^{m}\prod_{j=1}^{n_i} dx_{ij}$$

$$= \frac{(\pi)^{\frac{1}{2}(n_1+\cdots+n_m)}}{\prod_{i=1}^{m}\Gamma(\frac{1}{2}n_i)}\int_0^\infty\cdots\int_0^\infty f(u_1, \ldots, u_m)\left(\prod_1^m u_i^{\frac{1}{2}n_i-1}\,du_i\right).$$

2.5. Spherical Distributions

Multivariate normal distribution has been the central subject treated in multivariate analysis. Statisticians have been trying to extend the theory of multivariate analysis to more general cases. Recently, many statisticians have been interested in a class of distributions whose contours of equal density have the same elliptical shape as the normal and which contains long-tailed and short-tailed distributions (relative to the normal). This is called a class of elliptically contoured distributions, or simply, one of elliptical distributions.

Definition 2.5.1. If the c.f. of an n-dimensional random vector $\boldsymbol{x}$ has the form $\exp(i\boldsymbol{t}'\mu)\,\phi(\boldsymbol{t}'\Sigma\boldsymbol{t})$, where μ: $n \times 1$, Σ: $n \times n$, and $\Sigma \geqslant 0$, we say that $\boldsymbol{x}$ is *distributed according to an elliptically contoured distribution with parameters* μ, Σ, and ϕ, and we write $\boldsymbol{x} \sim EC_n(\mu, \Sigma, \phi)$. In particular, when $\mu = \boldsymbol{0}$ and $\Sigma = \boldsymbol{I}_n$, $EC_n(\boldsymbol{0}, \boldsymbol{I}_n, \phi)$ is called a *spherical distribution* and denoted by $S_n(\phi)$.

The class of spherical distributions is studied in this section and the class of elliptical distributions will be discussed in the next section.

2.5.1. Uniform Distribution and Its Stochastic Representation

Let $\boldsymbol{u}^{(n)}$ denote a random vector distributed uniformly on the unit sphere in R^n. We will show that $\boldsymbol{u}^{(n)}$ is distributed according to a spherical distribution. In order to show this fact, we need the following lemma:

Lemma 2.5.1. *Assume that $g(.)$ is a nonnegative Borel function and $\boldsymbol{a} = (a_1, \ldots, a_n)' = \boldsymbol{0}$. Then*

$$(2.5.1)\qquad \int\limits_{S:\boldsymbol{x}'\boldsymbol{x} = c^2} g(\boldsymbol{a}'\boldsymbol{x})\, ds = \frac{2c\pi^{(n-1)/2}}{\Gamma(\frac{1}{2}(n-1))} \int_{-c}^{c} g(\|\boldsymbol{a}\| y)(c^2 - y^2)^{\frac{1}{2}(n-3)} dy.$$

Proof. Let $\boldsymbol{T}$ be an $n \times n$ orthogonal matrix with the first row $(a_1 / \|\boldsymbol{a}\|, \ldots, a_n / \|\boldsymbol{a}\|)$ and take the transformation $\boldsymbol{y} = \boldsymbol{T}\boldsymbol{x}$. Then $\boldsymbol{y}'\boldsymbol{y} = \boldsymbol{x}'\boldsymbol{x} = c^2$, $y_1 = \boldsymbol{a}'\boldsymbol{x} / \|\boldsymbol{a}\|$. We can take $\boldsymbol{T}$ such that $|\boldsymbol{T}| = 1$, because we can interchange rows of $\boldsymbol{T}$ except the first row. Consider that $x_1, \ldots, x_{n-1}$ and $y_1, \ldots, y_{n-1}$ are independent variables. Hence $x_n = \pm (c^2 - x_1^2 - \cdots - x_{n-1}^2)^{\frac{1}{2}}$ and

$$\left|\frac{\partial(y_1, \ldots, y_{n-1})}{\partial(x_1, \ldots, x_{n-1})}\right| = \begin{vmatrix} t_{11} - t_{1n}x_1/x_n & \cdots & t_{1,n-1} - t_{1n}x_{n-1}/x_n \\ \vdots & & \vdots \\ t_{n-1,1} - t_{n-1,n}x_1/x_n & \cdots & t_{n-1,n-1} - t_{n-1,n}x_{n-1}/x_n \end{vmatrix}$$
$$= t_{nn} + t_{n1}x_1/x_n + \cdots + t_{n,n-1}x_{n-1}/x_n$$
$$= y_n / x_n$$

by using $|\boldsymbol{T}| = 1$. Therefore

$$\overbrace{\int\cdots\int}^{n}_{S:\boldsymbol{x}'\boldsymbol{x} = c^2} g(\boldsymbol{a}'\boldsymbol{x})\, ds = c \overbrace{\int\cdots\int}^{n-1}_{\boldsymbol{x}'\boldsymbol{x} = c^2} g(\boldsymbol{a}'\boldsymbol{x}) / |x_n|\, dx_1 \ldots dx_{n-1}$$

$$= 2c \overbrace{\int\cdots\int}^{n-1}_{y_1^2 + \ldots + y_{n-1}^2 \leqslant c^2} g(\|\boldsymbol{a}\| y_1) \frac{dy_1 \ldots dy_{n-1}}{\sqrt{c^2 - y_1^2 - \ldots - y_{n-1}^2}}$$

$$= 2c \int_{-c}^{c} g(\|\boldsymbol{a}\| y_1)\, dy_1 \overbrace{\int\cdots\int}^{n-2}_{y_2^2 + \ldots + y_{n-1}^2 \leqslant c^2 - y_1^2} \frac{dy_2 \ldots dy_{n-1}}{\sqrt{(c^2 - y_1^2) - y_2^2 - \cdots - y_{n-1}^2}}$$

$$= 2c \int_{-c}^{c} \frac{\pi^{\frac{1}{2}(n-1)}}{\Gamma((n-1)/2)} (c^2 - y_1^2)^{\frac{1}{2}(n-3)} g(\|\boldsymbol{a}\| y_1)\, dy_1$$

$$= \frac{2c\pi^{\frac{1}{2}(n-1)}}{\Gamma((n-1)/2)} \int_{-c}^{c} g(\|\boldsymbol{a}\| y)(c^2 - y^2)^{\frac{1}{2}(n-3)} dy \qquad \square$$

Theorem 2.5.1. *The c.f. of $\boldsymbol{u}^{(n)}$ is*

$$(2.5.2)\qquad \Omega_n(\|\boldsymbol{t}\|^2) \equiv \int_0^{\pi} \exp(i\,\|\boldsymbol{t}\| \cos\theta) \sin^{n-2}\theta\, d\theta \,/\, B(\frac{1}{2}(n-1), \frac{1}{2}).$$

Thus $\boldsymbol{u}^{(n)}$ is distributed according to a spherical distribution.

Proof. Let S_n denote the surface area of the unit sphere in R^n. Then

$$S_n = 2\pi^{n/2}/\Gamma(n/2)$$

from Exercise 2.9. By Lemma 2.5.1., we have

$$\begin{aligned}
\phi_{u^{(n)}}(t) &= \frac{1}{S_n}\int\cdots\int_{S:\boldsymbol{x}'\boldsymbol{x}=1} e^{it'x}\,ds \\
&= \frac{1}{S_n}\frac{2\pi^{\frac{1}{2}(n-1)}}{\Gamma((n-1)/2)}\int_{-1}^{1} e^{i\|\boldsymbol{t}\|x}(1-x^2)^{\frac{1}{2}(n-3)}\,dx \\
&= \frac{\Gamma(n/2)}{\pi^{\frac{1}{2}}\Gamma((n-1)/2)}\int_{-1}^{1} e^{i\|\boldsymbol{t}\|x}(1-x^2)^{\frac{1}{2}(n-3)}\,dx \ (\text{taking } x=\cos\theta) \\
&= \frac{\Gamma(n/2)}{\pi^{\frac{1}{2}}\Gamma((n-1)/2)}\int_{0}^{\pi} e^{i\|\boldsymbol{t}\|\cos\theta}\sin^{n-2}\theta\,d\theta.
\end{aligned}$$

Hence the theorem is proved. □

Denote

(2.5.3) $$\Phi_n = \{\phi(.)|\phi(t_1^2+\ldots+t_n^2) \text{ is a c. f.}\}.$$

It is easy to see that

(2.5.4) $$\Phi_1 \supset \Phi_2 \supset \Phi_3 \cdots\cdots.$$

Let

(2.5.5) $$\Phi_\infty = \bigcap_{n=1}^{\infty}\Phi_n.$$

Exercise 2.11 shows that ϕ_{n+1} is a proper subset of ϕ_n for $n \geqslant 1$.

Theorem 2.5.2. *A function $\phi(.)\in\Phi_k$ if and only if*

(2.5.6) $$\phi(x) = \int_0^\infty \Omega_k(xr^2)\,dF(r),$$

where $\Omega_k(.)$ is defined in Theorem 2.5.1 and $F(.)$ is a cdf over $[0, \infty)$.

Proof. Assume $\phi(.)\in\Phi_k$. Then $g(t_1, \ldots, t_k) \equiv \phi(\|\boldsymbol{t}\|^2)$ is a c.f. of some random vector $\boldsymbol{y}$ with cdf $G(\boldsymbol{y})$. Thus $g(t_1, \ldots t_k)$ is a symmetric function of $t_1, \ldots, t_k$. For every $\boldsymbol{x}$ with $\|\boldsymbol{x}\| = 1$, we have

$$g(t_1, \ldots, t_k) = \phi(\boldsymbol{t}'\boldsymbol{t}) = g(\|\boldsymbol{t}\|x_1, \ldots, \|\boldsymbol{t}\|x_k)$$

and

$$\phi(\boldsymbol{t}'\boldsymbol{t}) = \frac{1}{S_k}\int_{S:\|\boldsymbol{x}\|=1} g(\|\boldsymbol{t}\|x_1, \ldots, \|\boldsymbol{t}\|x_k)\,ds$$

$$= \frac{1}{S_k} \int\limits_{S:\|x\|=1} \left[\int\limits_{R^k} e^{i\|t\|x'y} dG(y) \right] ds$$

$$= \int\limits_{R^k} \left[\frac{1}{S_k} \int\limits_{S:\|x\|=1} e^{i\|t\|x'y} ds \right] dG(y) = \int\limits_{R^k} \Omega_k(\|t\|^2 \|y\|^2) dG(y).$$

Let

$$F(u) = \int\limits_{\|y\| \leqslant u} dG(y).$$

Then F(.) is a cdf over $[0, \infty)$ and

$$\phi(x) = \int_0^\infty \Omega_k(xu^2) dF(u).$$

Assume that $\phi(.)$ can be expressed in the form of (2.5.6), and make a random variable $R \sim F(x)$ independent of $u^{(k)}$. Then the c.f. of $Ru^{(k)}$ is

$$E(e^{it'Ru^{(k)}}) = \int_0^\infty E(e^{irt'u^{(k)}}) \, dF(r)$$

$$= \int_0^\infty \Omega_k(r^2 \|t\|^2) dF(r) = \phi(\|t\|^2),$$

i.e., $\phi(\|t\|^2)$ is the c.f. of $Ru^{(k)}$ and $\phi(.) \in \Phi_k$.

The two above theorems are due to Schoenberg (1938). The following important corollaries are the consequence of Theorem 2.5.2. .

Corollary 1. *Assume that the c.f. of a $k \times 1$ random vector x is $\phi(t't)$ and $\phi \in \Phi_k$. Then x has a stochastic representation*

(2.5.7) $$x \overset{d}{=} Ru^{(k)},$$

where $R \sim F(x)$ is related to ϕ as in (2.5.6) and is independent of $u^{(k)}$.

Let $O(n)$ denote the class of $n \times n$ orthogonal matrices.

Corollary 2. *An $n \times 1$ random vector $x \sim S_n(\phi)$ if and only if, for every $\Gamma \in O(n)$,*

(2.5.8) $$x \overset{d}{=} \Gamma x.$$

Proof. Assume $x \sim S_n(\phi)$. The c.f. of x is $\phi(t't) = \phi(\|t\|^2)$ for some $\phi \in \Phi_n$. Thus the c.f. of Γx is $\phi(t'\Gamma\ \Gamma't) = \phi(t't)$ and Formula (2.5.8) follows.

Assume that (2.5.8) holds for every $\Gamma \in O(n)$. The c.f. of x satisfies $\psi(t) = \psi(\Gamma t)$ for every $\Gamma \in O(n)$. Therefore $\psi(\cdot)$ must be of the form $\phi(\|t\|^2)$ for some $\phi \in \Phi_n$ by Theorem 1.7.1 and Example 1.7.1. □

Now we can summarize the above results as follows:

Theorem 2.5.3. *Assume that x is an $n \times 1$ random vector. Them the following statements are equivalent:*

(1) *The c.f. of x has the form $\phi(\|t\|^2)$, where $\phi \in \Phi_n$;*

(2) *x has a stochastic representation $x \overset{d}{=} Ru^{(n)}$, where $R \geqslant 0$ is independent of $u^{(n)}$;*

(3) *$x \overset{d}{=} \Gamma x$ for every $\Gamma \in O(n)$.*

Corollary 1. *Suppose $x \overset{d}{=} Ru^{(n)} \sim S_n(\phi)$ and $P(x = 0) = 0$. Then*

$$\|x\| \overset{d}{=} R, \qquad x/\|x\| \overset{d}{=} u^{(n)}, \tag{2.5.9}$$

and they are independent.

Proof. As $P(x = 0) = P(R = 0) = 0$, take $f_1(x) = (x'x)^{\frac{1}{2}}$ and $f_2(x) = x/\sqrt{x'x}$. Then

$$\begin{bmatrix} \|x\| \\ x/\|x\| \end{bmatrix} = \begin{bmatrix} f_1(x) \\ f_2(x) \end{bmatrix} \overset{d}{=} \begin{bmatrix} f_1(Ru^{(n)}) \\ f_2(Ru^{(n)}) \end{bmatrix} = \begin{bmatrix} R \\ u^{(n)} \end{bmatrix}$$

which completes the proof. □

From now on we denote $x \sim S_n^+(\phi)$ if $x \sim S_n(\phi)$ and $P(x = 0) = 0$.

The distribution of $x/\|x\|$ is independent of any special element of the class of $S_n^+(\phi)$'s. In particular, we can assume $x \sim N_n(0, I_n)$. This is a very important fact and we will use it many times. First, we obtain the following results.

Corollary 2.

$$\mathcal{E}(u^{(n)}) = 0, \quad \mathcal{D}(u^{(n)}) = \frac{1}{n} I_n. \tag{2.5.10}$$

Proof. Let $x \sim N_n(0, I_n)$. Form Corollary 1 we have $x \overset{d}{=} \|x\| u^{(n)}$, where $\|x\|$ is independent of $u^{(n)}$. It is a well-known fact that $\|x\|^2 \sim \chi_n^2$. The assertion follows from $\mathcal{E}(x) = 0$, $E\|x\| > 0$, $E\|x\|^2 = n$ *and* $\mathcal{D}(x) = I_n$. □

Denote $u^{(n)} = (u_1, \ldots, u_n)'$. As $u^{(n)'} u^{(n)} = 1$, $u^{(n)}$ does not have a density, but all the marginal distributions of $u^{(n)}$ will have densities. By (2.4.11) the marginal density of $u_1, \cdots, u_k$ is

$$\frac{\Gamma(\frac{1}{2}n)}{\Gamma(\frac{1}{2}(n-k))\pi^{\frac{1}{2}k}} \left(1 - \sum_1^k u_i^2\right)^{\frac{1}{2}(n-k)-1}, \qquad \text{if } \sum_1^k u_i^2 < 1. \tag{2.5.11}$$

Corollary 3. *Partition $u^{(n)}$ into m parts $u^{(n)} = (u'_{(1)}, \cdots, u'_{(m)})'$ with $n_1, \cdots, n_m$ components of $u^{(n)}$. Then*

$$\begin{bmatrix} u_{(1)} \\ \vdots \\ u_{(m)} \end{bmatrix} \overset{d}{=} \begin{bmatrix} d_1 u_1 \\ \vdots \\ d_m u_m \end{bmatrix}, \tag{2.5.12}$$

where $d_i \geqslant 0$, $i = 1, \ldots, m$; $(d_1^2, \ldots, d_m^2) \sim D_m(\frac{1}{2}n_1, \ldots, \frac{1}{2}n_m)$ is independent of $(u_1, \ldots, u_m)$; $u_1, \ldots, u_m$ are independent and $u_j \overset{d}{=} u^{(n_j)}$, $j = 1, \ldots, m$.

Proof. Let $\boldsymbol{x} \sim N_n(\mathbf{0}, \boldsymbol{I}_n)$ be partitioned as (2.4.5). By Corollary 1, we have

$$\begin{bmatrix} \boldsymbol{u}_{(1)} \\ \vdots \\ \boldsymbol{u}_{(m)} \end{bmatrix} \stackrel{d}{=} \begin{bmatrix} \boldsymbol{x}^{(1)} / \|\boldsymbol{x}\| \\ \vdots \\ \boldsymbol{x}^{(m)} / \|\boldsymbol{x}\| \end{bmatrix} = \begin{bmatrix} \boldsymbol{x}^{(1)} / \|\boldsymbol{x}^{(1)}\| \cdot \|\boldsymbol{x}^{(1)}\| / \|\boldsymbol{x}\| \\ \vdots \\ \boldsymbol{x}^{(m)} / \|\boldsymbol{x}^{(m)}\| \cdot \|\boldsymbol{x}^{(m)}\| / \|\boldsymbol{x}\| \end{bmatrix}.$$

Let $d_j = \|\boldsymbol{x}^{(j)}\| / \|\boldsymbol{x}\|$, $j = 1, \ldots, m$. *Clearly*, $(d_1, \ldots, d_m)$ satisfies the required condition (cf. Corollary 2 of Theorem 2.4.1). Applying Corollary 1 again, $\boldsymbol{u}_j = \boldsymbol{x}^{(j)} / \|\boldsymbol{x}^{(j)}\| \stackrel{d}{=} \boldsymbol{u}^{(n_j)}$, $j = 1, \ldots, m$. The corollary follows from the facts that $\{\boldsymbol{x}^{(j)}, j = 1, \ldots, m\}$ are independent, and $\|\boldsymbol{x}^{(j)}\|$ is independent of $\boldsymbol{x}^{(j)} / \|\boldsymbol{x}^{(j)}\|$. □

We now wish to show that the relationship (2.5.6) between $\phi \in \Phi_k$ and F is one-to-one.

Theorem 2.5.4. $\phi \in \Phi_k$ *if and only if* ϕ *is continuous and*

$$\int_0^\infty \phi(2sv)\, g_k(V)\, dv. \qquad s \geqslant 0 \tag{2.5.13}$$

is the Laplace transform of a nonnegative random variable whose distribution function is $F(\sqrt{\cdot})$ *when* ϕ *is given by* (2.5.6), *where* g_k *denotes the chi-squared density with k degrees of freedom* $(1 \leqslant k \leqslant \infty)$.

Proof. Since $N_k(\mathbf{0}, \boldsymbol{I}_k)$ is $S_k(\phi)$ with $\phi(u) = \exp(-u/2)$, (2.5.6) yields the identity

$$\exp(-u/2) = \int_0^\infty \Omega_k(uv)\, g_k(v)\, dv, \qquad u \geqslant 0, \tag{2.5.14}$$

where g_k is the density of the quadratic form $R^2 = \boldsymbol{x}'\boldsymbol{x}$ (cf. (2.5.9)) with $\boldsymbol{x} \sim N_k(\mathbf{0}, \boldsymbol{I}_k)$. When $\phi \in \Phi_k$, it follows immediately from (2.5.6) and (2.5.14), so that (2.5.13) is the Laplace transform of R^2 where R has the cdf F. Conversely, suppose (2.5.13) is the Laplace transform of R^2 and ϕ is continuous. Let F be the cdf of R, and let $\phi_0 \in \Phi_k$ be the function defined in (2.5.6). From the first part of the proof we have

$$\int_0^\infty \phi_0(2sv)\, g_k(v)\, dv = \int_0^\infty \phi(2sv)\, g_k(v)\, dv. \qquad s \geqslant 0$$

and since $g_k(v) = c v^{\frac{1}{2}k-1} \exp(-v/2)$ (cf. Section 2.4, Example 2.4.1.), the uniqueness of Laplace transform implies that $\phi_0 = \phi$. □

This theorem is from Cambians, Huang and Simons (1981). In fact, we can prove that the relationship between ϕ and F is one-to-one by using the operation $\stackrel{d}{=}$ (cf. Exercise 2.13).

If $\phi(.) \in \Phi_n$, then for each $1 \leqslant m \leqslant n$, $\phi \in \Phi_m$. As the relationship (2.5.6) between ϕ and F is one-to-one, there exist two cdf's F_n and F_m such that

$$\phi(x) = \int_0^\infty \Omega_n(xu^2)\,dF_n(u),$$

and

$$\phi(x) = \int_0^\infty \Omega_m(xu^2)\,dF_m(u).$$

The precise relationship between the various F_n and F_m is specified in the following

Corollary. *If R_n and R_m have cdfs F_n and F_m, $1 \leqslant m \leqslant n$, respectively, then $R_m \overset{d}{=} R_n b_{\frac{1}{2}m, \frac{1}{2}(n-m)}$, where $b_{\frac{1}{2}m, \frac{1}{2}(n-m)} \geqslant 0$ and $b^2_{\frac{1}{2}m, \frac{1}{2}(n-m)} \sim B(\frac{1}{2}m, \frac{1}{2}(n-m))$ is independent of R_n.*

Proof. Let $\boldsymbol{x} = (\boldsymbol{x}^{(1)\prime}, \boldsymbol{x}^{(2)\prime})' \sim S_n(\phi)$, where $\boldsymbol{x}^{(1)}: m \times 1$. Then the c.f. of $\boldsymbol{x}^{(1)}$ is $\phi(\|\boldsymbol{t}_{(1)}\|^2)$, where $\boldsymbol{t}^{(1)}: m \times 1$. On the other hand, by Corollary 3 of Theorem 2.5.3, $\boldsymbol{x}^{(1)} \overset{d}{=} R_n d_1 \boldsymbol{u}_1^{(m)}$ which completes the proof. □

The above relationship between ϕ and R_n (or R_m) can be written as $\phi \in \Phi_n \leftrightarrow R_n$ (or F_n) and $\phi \in \Phi_m \leftrightarrow R_m \overset{d}{=} R_n b_{\frac{1}{2}m, \frac{1}{2}(n-m)}$ (or F_m). It can be verified that R_m has a density over $(0, \infty)$, which is

$$(2.5.15) \qquad f_m(x) = \frac{2x^{m-1}\Gamma(\frac{1}{2}n)}{\Gamma(\frac{1}{2}m)\Gamma(\frac{1}{2}(n-m))} \int_x^\infty r^{-(n-2)}(r^2 - x^2)^{\frac{1}{2}(n-m)-1}\,dF_n(r).$$

2.5.2. Densities

If $\boldsymbol{x} \sim S_n(\phi)$, in general, it is not necessary that $\boldsymbol{x}$ has a density. It is easy to see from Theorem 2.5.3 that the density of $\boldsymbol{x}$ must have the form of $f(\boldsymbol{x}'\boldsymbol{x})$ for some nonnegative function f(.) if $\boldsymbol{x}$ has a density. In this case, by (2.4.15) we have

$$1 = \int f(\boldsymbol{x}'\boldsymbol{x})\,d\boldsymbol{x} = \frac{\pi^{\frac{1}{2}n}}{\Gamma(\frac{1}{2}n)} \int_0^\infty y^{\frac{1}{2}n-1} f(y)\,dy.$$

Hence, a function $f(.)$ can be used to define a density $cf(.)$ of spherical distribution if and only if

$$(2.5.16) \qquad \int_0^\infty y^{\frac{1}{2}n-1} f(y)\;dy < \infty.$$

In this case, we write $\boldsymbol{x} \sim S_n(cf)$ instead of $\boldsymbol{x} \sim S_n(\phi)$ if the density of $\boldsymbol{x}$ exists.

Theorem 2.5.5. *Let $\boldsymbol{x} \overset{d}{=} R\boldsymbol{u}^{(n)} \sim S_n(\phi)$. Then $\boldsymbol{x}$ has a density $f(.)$ if and only if R has a density $g(.)$. And there is a relationship between $f(.)$ and $g(.)$ as follows:*

$$(2.5.17) \qquad g(r) = \frac{2\pi^{\frac{1}{2}n}}{\Gamma(\frac{1}{2}n)} r^{n-1} f(r^2).$$

Proof. Assume that $\boldsymbol{x}$ has a density $f(\boldsymbol{x}'\boldsymbol{x})$. Let $h(.)$ be any nonnegative Borel function. By using (2.4.15) we have

$$E[h(R)] = \int h[(\boldsymbol{x}'\boldsymbol{x})^{\frac{1}{2}}] f(\boldsymbol{x}'\boldsymbol{x})\, d\boldsymbol{x}$$

$$= \frac{\pi^{\frac{1}{2}n}}{\Gamma(\frac{1}{2}n)} \int_0^\infty h(y^{\frac{1}{2}})\, y^{\frac{1}{2}n-1} f(y)\, dy \qquad (let\ r = y^{\frac{1}{2}})$$

$$= \frac{2\pi^{\frac{1}{2}n}}{\Gamma(\frac{1}{2}n)} \int_0^\infty h(r) r^{n-1} f(r^2)\, dr$$

which shows that R has a density $g(.)$ of the form (2.5.17). The converse is obvious. □

Theorem 2.5.6. *Assume* $\boldsymbol{x} \stackrel{d}{=} R\boldsymbol{u}^{(n)} \sim S_n^+(\phi)$. *Then all the marginal distributions of* $\boldsymbol{x}$ *will have densities. In particular, the marginal density of* $X_1, \ldots, X_k$ $(1 \leqslant k < n)$ *is*

$$(2.5.18) \qquad \frac{\Gamma(\frac{1}{2}n)}{\Gamma(\frac{1}{2}(n-k))\, \pi^{\frac{1}{2}k}} \int_{\left(\sum_1^k x_i^2\right)^{\frac{1}{2}}}^\infty r^{-(n-2)} \left(r^2 - \sum_1^k x_i^2\right)^{\frac{1}{2}(n-k)} dF(r),$$

where $F(r)$ *is the cdf of* R.

Proof. The first assertion is obvious. We only prove (2.5.18). By (2.5.11), the cdf of $X_1, \ldots, X_k$ is

$$\frac{\Gamma(\frac{1}{2}n)}{\Gamma(\frac{1}{2}(n-k))\, \pi^{\frac{1}{2}k}} \int_0^\infty dF(r) \int_{-1}^{x_1/r} \cdots \int_{-1}^{x_k/r} \left(1 - \sum_1^k u_i^2\right)^{\frac{1}{2}(n-k)-1} I_A(u_1, \ldots, u_k)\, du_1 \ldots du_k,$$

where

$$I_A(u_1, \ldots, u_k) = \begin{cases} 1 & \sum_1^k u_i^2 < 1 \\ 0 & \text{otherwise.} \end{cases}$$

By differentiating the above formula with respect to $x_1, \ldots, x_k$ respectively. Formula (2.5.18) follows. □

If $\boldsymbol{x} \sim S_n(\phi)$ has a density, there exist all the marginal densities. Denote the density of $(X_1, \ldots, X_m)$ by $g_m(x_1^2 + \ldots + x_m^2)$, $1 \leqslant m \leqslant n$. *Then, for* $m \geqslant 3$,

$$g_{m-2}(u) = \int_{-\infty}^\infty \int_{-\infty}^\infty g_m(u^2 + x_{m-1}^2 + x_m^2)\, dx_{m-1}\, dx_m$$

$$= \int_0^{2\pi} \int_0^\infty r g_m(u^2 + r^2)\, dr d\theta = 2\pi \int_{|u|}^\infty y g_m(y^2)\, dy.$$

From the above formula we immediately obtain the following facts:

(1) If $\boldsymbol{x} \sim S_n(\phi)$ has a density, the marginal densities of dimension less than or equal to n$-$2 are continuous and the marginal densities of dimension less than or equal

to $n-4$ are differentiable (except possibly at the origin in both cases).

(2) Univariate marginal densities for $n \geqslant 3$ are non-decreasing on $(-\infty, 0)$ and non-increasing on $(0, \infty)$.

(3) For $1 \leqslant m < n-1$, we have

$$g_{m+2}(x) = -(1/\pi)g_m'(x) \qquad \text{for a.e. } x > 0 \tag{2.5.19}$$

(4) Equation (2.5.19) enables one to construct all of the marginal densities if only the univariate marginal density is known.

The material in this paragraph is from Kelker (1970) and Cambanis, Huang and Simons (1981).

2.5.3. The Class Φ_∞

The class Φ_∞ is defined by (2.5.5). Parallel to Theorem 2.5.2. we can obtain the following theorem.

Theorem 2.5.7. *A function* $\phi \in \Phi_\infty$ *if and only if*

$$\phi(x) = \int_0^\infty e^{-xr2}\, dF_\infty(r), \tag{2.5.20}$$

where $F_\infty(.)$ *is a cdf over* $[0, \infty)$.

Proof. If π is any permutation of $\{1, ..., n\}$, the linear mapping of R^n into itself which sends $(X_1, ..., X_n)$ into $(X_{\pi 1}, ..., X_{\pi n})$ is a rotation, and spherical symmetry therefore requires $(X_1, ..., X_n) \overset{d}{=} (X_{\pi 1}, ..., X_{\pi n})$. Thus the sequence (X_n) is exchangeable, and de Finetti's theorem (Feller, 1971, Chapter VII, 4, and references cited there) shows that there is a σ-field $\mathfrak{F}$ of events conditional upon whicthe X_n', $n = 1$, 2, ..., are independent and have the same distribution function F, say. Write

$$\Phi(t) = \int_{-\infty}^{\infty} e^{itx}\, dF(x) = E(e^{itX_n} | \mathfrak{F}), \qquad n = 1, 2, ... \tag{2.5.21}$$

so that Φ is a random, but $\mathfrak{F}$–measurable, continuous function and

$$\Phi(-t) = \overline{\Phi(t)}, \qquad |\Phi(t)| \leqslant 1, \qquad \Phi(0) = 1. \tag{2.5.22}$$

The conditional independence means that, for real $t_1, ..., t_n$,

$$E\left\{\exp\left(i \sum_{r=1}^n t_r X_r\right) \middle| \mathfrak{F}\right\} = \prod_{r=1}^n \Phi(t_r), \tag{2.5.23}$$

so that

$$\Phi(t_1^2 + ... + t_n^2) = E\left\{\exp\left(i \sum_{r=1}^n t_r X_r\right)\right\} = E\left\{\prod_{r=1}^n \Phi(t_r)\right\}. \tag{2.5.24}$$

Spherical symmetry implies that the left-hand side of (2.5.24), and thus also the right–hand side, depends only on $t_1^2 + ... + t_n^2$. For any real u and v, write $t = (u^2 + v^2)^{\frac{1}{2}}$, and use (2.5.22) and (2.5.24) to compute

$$E\{|\Phi(t) - \Phi(u)\Phi(v)|^2\} = E[\{\Phi(t) - \Phi(u)\Phi(v)\}\{\Phi(-t) - \Phi(-u)\Phi(-v)\}]$$
$$= E\{\Phi(t)\Phi(-t)\} - E\{\Phi(t)\Phi(-u)\Phi(-v)\} - E\{\Phi(u)\Phi(v)\Phi(-t)\}$$

$$+ E\{\Phi(u)\Phi(v)\Phi(-u)\Phi(-v)\}.$$

The four terms in this final expression are all of the form of the right-hand side of (2.5.24), and, because of $t = (u^2 + v^2)^{\frac{1}{2}}$,

$$E\{\Phi(t)\Phi(-t)\} = \psi(2t^2)$$
$$E\{\Phi(t)\Phi(-u)\Phi(-v)\} = \psi(t^2 + u^2 + v^2) = \psi(2t^2)$$
$$E\{\Phi(u)\Phi(v)\Phi(-t)\} = \psi(u^2 + v^2 + t^2) = \psi(2t^2)$$
$$E\{\Phi(u)\Phi(v)\Phi(-u)\Phi(-v)\} = \psi(2u^2 + 2v^2) = \psi(2t^2).$$

Hence they are all equal, and

$$E\{|\Phi(t) - \Phi(u)\Phi(v)|^2\} = 0$$

or

$$\Phi(t) = \Phi(u)\Phi(v)$$

with probability one. Hence Φ satisfies, with probability one, the functional equation

$$\Phi\{(u^2 + v^2)^{\frac{1}{2}}\} = \Phi(u)\Phi(v)$$

for all rational u and v, and hence by continuity for all real u and v. Hence, (Feller, 1971, Chapter III.4) $\Phi(t) = \exp(-\frac{1}{2}At^2)$ for some complex A, and (2.5.22) shows that A is real and nonnegative.

Since $A = -2\log\Phi(1)$, A is an $\mathfrak{F}$-measurable random variable, and so

$$E(Z|A) = E\{E(Z|\mathfrak{F})|A\}$$

for any random variable Z (cf. Loève (1960), p.350). Taking

$$Z = \exp\left(i\sum_{r=1}^{n} t_r X_r\right)$$

and using (2.5.23), we have

$$E\left\{\exp\left(i\sum_{r=1}^{n} t_r X_r\right)\middle| A\right\} = E\left(\prod_{r=1}^{n} e^{-\frac{1}{2}At_r^2}\middle| A\right) = \prod_{r=1}^{n} e^{-\frac{1}{2}At_r^2} = \exp\left(-\tfrac{1}{2}A\sum_{1}^{n} t_r^2\right).$$

Hence, conditional on A, the X_r's are independent, with distribution $N(0, A)$. Let $F(.)$ be the cdf of A, *then*

$$E\left\{\exp\left(i\sum_{1}^{n} t_r X_r\right)\right\} = \int_0^\infty \exp\left(-\tfrac{1}{2}a\sum_{1}^{n} t_r^2\right) dF(a),$$

which completes the proof with $F_\infty(.) = F(.)$. □

The fact in Theorem 2.5.7 was first proved by Schoenberg (1938). Here the proof is from Kingman (1972).

Corollary 1. $\boldsymbol{x} \sim S_n(\phi)$ *with* $\phi \in \Phi_\infty$ *if and only if*

(2.5.25)
$$\boldsymbol{x} \overset{d}{=} R\boldsymbol{z}.$$

where $\boldsymbol{z} \sim N_n(\boldsymbol{0}, \boldsymbol{I}_n)$ *is independent of* $R \geqslant 0$. *This means that the distribution of* $\boldsymbol{x}$ *is a mixture of normal distributions. If* $\boldsymbol{x}$ *has a density, it is of the form*

(2.5.26)
$$\int_0^\infty \frac{1}{(2\pi r^2)^{n/2}} \exp(-\tfrac{1}{2}\boldsymbol{x}'\boldsymbol{x}/r^2)\, dF_\infty(r),$$

where $F_\infty(.)$ is the cdf of R.

Obviously, many results in Φ_∞ are much more easily obtained than in Φ_n with $n < \infty$. This is the reason why many facts are first established in Φ_∞.

2.5.4. Invariant Distribution

Assume that $\boldsymbol{x} = (X_1, ..., X_n)'$ is a random vector. Let

$$\overline{X} = \frac{1}{n}\sum_{i=1}^{n} X_i = \frac{1}{n}\mathbf{1}_n'\boldsymbol{x}, \quad s^2 = \frac{1}{n-1}\sum_{i=1}^{n}(X_i - \overline{X})^2 = \frac{1}{n-1}\boldsymbol{x}'\boldsymbol{D}\boldsymbol{x}$$

and

$$\boldsymbol{D} = \boldsymbol{I}_n - \frac{1}{n}\mathbf{1}_n\mathbf{1}_n'$$

where $\mathbf{1}_n = (1, ..., 1)'$. There are some very useful statistics in univariate analysis. For example,

(2.5.27)
$$t = \sqrt{n}\frac{\overline{X}}{s}$$

and

(2.5.28)
$$F = \frac{\boldsymbol{x}'\boldsymbol{C}_1\boldsymbol{x}/r}{\boldsymbol{x}'\boldsymbol{C}_2\boldsymbol{x}/k}$$

where $\boldsymbol{C}_1$ and $\boldsymbol{C}_2$ are projection matrices with $\mathrm{rk}(C_1) = r$ and $\mathrm{rk}(C_2) = k$ respectively, such that $\boldsymbol{C}_1\boldsymbol{C}_2 = \boldsymbol{O}$. It is a well-known fact that $t \sim t_{n-1}$ and $F \sim F(r, k)$ if $\boldsymbol{x} \sim N_n(\mathbf{0}, \boldsymbol{I}_n)$, where t_r denotes the t-distribution with r degrees of freedom and $F(r, k)$ denotes the F–distribution with r and k degrees of freedom. We want to show that this fact is true whenever $\boldsymbol{x} \sim S_n^+(\Phi)$. In fact, let $f(.)$ be defined by

$$f(\boldsymbol{x}) = \sqrt{n}\frac{\frac{1}{n}\mathbf{1}_n'\boldsymbol{x}}{\left(\frac{1}{n-1}\boldsymbol{x}'\boldsymbol{D}\boldsymbol{x}\right)^{\frac{1}{2}}}.$$

Then

$$t = f(\boldsymbol{x}) \overset{d}{=} f(R\boldsymbol{u}^{(n)}) = \sqrt{n}\frac{\frac{R}{n}\mathbf{1}_n'\boldsymbol{u}^{(n)}}{\left(\frac{R^2}{n-1}\boldsymbol{u}^{(n)\prime\prime}\boldsymbol{D}\boldsymbol{u}^{(n)}\right)^{\frac{1}{2}}} = \sqrt{n}\frac{\frac{1}{n}\mathbf{1}_n'\boldsymbol{u}^{(n)}}{\left(\frac{1}{n-1}\boldsymbol{u}^{(n)\prime}\boldsymbol{D}\boldsymbol{u}^{(n)}\right)^{\frac{1}{2}}}$$

whose distribution is independent of R. As the normal distribution $N_n(\mathbf{0}, \mathbf{I}_n)$ is $S_n^+(\phi)$ with $\phi(u) = \exp(-u/2)$, therefore t has the same distribution t_{n-1}, as in the normal case, for the whole class $\{S_n^+(\phi)\}$. Similarly, F has the same distribution $F(r, k)$ in the class $\{S_n^+(\phi)\}$. In general we have the following theorem:

Theorem 2.5.8. *A statistic $t(\boldsymbol{x})$'s distribution remains the same whenever $\boldsymbol{x} \sim S_n^+(\phi)$ if*

(2.5.29) $t(\alpha \boldsymbol{x}) \stackrel{d}{=} t(\boldsymbol{x})$ for each $\alpha > 0$ and each $\boldsymbol{x} \sim S_n^+(.)$.

Proof. From the assumption $\boldsymbol{x} \stackrel{d}{=} R\boldsymbol{u}^{(n)}$. $R \sim F(r)$ is independent of $\boldsymbol{u}^{(n)}$ and we have

$$E(e^{ist(\boldsymbol{x})}) = E(e^{ist(R\boldsymbol{u}^{(n)})}) = \int_{(0,\infty)} E(e^{ist(r\boldsymbol{u}^{(n)})} | R = r)\, dF(r)$$

$$= \int_{(0,\infty)} E(e^{ist(\boldsymbol{u}^{(n)})})\, dF(r) = E(e^{ist(\boldsymbol{u}^{(n)})})$$

which is independent of R (or ϕ). ∎

More generally, if $\boldsymbol{x} \sim EC_n(\mu, \Sigma, \phi)$ with $\Sigma > 0$ and $P(\boldsymbol{x} = \boldsymbol{0}) = 0$ we have the following corollary.

Corollary 1. *Suppose that $\Omega_m \subset R^n$ and Ω_c is given where $\boldsymbol{0} \in \Omega_m$, $\boldsymbol{I}_n \in \Omega_c$, and Ω_c is a subset of $n \times n$ positive definite matrices such that $\boldsymbol{A} \in \Omega_c$ implies $\alpha\, \boldsymbol{A} \in \Omega_c$ for each $\alpha > 0$. The distribution of $t(\boldsymbol{x})$ remains the same in the class that $\boldsymbol{x} \sim EC_n(\mu, \Sigma, \phi)$ with $(\mu, \Sigma) \in \Omega_m \times \Omega_c$ and $P(\boldsymbol{x} = \mu) = 0$ if the following condition holds:*

(2.5.30)

$t(\Sigma^{-\frac{1}{2}}(\boldsymbol{x} - \mu)) \stackrel{d}{=} t(\boldsymbol{x})$ for each $(\mu, \Sigma) \in \Omega_m \times \Omega_c$ and each $\boldsymbol{x}$ in the class.

Proof. Taking $\Sigma = \sqrt{\alpha}\, \boldsymbol{I}_n$ in (2.5.30) implies (2.5.29) and the assertion follows. ∎

2.6. Elliptically Contoured Distributions

The definition of elliptically contoured distributions (ECD) was given in the preceding section. In this section we will discuss their properties. Readers will find that many properties of elliptically contoured distributions are similar to those of normal distributions.

2.6.1. The Stochastic Representation

Theorem 2.6.1. *A univariate function $\phi(.)$ can be used to define an ECD $DC_n(\mu, \Sigma. \phi)$ for every $\mu \in R^n$ and $\Sigma \geqslant 0$ with $rk\ \Sigma = k$ if and only if $\phi \in \Phi_k$.*

Proof. By definition of $\boldsymbol{\Phi}_k$, the "only if" part follows from taking $\mu = \boldsymbol{0}$ *and* $\Sigma = \begin{pmatrix} \boldsymbol{I}_k & \boldsymbol{0} \\ \boldsymbol{0} & \boldsymbol{0} \end{pmatrix}$.

Assume $\phi \in \Phi_k$. For any given μ and $\Sigma \geqslant 0$ with rk $\Sigma = k$, there exists a $k \times n$ matrix $\boldsymbol{A}$ such that $\boldsymbol{A}'\boldsymbol{A} \neq \Sigma$. Let $\boldsymbol{x}$ be a $k \times 1$ random vector with c.f. $\phi(\boldsymbol{t}'\boldsymbol{t})$. Define $\boldsymbol{y} = \mu + \boldsymbol{A}'\boldsymbol{x}$. Then the c.f. of $\boldsymbol{y}$ is

$$e^{it'\mu}\phi(\boldsymbol{t}'\boldsymbol{A}'\boldsymbol{A}\boldsymbol{t}) = e^{it'\mu}\phi(\boldsymbol{t}'\Sigma\, \boldsymbol{t}).$$ ∎

Corollary 1. $\boldsymbol{x} \sim EC_n(\mu, \Sigma, \phi)$ *with* rk $\Sigma = k$ *if and only if*

$$\boldsymbol{x} \overset{d}{=} \mu + R\boldsymbol{A}'\boldsymbol{u}^{(k)}, \tag{2.6.1}$$

where $R \geqslant 0$ *is independent of* $\boldsymbol{u}^{(k)}$, $R \leftrightarrow \phi \in \Phi_k$, and $\boldsymbol{A}$ *is a* $k \times n$ *matrix such that* $\boldsymbol{A}'\boldsymbol{A} = \Sigma$.

Proof. The "if" part follows from the proof of Theorem 2.6.1. Assume $\boldsymbol{x} \sim EC_n(\mu, \Sigma, \phi)$ with rk $\Sigma = k$, i.e., the c.f. of $\boldsymbol{x}$ is $\exp(i\boldsymbol{t}'\mu)\,\phi\,(\boldsymbol{t}'\Sigma\boldsymbol{t})$. As the c.f. of $\mu + R\boldsymbol{A}'\boldsymbol{u}^{(k)}$ defined in the corollary is the same with $\boldsymbol{x}$, the assertion follows. □

Corollary 2. *Assume* $\boldsymbol{x} = \mu + R\boldsymbol{A}'\boldsymbol{u}^{(k)} \sim EC_n(\mu, \Sigma, \phi)$ *with rk* $\Sigma = k$. *Then*

$$Q(\boldsymbol{x}) = (\boldsymbol{x} - \mu)'\,\Sigma^-\,(\boldsymbol{x} - \mu) \overset{d}{=} R^2, \tag{2.6.2}$$

where Σ^- *is a generalized inverse of* Σ (*cf. Section 1.3*).

Proof. By (2.6.1)

$$Q(\boldsymbol{x}) \overset{d}{=} R^2\boldsymbol{u}^{(k)\prime}\,\boldsymbol{A}\,(\boldsymbol{A}'\boldsymbol{A})^-\,\boldsymbol{A}'\boldsymbol{u}^{(k)} = R^2\boldsymbol{u}^{(k)\prime}\,\boldsymbol{u}^{(k)} = R^2,$$

because $\boldsymbol{A}(\boldsymbol{A}'\boldsymbol{A})^-\boldsymbol{A}'$ is a project matrix with rank k. □

We now show that the stochastic representations, as well as the parametric ones of nondegenerate ECD are essentially unique, i.e., up to only obvious redundancies.

Theorem 2.6.2. *Let* $\boldsymbol{x}$ *be nondegenerate.*

(i) *If* $\boldsymbol{x} \sim EC_n(\mu, \Sigma, \phi)$ *and* $\boldsymbol{x} \sim EC_n(\mu^*, \Sigma^*, \phi^*)$, *then there is* $c > 0$ *such that*

$$\mu^* = \mu, \qquad \Sigma^* = c\Sigma, \qquad \phi^*(.) = \phi(c^{-1}.). \tag{2.6.3}$$

(ii) *If* $\boldsymbol{x} \overset{d}{=} \mu + R\boldsymbol{A}'\boldsymbol{u}^{(r)} \overset{d}{=} \mu^* + R^*\boldsymbol{A}^{*\prime}\boldsymbol{u}^{(r^*)}$, *where* $r \geqslant r^*$, *then there is* $c > 0$ *such that*

$$\mu^* = \mu, \qquad \boldsymbol{A}^{*\prime}\boldsymbol{A}^* = c\boldsymbol{A}'\boldsymbol{A}, \qquad R^* \overset{d}{=} c^{-\frac{1}{2}}Rb_{r^*/2,(r-r^*)/2}, \tag{2.6.4}$$

where $b_{r^*/2,(r-r^*)/2} \geqslant 0$ *is independent of* R *and* $b^2_{r^*/2,(r-r^*)/2} \sim B(r^*/2, (r - r^*)/2)$ *if* $r > r^*$, *and* $b_{r^*/2,(r-r^*)/2} \equiv 1$ *if* $r = r^*$.

Proof. (i) Since $\boldsymbol{x} - \mu$ and $\boldsymbol{x} - \mu^*$ are symmetric about 0, we have

$$\boldsymbol{x} - \mu \overset{d}{=} (-\boldsymbol{x} - \mu) \overset{d}{=} (\mu - \mu^*) - (\boldsymbol{x} - \mu^*) \overset{d}{=} (\mu - \mu^*) + (\boldsymbol{x} - \mu^*) = \boldsymbol{x} - (2\mu^* - \mu),$$

which implies $\mu^* = \mu$. Write $\Sigma = (\sigma_{ij})$, $\Sigma^* = (\sigma^*_{ij})$, and $\mu = (\mu_i)$. Since $\boldsymbol{x}$ is nondegenerate, one of its components X_j is nondegenerate, and the c.f. of $X_j - \mu_j$ is given by $\phi(\sigma_{jj}u^2) = \phi^*(\sigma^*_{jj}u^2)$, $u \in R^1$, with $\sigma_{jj}, \sigma^*_{jj} > 0$, which establishes $\phi^*(.) = \phi(c^{-1}.)$ with $c = \sigma^*_{jj}/\sigma_{jj}$. Then the c.f. of $\boldsymbol{x} - \mu$ is given by $\phi(\boldsymbol{t}'\Sigma\boldsymbol{t}) = \phi^*(\boldsymbol{t}'\Sigma^*\boldsymbol{t}) = \phi(c^{-1}\boldsymbol{t}'\Sigma^*\boldsymbol{t})$, $\boldsymbol{t} \in R^n$. Now if $\Sigma^* \neq c\Sigma$ then for some $\boldsymbol{t}_0 \in R^n$ we have $a^2 = \boldsymbol{t}_0'\Sigma\boldsymbol{t}_0 \neq c^{-1}\boldsymbol{t}_0'\Sigma^*\boldsymbol{t}_0 = b^2$; and putting $\boldsymbol{t} = u\boldsymbol{t}_0$ we obtain $\phi(a^2u^2) = \phi(b^2u^2)$, $u \in R^1$. But this implies through recursion $\phi(u^2) = \phi(0) = 1$, $u \in R^1$, which contradicts the nondegeneracy of $\boldsymbol{x}$; and thus $\Sigma^* = c\Sigma$ follows.

(ii) The assumptions imply that $\boldsymbol{x} \sim EC_n(\mu, \boldsymbol{A}'\boldsymbol{A}, \phi)$ and $\boldsymbol{x} \sim EC_n(\mu^*, \boldsymbol{A}^{*\prime}\boldsymbol{A}^*, \phi^*)$; so by (i), $\mu^* = \mu$, $\boldsymbol{A}^{*\prime}\boldsymbol{A}^* = c\boldsymbol{A}'\boldsymbol{A}$ and $\phi^*(\cdot) = \phi(c^{-1}\cdot)$. Thus, putting $R_r = R$, $R_{r^*} = c^{\frac{1}{2}}R^*$, $\Sigma = \boldsymbol{A}'\boldsymbol{A}$, we have

$$x - \mu \overset{d}{=} R_r A' u^{(1)} \sim EC_n(\mathbf{0}, \Sigma, \phi), \qquad \phi \in \Phi_r,$$

$$x - \mu \overset{d}{=} R r_*(c^{-\frac{1}{2}} A^*)' u^{(r^*)} \sim EC_n(\mathbf{0}, \Sigma, \phi), \qquad \phi \in \Phi_{r^*}.$$

When $n^* < n$, $R_r^* \overset{d}{=} R_r b_{r^*/2,(r-r^*)/2}$ from Corollary of Theorem 2.5.4 with $m = r^*$ *and* $n = r$. *This* $R^* \overset{d}{=} c^{-\frac{1}{2}} R b_{r^*/2,(r-r^*)/2}$. When $r^* = r$, the one-to-one correspondence between ϕ and F implies $R_{n^*} \overset{d}{=} R_r$. Thus $R^* \overset{d}{=} c^{-\frac{1}{2}} R$. □

2.6.2. Combination and Marginal Distributions

Theorem 2.6.3. *Assume that* $x \sim EC_n(\mu, \Sigma, \phi)$ *with* $\mathrm{rk}\,\Sigma = k$, B *is an* $n \times m$ *matrix and* γ *is an* $m \times 1$ *vector. Then*

$$\gamma + B'x \sim EC_m(B'\mu + \gamma, B'\Sigma B, \phi). \tag{2.6.5}$$

Proof. The theorem follows from

$$\gamma + B'x \overset{d}{=} (B'\mu + \gamma) + R(AB)'u^{(k)}. \quad □$$

Partition x, μ and Σ into

$$x = \begin{bmatrix} x^{(1)} \\ x^{(2)} \end{bmatrix}, \quad \mu = \begin{bmatrix} \mu^{(1)} \\ \mu^{(2)} \end{bmatrix} \quad \text{and } \Sigma = \begin{bmatrix} \Sigma_{11} & \Sigma_{12} \\ \Sigma_{21} & \Sigma_{22} \end{bmatrix}, \tag{2.6.6}$$

where $x^{(1)}$: $m \times 1$, and Σ_{11}: $m \times m$. Similar to the case of multivariate normal distribution, marginal distributions of ECD can be found. □

Corollary 1. *Assume* $x \sim EC_n(\mu, \Sigma, \phi)$. *Then* $x^{(1)} \sim ES_m(\mu^{(1)}, \Sigma_{11}, \phi)$ *and* $x^{(2)} \sim EC_{n-m}(\mu^{(2)}, \Sigma_{22}, \phi)$.

2.6.3. Moments

If $x \sim N_n(\mu, \Sigma)$, we know that $\mathcal{E}(x) = \mu$ and $\mathcal{D}(x) = \Sigma$ in Section 2.3. In the case of ECD, it is not always true that there exist moments. Suppose $x \sim EC_n(\mu, \Sigma, \phi)$. From (2.6.1) it can be verified that there exists $\mathcal{E}(x)$ ($\mathcal{D}(x)$) if and only if $E(R) < \infty$ ($E(R^2) < \infty$).

Theorem 2.6.4. *Assume* $x \sim EC_n(\mu, \Sigma, \phi)$ *and* $E(R^2) < \infty$. *Then*

$$\mathcal{E}(x) = \mu \quad \text{and} \quad \mathcal{D}(x) = (ER^2/rk\,\Sigma)\Sigma. \tag{2.6.7}$$

Proof. Denote $k = \mathrm{rk}\Sigma$. We have $x \overset{d}{=} \mu + RA'u^{(k)}$. By Corollary 2 of Theorem 2.5.3, we see

$$\mathcal{E}(x) = \mu + RA'\,\mathcal{E}(u^{(k)}) = \mu$$

and

$$\mathcal{D}(x) = \mathcal{D}(RA'u^{(k)}) = ER^2 \cdot A'\,\mathcal{D}(u^{(k)})A = ER^2 \cdot \frac{1}{k} A' I_K A = \frac{1}{\mathrm{k}} ER^2 \cdot \Sigma,$$

which completes the proof. □

Here the covariance matrix of $\boldsymbol{x}$ is proportional to Σ, and genenally it is not necessarily equal to Σ.

An alternative characterization and expression of moments, in terms of ϕ, is given as follows:

Theorem 2.6.5. *Let $\boldsymbol{x} \sim EC_n(\mu, \Sigma, \phi)$ be nondegenerate. Then*
(a) $\Gamma_1(\boldsymbol{x}) = \mu$;
(b) $\Gamma_2(\boldsymbol{x}) = \mu\mu' - 2\phi'(0)\Sigma$ and $\mathcal{D}(\boldsymbol{x}) = -2\phi'(0)\Sigma$;
(c) $\Gamma_3(\boldsymbol{x}) = \mu \otimes \mu' \otimes \mu - 2\phi'(0)[\mu \otimes \Sigma + \Sigma \otimes \mu + \mathrm{vec}(\Sigma)\mu']$

If they exist, where $\Gamma_k(\boldsymbol{x})$ $(k = 1, 2, \ldots)$ were defined in (2.2.18), *and the operator "vec" was discussed in Section* 1.4.

Proof. By using (2.2.20) the theorem follows from direct calculation For example, let $\phi_{\boldsymbol{x}}(\boldsymbol{t}) = \exp(i\boldsymbol{t}'\mu)\phi(\boldsymbol{t}'\Sigma\boldsymbol{t})$ be the c.f. of $\boldsymbol{x}$. Then

$$\frac{\partial \phi_{\boldsymbol{x}}(\boldsymbol{t})}{\partial \boldsymbol{t}} = ie^{it'\mu}\phi(\boldsymbol{t}'\Sigma\boldsymbol{t})\mu + 2e^{it'\mu}\phi'(\boldsymbol{t}'\Sigma\boldsymbol{t})\Sigma\boldsymbol{t},$$

$$\Gamma_1(\boldsymbol{x}) = \frac{1}{i}\frac{\partial\phi_{\boldsymbol{x}}(\boldsymbol{t})}{\partial\boldsymbol{t}}\bigg|_{\boldsymbol{t}=\boldsymbol{0}} = \mu,$$

$$\frac{\partial^2\phi_{\boldsymbol{x}}(\boldsymbol{t})}{\partial\boldsymbol{t}\,\partial\boldsymbol{t}'} = i^2 e^{it'\mu}\phi(\boldsymbol{t}'\Sigma\boldsymbol{t})\mu\mu' + 4e^{it'\mu}\phi''(\boldsymbol{t}'\Sigma\boldsymbol{t})[\Sigma\boldsymbol{t}\otimes\boldsymbol{t}'\Sigma]$$

$$+2ie^{it'\mu}\phi'(\boldsymbol{t}'\Sigma\boldsymbol{t})[\mu\otimes\boldsymbol{t}'\Sigma + \mu'\otimes\Sigma\boldsymbol{t} - i\Sigma],$$

and

$$\Gamma_2(\boldsymbol{x}) = \frac{1}{i^2}\frac{\partial^2\phi_{\boldsymbol{x}}(\boldsymbol{t})}{\partial\boldsymbol{t}\,\partial\boldsymbol{t}'}\bigg|_{\boldsymbol{t}=\boldsymbol{0}} = \mu\mu' - 2\phi'(\boldsymbol{0})\Sigma.$$

The calculation of $\Gamma_3(\boldsymbol{x})$ is left to the reader. ∎

When the covariance matrix Σ_0 of $\boldsymbol{x} \sim EC_n(\mu, \Sigma, \phi)$ exists, choosing Σ to be Σ_0 has special appeal. This occurs if and only if ϕ is chosen so as to make $-2\phi'(\boldsymbol{0}) = 1$.

2.6.4. Conditional Distributions

Let $\boldsymbol{x}$ be an $n \times 1$ random vector and $\boldsymbol{x} = (\boldsymbol{x}^{(1)\prime}, \boldsymbol{x}^{(2)\prime})'$, where $\boldsymbol{x}^{(1)}$: $m \times 1$, $0 < m < n$. We consider the conditional distribution of $\boldsymbol{x}^{(1)}$ given $\boldsymbol{x}^{(2)} = \boldsymbol{x}_0^{(2)}$.

Theorem 2.6.6. *Let $\boldsymbol{x} = R\boldsymbol{u}^{(n)} \sim S_n(\phi)$. Then the conditional distribution of $\boldsymbol{x}^{(1)}$ given $\boldsymbol{x}^{(2)} = \boldsymbol{x}_0^{(2)}$ is given by*

$$(\boldsymbol{x}^{(1)} \mid \boldsymbol{x}^{(2)} = \boldsymbol{x}_0^{(2)}) \sim EC_m(\boldsymbol{0}, \boldsymbol{I}_m, \phi_{\|\boldsymbol{x}_0^{(2)}\|^2}) \tag{2.6.8}$$

with stochastic representation

$$(\boldsymbol{x}^{(1)} \mid \boldsymbol{x}^{(2)} = \boldsymbol{x}_0^{(2)}) \overset{\mathrm{d}}{=} R(\|\boldsymbol{x}_0^{(2)}\|^2)\boldsymbol{u}^{(m)}, \tag{2.6.9}$$

where for each $a^2 \geqslant 0$, $R(a^2)$ and $\boldsymbol{u}^{(m)}$ are independent, and

$$R(\|\boldsymbol{x}_0^{(2)}\|^2) \overset{\mathrm{d}}{=} ((R^2 - \|\boldsymbol{x}_0^{(2)}\|^2)^{\frac{1}{2}} \mid \boldsymbol{x}^{(2)} = \boldsymbol{x}_0^{(2)}), \tag{2.6.10}$$

and the function ϕ_{a^2} *is given by* (2.5.6) *with* $k = m$ *and* F *replaced by the distribution function of* $R(a^2)$.

Proof. From Corollary 3 of Theorem 2.5.3 we have

$$\begin{bmatrix} \boldsymbol{x}^{(1)} \\ \boldsymbol{x}^{(2)} \end{bmatrix} \stackrel{d}{=} R \begin{bmatrix} \boldsymbol{u}_{(1)} \\ \boldsymbol{u}_{(2)} \end{bmatrix} \stackrel{d}{=} \begin{bmatrix} R d_1 \boldsymbol{u}^{(m)} \\ R d_2 \boldsymbol{u}^{(n-m)} \end{bmatrix},$$

where R, d_1 (or d_2), $\boldsymbol{u}^{(m)}$ and $\boldsymbol{u}^{(n-m)}$ are independent, $d_1^2 + d_2^2 = 1$, and $d_1^2 \sim B(\frac{1}{2}m, \frac{1}{2}(n-m))$. When $R d_2 \boldsymbol{u}^{(n-m)} = \boldsymbol{x}_0^{(2)}$, it is easy to see $R d_1 = (R^2 - \|\boldsymbol{x}_0^{(2)}\|^2)^{\frac{1}{2}}$, and (2.6.10) and (2.6.9) follow. That $R(\|\boldsymbol{x}_0^{(2)}\|^2)$, as defined in (2.6.10), depends upon $\boldsymbol{x}_0^{(2)}$ only through $\|\boldsymbol{x}_0^{(2)}\|^2$, as the notation indicates, is obvious when $\boldsymbol{x}_0^{(2)} = \boldsymbol{0}$; and when $\boldsymbol{x}_0^{(2)} \neq \boldsymbol{0}$, it follows from

$$\begin{aligned} ((R^2 - \|\boldsymbol{x}_0^{(2)}\|^2)^{\frac{1}{2}} | \boldsymbol{x}^{(2)} = \boldsymbol{x}_0^{(2)}) &\stackrel{d}{=} ((R^2 - \|\boldsymbol{x}_0^{(2)}\|^2)^{\frac{1}{2}} | R d_2 \boldsymbol{u}^{(n-m)} = \boldsymbol{x}_0^{(2)}) \\ &\stackrel{d}{=} ((R^2 - \|\boldsymbol{x}_0^{(2)}\|^2)^{\frac{1}{2}} | R^2 d_2^2 = \|\boldsymbol{x}_0^{(2)}\|^2) \end{aligned}$$

since $\boldsymbol{u}^{(n-m)}$ is independent of R and d_1. Thus (2.6.8) follows from

$$(\boldsymbol{x}^{(1)} | \boldsymbol{x}^{(2)} = \boldsymbol{x}_0^{(2)}) \stackrel{d}{=} (R d_1 \boldsymbol{u}^{(m)} | \boldsymbol{x}^{(2)} = \boldsymbol{x}_0^{(2)}) \stackrel{d}{=} ((R^2 - \|\boldsymbol{x}_0^{(2)}\|^2)^{\frac{1}{2}} \boldsymbol{u}^{(m)} | \boldsymbol{x}^{(2)} = \boldsymbol{x}_0^{(2)})$$

$$\sim EC_m(\boldsymbol{0}, \boldsymbol{I}_m, \phi_{\|\boldsymbol{x}_0^{(2)}\|^2}). \qquad \square$$

In order to obtain the distribution of R_{a^2}, we need the following lemma whose proof is straightforward and therefore omitted.

Lemma 2.6.1. *Suppose that nonnegative random variables* R *and* S *are independent,* R *has cdf* F, *and* S *is absolutely continuous with density* g. *Then* $T = RS$ *has an atom of size* $F(0)$ *at zero if* $F(0) > 0$. *It is absolutely continuous on* $(0, \infty)$ *with density* h *given by*

$$h(t) = \int_{(0,\infty)} r^{-1} g(t/r)\, dF(r).$$

And a regular version of the conditional distribution of R *given* $T = t$ *can be expressed as*

$$P(R \leqslant \rho | T = t) = \begin{cases} 0 & \text{for } \rho < 0; \\ 1 & \text{for } t = 0 \text{ or } t > 0 \text{ with } h(t) = 0,\ \rho \geqslant 0; \\ \dfrac{1}{h(t)} \displaystyle\int_{(0,\rho]} r^{-1} g(t/r)\, dF(r) & \text{for } t > 0 \text{ with } h(t) \neq 0,\ \rho \geqslant 0. \end{cases}$$

Corollary 1. *Under the notation of Theorem 2.6.6. we have*

$$R_a{}^2 = 0 \text{ a.s. when } a = 0 \text{ or } F(a) = 1,$$

$$P(R_a^2 \leqslant \rho) = \frac{\displaystyle\int_{(a,(\rho^2+a^2)^{\frac{1}{2}}]} (r^2 - a^2)^{m/2-1} r^{-(n-2)}\, dF(r)}{\displaystyle\int_{(a,\infty)} (r^2 - a^2)^{m/2-1} r^{-(n-2)}\, dF(r)} \tag{2.6.11}$$

for $\rho \geqslant 0$, $a > 0$ *and* $F(a) < 1$. *Here* $R_a{}^2 = R(a^2)$ *for simplicity.*

Proof. By the definition of $R_a{}^2$,

$$R_a{}^2 = (\sqrt{R^2 - a^2} \mid \|\boldsymbol{x}_2\|^2 = a^2) = (\sqrt{R^2 - a^2} \mid R^2 d_2^2 = a^2).$$

Then

$$P(R_a^2 \leqslant \rho) = P(R \leqslant \sqrt{a^2 + \rho^2} \mid Rd_2 = a).$$

Taking $T = \|\boldsymbol{x}_2\|$, $R = R$ and $S = d_2$ in Lemma 2.6.1 and noting $d_2^2 \sim B((n-m)/2, m/2)$ and

$$g(t) = \frac{2\Gamma(m/2)\Gamma((n-m)/2)}{\Gamma(n/2)} (1 - t^2)^{m/2-1} t^{n-m-1}, \quad 0 < t < 1,$$

where g(t) is the density of d_2, (2.6.11) follows from Lemma 2.6.1. □

As (2.6.11) shows how the distribution function $F_a{}_2$ of $R_a{}_2$ is determined by F (together with m, n and a), for the conversne case we have

Corollary 2. *Denoting the denominator in* (2.6.11) *by* $C_a{}_2$, *we have*

$$1 - F(r) = C_a{}_2 \int_{((r^2-a^2)^{\frac{1}{2}},\, \infty)} (\rho^2 + a^2)^{n/2-1} \rho^{-(m-2)}\, dF_a{}_2(\rho),\ r \geqslant a > 0. \tag{2.6.12}$$

Proof. Differentiating both sides of (2.6.11) with respect to ρ, then(2.6.12) follows. □

Thus, when a > 0, $F_a{}_2$ determines F on the interval $[a, \infty)$ up to an unknown multiplicative factor $C_a{}_2 \geqslant 0$, and, of course, this gives no information about the values of F on $[0, a)$.

Now we can extend Theorem 2.6.6 to the case of $\boldsymbol{x} \sim EC_n(\mu, \Sigma, \phi)$.

Corollary 3. *Let* $\boldsymbol{x} \overset{d}{=} \mu + RA'\boldsymbol{u}^{(n)} \sim EC_n(\mu, \Sigma, \phi)$ *with* $\Sigma > 0$. *Let*

$$\boldsymbol{x} = \begin{bmatrix} \boldsymbol{x}^{(1)} \\ \boldsymbol{x}^{(2)} \end{bmatrix}, \quad \mu = \begin{bmatrix} \mu^{(1)} \\ \mu^{(2)} \end{bmatrix}, \quad \Sigma = \begin{bmatrix} \Sigma_{11} & \Sigma_{12} \\ \Sigma_{21} & \Sigma_{22} \end{bmatrix},$$

where $\boldsymbol{x}^{(1)}$: $m \times 1$, $\mu(1)$: $m \times 1$, Σ_{11}: $m \times m$ *and* $0 < m < n$. *Then*

$$(\boldsymbol{x}^{(1)} | \boldsymbol{x}^{(2)} = \boldsymbol{x}_0^{(2)}) \overset{d}{=} \mu_{1.2} + R_{q_{(\boldsymbol{x}_0^{(2)})}} A'_{11.2}\, \boldsymbol{u}^{(n)} \sim EC_m(\mu_{1.2}, \Sigma_{11.2}, \phi_{q_{(\boldsymbol{x}_0^{(2)})}}) \tag{2.6.13}$$

where

$$\begin{cases} \mu_{1.2} = \mu^{(1)} + \Sigma_{12}\Sigma_{22}^{-1}(\boldsymbol{x}_0^{(2)} - \mu^{(2)}), \\ \Sigma_{11.2} = \Sigma_{11} - \Sigma_{12}\Sigma_{22}^{-1}\Sigma_{21}, \\ q(\boldsymbol{x}_0^{(2)}) = (\boldsymbol{x}_0^{(2)} - \mu^{(2)})'\Sigma_{22}^{-1}(\boldsymbol{x}_0^{(2)} - \mu^{(2)}), \end{cases} \tag{2.6.14}$$

and $\Sigma_{11.2} = A'_{11.2} A_{11.2}$. *Moreover, for each* $a \geqslant 0$, $R_a{}^2$ *is independent of* $u^{(m)}$ *and its distribution is given by* (2.6.11);

$$R_{q_{(x_0^{(2)})}} \overset{d}{=} ((R^2 - q(x_0^{(2)}))^{\frac{1}{2}} | x^{(2)} = x_0^{(2)}),$$

and the function $\phi_a{}^2$ *is given by* (2.5.6) *with* $k = m$ *and* F *being replaced by cdf of* $R_a{}^2$.

Proof. Let $A'A = \Sigma$, $A'_2 A_2 = \Sigma_{22}$, $A_2 > 0$ and

$$\Sigma = A' A = \begin{bmatrix} I & \Sigma_{12}\Sigma_{22}^{-1} \\ 0 & I \end{bmatrix} \begin{bmatrix} \Sigma_{11.2} & 0 \\ 0 & \Sigma_{22} \end{bmatrix} \begin{bmatrix} I & 0 \\ \Sigma_{22}^{-1}\Sigma_{21} & I \end{bmatrix} = FF$$

where

$$F = \begin{bmatrix} A_{11.2} & 0 \\ 0 & A_2 \end{bmatrix} \begin{bmatrix} I & 0 \\ \Sigma_{22}^{-1}\Sigma_{21} & \mathrm{I} \end{bmatrix} = \begin{bmatrix} A_{11.2} & 0 \\ A_2\Sigma_{22}^{-1}\Sigma_{21} & A_2 \end{bmatrix}.$$

Let $y \sim EC_n(0, I_n, \phi)$. We have

$$x \overset{d}{=} \mu + A' y \overset{d}{=} \mu + F' y = \begin{bmatrix} \mu^{(1)} \\ \mu^{(2)} \end{bmatrix} + \begin{bmatrix} A'_{11.2} y^{(1)} + \Sigma_{12}\Sigma_{22}^{-1} A'_2 y^{(2)} \\ A'_2 y^{(2)} \end{bmatrix}$$

and

$$\begin{aligned}(x^{(1)} | x^{(2)}) &= (A'_{11.2} y^{(1)} + \Sigma_{12}\Sigma_{22}^{-1} A'_2 y^{(2)} + \mu^{(1)} | \mu^{(2)} + A'_2 y^{(2)} = x^{(2)}) \\ &= \mu^{(1)} + \Sigma_{12}\Sigma_{22}^{-1}(x^{(2)} - \mu^{(2)}) + A'_{11.2}(y^{(1)} | y^{(2)} = A_2^{-1}(x^{(2)} - \mu^{(2)})).\end{aligned}$$

Now we see that the Corollary is a consequence of Theorem 2.6.6. ∎

This corollary can be extended to the case of Σ being singular (cf. Cambanis, Huang, and Simons (1981)). Also, the material in this paragraph is from their paper.

2.6.5. Densities

Suppise $x \sim EC_n(\mu, \Sigma, \phi)$. In general, it is not necessary that x has a density (cf. the preceding section). Now we consider two situations: (1) x has a density; (2) $\Sigma > 0$ and $\mathrm{p}(x = 0) = 0$.

Clearly, the necessary condition that $x \sim EC_n(\mu, \Sigma, \phi)$ has a density is $\mathrm{rk}\,\Sigma = n$. In this case, the stochastic representation becomes

$$x \overset{d}{=} \mu + RA' u^{(n)},$$

where A is a nonsingular matrix. Let $y = (A')^{-1}(x - \mu)$. Then the c.f. of y is $\phi(t' A'^{-1} \Sigma A^{-1} t) = \phi(t' t)$, i.e., $y \sim \mathrm{S}_n(\phi)$ and has a density.

In the preceding section we have seen that the density of y has the form of $\mathrm{f}(y' y)$. As $x = \mu + A' y$, the density of x has the form of

$$|\Sigma|^{-\frac{1}{2}} f((x - \mu)' \Sigma^{-1} (x - \mu)) \tag{2.6.15}$$

and R has a density (2.5.7).

If x does not have a density and $\mathrm{P}(R = 0) = 0$ and $\Sigma > 0$, then, often making the same transformation $x = \mu + A' y$, $\mathrm{P}(y = 0) = 0$ and y has all the marginal density and so does x. In this case, by (2.5.18) the marginal density of $x_{(k)} = (X_1, ..., X_k)' - \mu_{(k)}$, $1 \leqslant k < n$, where $\mu_{(k)} = (\mu_1, ..., \mu_k)'$, is

(2.6.16) $$\frac{\Gamma\left(\frac{1}{2}n\right)}{\Gamma\left(\frac{1}{2}(n-k)\right)\pi^{\frac{1}{2}k}} \int_{(\boldsymbol{x}'_{(k)}\Sigma_k^{-1}\boldsymbol{x}_{(k)})^{\frac{1}{2}}}^{\infty} r^{-(n-2)}\left(r^2 - \boldsymbol{x}'_{(k)}\Sigma_k^{-1}\boldsymbol{x}_{(k)}\right)^{\frac{1}{2}(n-k)-1} dF(r),$$

where Σ_k is the first principal minor of order k *of* Σ.

Every function f(.) which satisfies (2.5.16) can define a density (2.6.15) of ECD with a normalizing constant c_n, where

(2.6.17) $$c_n = \frac{\Gamma(n/2)}{2\pi^{n/2}\int_0^{\infty} r^{n-1} f(r^2)\, dr}$$

Here are some examples:

Example 2.6.1. For $r, s > 0$, $2N + n > 2$ let

(2.6.18) $$f(t) = t^{N-1}\exp(-rt^s).$$

From (2.6.17),

(2.6.19) $$c_n = s\pi^{-n/2} r^{(2N+n-2)/(2s)}\Gamma(n/2)/\Gamma((2N+n-2)/(2s)).$$

The multivariate normal distribution is the special case $N = 1$, $s = 1$, $r = \frac{1}{2}$. The case $s = 1$ was introduced by Kotz (1975).

Example 2.6.2. The Pearson type VII distributions. Now

(2.6.20) $$f(t) = (1 + t/s)^{-N}, \qquad N > n/2,\ s > 0,$$

(2.6.21) $$c_n = (\pi s)^{-n/2}\Gamma(n)/\Gamma(N - n/2).$$

This family includes multivariate t–distributions. Let $\boldsymbol{y} \sim N_n(\boldsymbol{0}, \Sigma)$ $(\Sigma > 0)$ and $S \sim X_m$ be independet. The distribution of $\boldsymbol{x} = m^{\frac{1}{2}}\boldsymbol{y}/S$ is called the *multivariate t–distribution* (cf. Johnson and Kotz (1972)). This is the special case $N = (n + m)/2$, $s = m$ of (2.6.20). When $m = 1$, the corresponding distribution is called the *multivariate Cauchy distribution.*

Example 2.6.3. Uniform distribution in the unit sphere in R^n. Let $\boldsymbol{x}$ be distributed according to the uniform distribution in the unit sphere in R^n. From Example 2.4.2 its density is

$$p_x(\boldsymbol{x}) = \begin{cases} \dfrac{\Gamma(\frac{1}{2}n)}{\pi^{\frac{1}{2}n}} & \text{if } \sum_1^n x_i^2 \leqslant 1, \\ 0 & \text{otherwise,} \end{cases}$$

Clearly, $\Gamma\boldsymbol{x} \overset{d}{=} \boldsymbol{x}$ for every $\Gamma \in O(n)$ and the uniform distribution in the unit sphere belongs to the class of ECD. The $\boldsymbol{x}$ can be expressed as $\boldsymbol{x} \overset{d}{=} R\boldsymbol{u}^{(n)}$. From (2.5.17) the density of $R = \|\boldsymbol{x}\|$ is

$$g_R(r) = \begin{cases} \dfrac{2\pi^{\frac{1}{2}n}\,\Gamma(\frac{1}{2}(n+2))}{\Gamma(\frac{1}{2}n)\,\pi^{\frac{1}{2}n}} r^{n-1} = nr^{n-1}, & \text{if } 0 \leqslant r \leqslant 1 \\ 0, & \text{otherwise.} \end{cases}$$

Example 2.6.4. The generalized Laplace or Bessel distribution. By taking

$$f(t) = (t^{\frac{1}{2}}/\beta)^a \; K_a(t^{\frac{1}{2}}/\beta), \qquad a > -n/2,\ \beta > 0. \tag{2.6.22}$$

the normalizing constant is

$$c_n^{-1} = 2^{a+n-1}\pi^{n/2}\beta^n\Gamma(a+n/2). \tag{2.6.23}$$

Here $K_a(\cdot)$ denotes the modified Bessel function of the third kind, i.e.,

$$K_a(z) = \frac{\pi}{2}\,\frac{I_{-a}(z) - I_a(z)}{\sin(a\pi)}, \qquad |\arg z| < \pi,\ a = 0,\ \pm 1,\ \pm 2,\ \ldots,$$

$$I_a(z) = \sum_{k=0}^{\infty} \frac{1}{k!\,\Gamma(k+a+1)}(z/2)^{a+2k}, \qquad |z| < \infty,\ |\arg z| < \pi,$$

which is of great importance in the special case $a = n/2 - 1$ (cf. Laurent (1974)).

2.7. Characterizations of Normality

Normal distributions are elements of the class of ECD. In this section we focus our attention on several properties of normal distributions which cannot be extended to other ECD. The main results in this section are from Kelker (1970) and Cambanis, Huang and Simons (1981). For instance, assume $\boldsymbol{x} \sim EC_n(\mu, \Sigma, \phi)$ with $\text{rk}\Sigma = k$. Then $\boldsymbol{x}$ has a normal distribution if and only if $Q(\boldsymbol{x}) = (\boldsymbol{x} - \mu)'\Sigma^{-}(\boldsymbol{x} - \mu) \sim \chi_k^2$, because the relationship between ϕ and F is one-to-one.

Theorem 2.7.1. *Assume $\boldsymbol{x} \sim EC_n(\mu, \Sigma, \phi)$. Then any marginal distribution is normal if and only if $\boldsymbol{x}$ is normally distributed.*

Proof. Assume that $\boldsymbol{x}^{(1)}$, where $\boldsymbol{x} = \begin{pmatrix} \boldsymbol{x}^{(1)} \\ \boldsymbol{x}^{(2)} \end{pmatrix}$ has a normal distribution. As the c.f. of $\boldsymbol{x}^{(1)}$ is $\phi(\boldsymbol{t}'\,\boldsymbol{t})$ where $\boldsymbol{t}$ has the same dimension as $\boldsymbol{x}^{(1)}$. we have $\phi(\boldsymbol{u}) = \exp(-u/2)$ from the normality of $\boldsymbol{x}^{(1)}$. Thus $\boldsymbol{x}$ has a normal distribution. The inverse is obvious (cf. Section 2.3). □

Theorem 2.7.2. *Assume $\boldsymbol{x} \sim EC_n(\mu, \Sigma, \phi)$ with $\Sigma = \text{diag}(\sigma_{11}, \ldots, \sigma_{nn})$. Then the following statements are equivalent.*

(a) *$\boldsymbol{x}$ is normally distributed;*

(b) *Components of $\boldsymbol{x}$ are independent;*

(c) *X_i and X_j $(1 \leqslant i < j \leqslant n)$ are independent.*

Proof. (a) ⇒ (b) ⇒ (c) are trivial. Now we prove (c) ⇒ (a). From assumption we have

$$\phi(t_i^2\sigma_{ii} + t_j^2\sigma_{jj}) = \phi(t_i^2\sigma_{ii})\phi(t_j^2\sigma_{jj}) \qquad (\text{Let} \quad u_k = t_k\sigma_{kk}^{\frac{1}{2}}\ ,\ k = i,\ j).$$

Thus

$$\phi(u_i + u_j) = \phi(u_i)\,\phi(u_j).$$

This equation, known as Hamel's equation, has the solution $\phi(u) = e^{au}$ for some constant a (Although $\phi(u) \equiv 1$ is a solution of the equation, the corresponding distribution, however, is singular). As $\phi(u)$ is a c.f., the constant a must be negative which completes the proof. □

Theorem 2.7.2 shows that the components of $\boldsymbol{x} \sim S_n(\phi)$ are uncorrelative, but are dependent except in the case of normal distribution. If $\boldsymbol{x} = \begin{pmatrix} \boldsymbol{x}^{(1)} \\ \boldsymbol{x}^{(2)} \end{pmatrix}$ is normally distributed, then, of course, the conditional distribution of $\boldsymbol{x}^{(1)}$, given $\boldsymbol{x}^{(2)}$, is normal and the function $\phi_{q(x^{(2)})}$ in Corollary 3 of Theorem 2.6.4 assumes the form $\phi_{q(x^{(2)})} = \exp(-cu/2)$, where $c \geqslant 0$ is independent of $x^{(2)}$. The failure of $\phi_{q(x^{(2)})}$ to depend upon the value of $q(\boldsymbol{x}^{(2)})$ characterizes normality.

Theorem 2.7.3. *Assume* $\boldsymbol{x} = \begin{pmatrix} \boldsymbol{x}^{(1)} \\ \boldsymbol{x}^{(2)} \end{pmatrix} \sim EC_n(\mu, \Sigma, \phi)$ *with* $\Sigma > 0$ *and* $\boldsymbol{x}^{(1)}$*:* $m \times 1$, $0 < m < n$. *Then* $\phi_q(\boldsymbol{x}^{(2)})$ *does not depend upon the value of* $\boldsymbol{x}^{(2)}$ *if and only if* $\boldsymbol{x}$ *is normally distributed.*

Proof. It is sufficient to show the "only if" part. By Corollary 3 of Lemma 2.6.1, for all $\boldsymbol{t} = \begin{pmatrix} \boldsymbol{t}^{(1)} \\ \boldsymbol{t}^{(2)} \end{pmatrix} \in R^n$, $\boldsymbol{t}^{(1)} \in R^m$,

$$\begin{aligned} \phi(\boldsymbol{t}'\Sigma\boldsymbol{t}) &= E[\exp(i\boldsymbol{t}'(\boldsymbol{x}-\mu))] \\ &= E\{\exp[i\boldsymbol{t}^{(2)\prime}(\boldsymbol{x}^{(2)}-\mu^{(2)})]\,E(\exp[i\boldsymbol{t}^{(1)\prime}(\boldsymbol{x}^{(1)}-\mu^{(1)})]|\boldsymbol{x}^{(2)})\} \\ &= E\{\exp[i\boldsymbol{t}^{(2)\prime}(\boldsymbol{x}^{(2)}-\mu^{(2)}) + i\boldsymbol{t}^{(1)\prime}(\mu_{1.2}-\mu^{(1)})]\,\phi_{q(x^{(2)})}(\boldsymbol{t}^{(1)\prime}\Sigma_{11.2}\boldsymbol{t}^{(1)})\}\ . \end{aligned}$$

From the assumption, putting, for each $u \geqslant 0$, $\psi(u) = \phi_{q(x^{(2)})}(u)$ a.s., it follows that

$$\begin{aligned} \phi(\boldsymbol{t}'\Sigma\boldsymbol{t}) &= \psi(\boldsymbol{t}^{(1)\prime}\Sigma_{11.2}\boldsymbol{t}^{(1)})\,E\exp[i(\boldsymbol{t}^{(2)}+\Sigma_{22}^{-1}\Sigma_{21}\boldsymbol{t}^{(1)})'(\boldsymbol{x}^{(2)}-\mu^{(2)})] \\ &= \psi(\boldsymbol{t}^{(1)\prime}\Sigma_{11.2}\boldsymbol{t}^{(1)})\,\phi\,\{(\boldsymbol{t}^{(2)}+\Sigma_{22}^{-1}\Sigma_{21}\boldsymbol{t}^{(1)})'\Sigma_{22}(\boldsymbol{t}^{(2)}+\Sigma_{22}^{-1}\Sigma_{21}\boldsymbol{t}^{(1)})\}. \end{aligned}$$

And since $\boldsymbol{t}'\Sigma\boldsymbol{t} = \boldsymbol{t}^{(1)\prime}\Sigma_{11.2}\boldsymbol{t}^{(1)} + (\boldsymbol{t}^{(2)}+\Sigma_{22}^{-1}\Sigma_{21}\boldsymbol{t}^{(1)\prime}\Sigma_{22}(\boldsymbol{t}^{(2)}+\Sigma_{22}^{-1}\Sigma_{21}\boldsymbol{t}^{(1)})$, it follows that $\phi(u+v) = \psi(u)\ \phi(v)$, $u, v \geqslant 0$. Setting $v = 0$ yields $\phi = \psi$ which implies $\phi(u) = \exp(-cu/2)$ for some $c > 0$. □

Further, we have the following

Theorem 2.7.4. *Assume* $\boldsymbol{x} = \begin{bmatrix} \boldsymbol{x}^{(1)} \\ \boldsymbol{x}^{(2)} \end{bmatrix} \sim EC_n(\mu, \Sigma, \phi)$ *with* $\Sigma > 0$ *and* $\boldsymbol{x}^{(1)}$*:* $m \times 1$, $0 < m < n$. *Then* $(\boldsymbol{x}^{(1)}\,|\,\boldsymbol{x}^{(2)})$ *is normally distributed with probability one if and only if* $\boldsymbol{x}$ *is normally distributed.*

Proof. Assume that $(\boldsymbol{x}^{(1)}\,|\,\boldsymbol{x}^{(2)})$ is normally distributed with probability one. We have $\phi_{q(x^{(2)})}(u) = \exp(-c(q(\boldsymbol{x}^{(2)}))u/2)$, $u \geqslant 0$, a.s. for some function c: $(0, \infty) \to (0, \infty)$. If it is shown that $c(q(\boldsymbol{x}^{(2)}))$ is a degenerate random variable, then the normality of $\boldsymbol{x}$ will follow from Theorem 2.7.3. Let A be the set of all $a > 0$ such that

$$\phi_{a^2}(u) = \exp(-c(a^2)u/2), \qquad u \geqslant 0. \tag{2.7.1}$$

Note $P(q(\boldsymbol{x}^{(2)})^{\frac{1}{2}} \in A) = 1$. Now (2.7.1) implies that the cdf F_a^2 of R_a^2, $a \in A$, is that of a chi–variable with k_1 degrees of freedom times $c(a^2)^{\frac{1}{2}}$. Combining this with (2.6.12), we have, for $a \in A$.

$$\begin{aligned} 1 - F(r) &= (\text{const}) \int_r^{\infty} s^{n-1}\ \exp(-s^2/2c(a^2))ds, \qquad r \geqslant a, \quad \text{i.e.,} \\ F(r) &= (\text{const})\ r^{n-1}\ \exp(-r^2/2c(a^2))ds, \qquad r \geqslant a. \end{aligned} \tag{2.7.2}$$

It is obvious from (2.7.2) that $d(a^2)$ is a constant for all $a \in A$, and the degeneracy of $d(q(\boldsymbol{x}^{(2)}))$ follows. □

Theorem 2.7.5. *$EC_n(\mu, \Sigma, \phi)$ with $\Sigma > 0$ is a normal distribution if and only if two marginal densities of different dimensions exist and have functional forms which agree up to a positive multiple.*

Proof. We shall only show the "if" part. Suppose $\boldsymbol{x} \sim EC_n(\mu, \Sigma, \phi)$ has marginal densities of dimensions p and $p+q$ with functional forms g_p and g_{p+q}, and

$$g_{p+q}(u) = c g_p(u), \qquad c\text{: constant}, \qquad u \geqslant 0. \tag{2.7.3}$$

(Here and below, c stands for a positive constant, not always the same.) Without loss of generality we assume $\mu = 0$ and $\Sigma = \boldsymbol{I}_n$ (Otherwise, we could consider $\boldsymbol{y} = \Sigma^{-\frac{1}{2}}(\boldsymbol{x} - \mu)$). Then

$$\begin{aligned} g_p(x_1^2 + \cdots + x_p^2) &= \int g_{p+q}(x_1^2 + \cdots + x_{p+q}^2)dx_{p+1} \cdots dx_{p+q} \\ &= c\int g_p(x_1^2 + \cdots + x_{p+q}^2)dx_{p+1} \cdots dx_{p+q}, \end{aligned}$$

which implies

$$g_p(u) = c\int g_p(u + z_1^2 + \cdots + z_q^2)dz_1 \cdots dz_q, \qquad u \geqslant 0.$$

It follows that

$$g_{p+q}(x_1^2 + \cdots + x_{p+q}^2) = c\int g_p(x_1^2 + \cdots + x_{p+2q}^2)dx_{p+q+1} \cdots dx_{p+2q}.$$

Thus, a multiple of $g_p(x_1^2 + \cdots + x_{p+2q}^2)$ is a density of some $\boldsymbol{z} \sim EC_{p+2q}(\boldsymbol{0}, \boldsymbol{I}_{p+2q}, \psi)$, and $\boldsymbol{z}$ has the $(p+q)$-dimensional marginal density $g_{p+q}(x_1^2 + \cdots + x_{p+q}^2)$. Consequently, $\phi = \psi \in \Phi_{p+2q}$. Similarly, it follows that $\phi \in \Phi_{p+jq}$ for all $j = 1, 2, \ldots\ldots$ Hence $\phi \in \Phi_\infty$, and there exists a cdf F_∞ on $[0, \infty)$ such that both g_p and g_{p+q} satisfy (2.5.26). By the uniqueness of Laplace transform and (2.7.3), we obtain $r^{-p}dF_\infty(r) = cr^{-(p+q)}dF_\infty(r)$, from which it follows that F_∞ is degenerate at some point $\sigma > 0$. Thus, $g_p(u) = \exp[-u/(2\sigma^2)]$ and $\boldsymbol{x}$ is normal. □

The reader can find more interesting results in Cambanis, Huang and Simons' papers, or in other sections of this book.

2.8. Distributions of Quadratic Forms and Cochran's Theorems

Distributions of quadratic forms and Cochran's theorems play an important role in the linear model. In this section, we will give distributions of quadratic forms and discuss Cochran's Theorem in both the normal and spherical distributions. The main results about ECD in this section are from Anderson and Fang (1982a) and (1984).

2.8.1. Distributions of Quadratic Forms

Suppose $\boldsymbol{x} \stackrel{d}{=} R\boldsymbol{u}^{(n)} \sim S_n(\phi)$ and $R \sim F(\cdot)$. Partition $\boldsymbol{x}$ into $\boldsymbol{x} = (\boldsymbol{x}^{(1)\prime}, \ldots, \boldsymbol{x}^{(m)\prime})'$, where $\boldsymbol{x}^{(1)}, \ldots, \boldsymbol{x}^{(m)}$ have $n_1, \ldots, n_m$ components of $\boldsymbol{x}$. From (2.5.12), we have

$$x \stackrel{d}{=} (Rd_1 u_1', \ldots, Rd_m u_m')', \tag{2.8.1}$$

where $d_1, \ldots, d_m \geqslant 0$, $(d_1^2, \ldots, d_m^2) \sim D_m(n_1/2, \ldots, n_m/2)$; R, $(d_1, \ldots, d_m)$, $u_1, \ldots, u_{m-1}$ and u_m are independent, and $u_j \stackrel{d}{=} u^{(n_j)}$, $j = 1, \ldots, m$. Thus

$$(x^{(1)\prime} x^{(1)}, \ldots, x^{(m)\prime} x^{(m)}) \stackrel{d}{=} R^2(d_1^2, \ldots, d_m^2). \tag{2.8.2}$$

We write $(x^{(1)\prime}x^{(1)}, \ldots, x^{(m)\prime}x^{(m)}) \sim G_m(n_1/2, \ldots, n_m/2; \phi)$ and $(x^{(1)\prime}x^{(1)}, \ldots, x^{(m-1)\prime}x^{(m-1)}) \sim G_m(n_1/2, \ldots, n_{m-1}/2; n_m/2; \phi)$. If $(y_1, \ldots, y_m) \sim G(n_1/2, \ldots, n_m/2; \phi)$, it is easily seen that $(y_1, \ldots, y_k) \sim G_{k+1}(n_1/2, \ldots, n_k/2; (n_{k+1} + \ldots + n_m)/2; \phi)$ for $1 \leqslant k < m$.

Lemma 2.8.1. *Suppose* $x \sim S_n(\phi)$.

(1) *If* $P(x = 0) = 0$ *and* $n_m \geqslant 1$, *then* $(Y_1, \ldots, Y_k) \equiv (x^{(1)\prime}x^{(1)}, \ldots, x^{(k)\prime}x^{(k)})$ $(1 \leqslant k < m)$ *has the density*

$$\frac{\Gamma(n/2)}{(n^*/2)\prod_1^k \Gamma(n_i/2)} \prod_1^k y_i^{n_i/2-1} \int_{(\sum_1^k y_i)^{\frac{1}{2}}}^{\infty} r^{2-n}(r^2 - \sum_1^k y_i)^{n^*/2-1} dF(r). \tag{2.8.3}$$

where $n^* = n_{k+1} + \cdots + n_m$.

(2) *If* x *has a density* $f(x'x)$, *then the joint density of* $Y_1, \ldots, Y_m$ *is*

$$\frac{\pi^{\frac{1}{2}n}}{\prod_1^m \Gamma(n_i/2)} \prod_1^m y_i^{n_i/2-1} f(\sum_1^m y_i). \tag{2.8.4}$$

Proof. (2.8.3) can be obtained from (2.8.2) and (2.5.18) directly. As the density function of $x^{(1)}, \ldots, x^{(m)}$ is $f(x'x) = f(\sum_1^m x^{(i)\prime} x^{(i)})$, for each nonnegative Borel function $h(\cdot)$ we have

$$Eh(Y_1, \ldots, Y_m) = \int h(x^{(1)\prime}x^{(1)}, \ldots, x^{(m)\prime}x^{(m)}) f(\sum_1^m x^{(i)\prime}x^{(i)}) \mathrm{d}x^{(1)} \ldots \mathrm{d}x^{(m)}$$

$$= \frac{\pi^{\frac{1}{2}n}}{\prod_1^m \Gamma(n_i/2)} \int h(y_1, \ldots, y_m) f(\sum_1^m y_i) \prod_1^m (y_i^{n_i/2-1} \, \mathrm{d}y_i)$$

by using (2.4.17) which implies (2.8.4). □

Example 2.8.1. Let x have a uniform distribution in the unit sphere in R^n. (cf. Example 2.6.3). Then, by (2.8.4) the density of $(Y_1, \ldots, Y_m)$ is

$$\frac{\Gamma(\frac{1}{2}n + 1)}{\prod_1^m \Gamma(n_i/2)} \prod_1^m y_i^{n_i/2-1} \qquad y_i > 0,\ i = 1, \ldots, m;\ \sum_1^m y_i \leqslant 1.$$

From (2.8.3) we obtain the interesting result that the marginal density of $Y_1, \ldots, Y_k$ $(1 \leqslant k < m)$ is $D_{k+1}(n_1/2, \ldots, n_k/2; (n^* + 2)/2)$ with $n^* = n_{k+1} + \ldots + n_m$.

If $\boldsymbol{x} \sim N_n(\boldsymbol{0}, \boldsymbol{I}_n)$, then $\boldsymbol{x}^{(1)\prime}\boldsymbol{x}^{(1)}, \ldots, \boldsymbol{x}^{(m)\prime}\boldsymbol{x}^{(m)}$ are independently distributed as chi–squared distribution with $n_1, \ldots, n_m$ degrees of freedom respectively. Conversely, if $\boldsymbol{x}^{(1)\prime}\boldsymbol{x}^{(1)}$ and $\boldsymbol{x}^{(2)\prime}\boldsymbol{x}^{(2)}$ are independent, we shall show that $\boldsymbol{x}$ must be normal.

Theorem 2.8.1. *Assume* $\boldsymbol{x} = \begin{bmatrix} \boldsymbol{x}^{(1)} \\ \boldsymbol{x}^{(2)} \end{bmatrix} \overset{d}{=} R\boldsymbol{u}^{(n)} \sim S_n^+(\phi)$, *where* $\boldsymbol{x}^{(1)}$: $m \times 1$ *and* $0 < m < n$. *Then* $\boldsymbol{x}^{(1)\prime}\boldsymbol{x}^{(1)}$ *and* $\boldsymbol{x}^{(2)\prime}\boldsymbol{x}^{(2)}$ *are independent if and only if* $\boldsymbol{x}$ *is normal.*

Proof. The "If" part is well known. Now assume that $\boldsymbol{x}^{(1)\prime}\boldsymbol{x}^{(1)}$ and $\boldsymbol{x}^{(2)\prime}\boldsymbol{x}^{(2)}$ are independent. We have $P(R = 0) = 0$. As $(y_1, y_2) = (\boldsymbol{x}^{(1)\prime}\boldsymbol{x}^{(1)}, \boldsymbol{x}^{(2)\prime}\boldsymbol{x}^{(2)}) \overset{d}{=} (R^2 d_1^2, R^2 d_2^2)$, the distribution of $Y_1/Y_2 \overset{d}{=} R^2 d_1^2/(R^2 d_2^2) = d_1^2/d_2^2$ is independent of $Y_1 + Y_2 = \boldsymbol{x}'\boldsymbol{x} \overset{d}{=} R^2$. Hence Y_1 and Y_2 have gamma distributions with the same scale parameter (cf. Lukacs (1956), p. 208), and R^2 has a gamma distribution with that scale parameter. Hence, $\boldsymbol{x}$ has a density $f(\boldsymbol{x}'\boldsymbol{x})$ and the density of Y_1 and Y_2 is

$$\text{const } y_1^{a-1} y_2^{b-1} e^{-(y_1+y_2)/2c} = \text{const } y_1^{m/2-1} y_2^{(n-m)/2-1} f(y_1 + y_2)$$

from (2.8.4). That is,

$$f(y_1 + y_2) = \text{const } y_1^{a-m/2} y_2^{b-(n-m)/2} e^{-(y_1+y_2)/2c}.$$

This can hold identically for $y_1 \geqslant 0$ and $y_2 \geqslant 0$ only if $a = m/2$ and $b = (n-m)/2$. Hence $R^2 = y_1 + y_2 \sim \text{const } \chi_n^2$. Then $x \sim N_n(\boldsymbol{0}, \sigma^2\boldsymbol{I}_n)$ for Some $\sigma > 0$ by Theorem 2.5.4. □

Example 2.8.2. Suppose that $\boldsymbol{x} = (X_1, \ldots, X_n)' \sim S_n(\phi)$ and has a density $f(\boldsymbol{x}'\boldsymbol{x})$. Let

$$\bar{X} = \frac{1}{n}\sum_1^n X_i, \qquad S = \sum_1^n (X_i - \bar{X})^2$$

Clearly, $\bar{X} = \boldsymbol{1}'_n\boldsymbol{x}/n$ and $S = \boldsymbol{x}'\boldsymbol{D}\boldsymbol{x}$ where $\boldsymbol{D} = \boldsymbol{I}_n - \frac{1}{n}\boldsymbol{1}_n\boldsymbol{1}'_n$. Let Γ be an $n \times n$ orthogonal matrix with the last row $(1/n^{\frac{1}{2}}, \ldots, 1/n^{\frac{1}{2}})$ and $\boldsymbol{y} = (Y_1, \ldots, Y_n)' = \Gamma\boldsymbol{x}$. Then $\boldsymbol{y} \sim S_n(\phi)$ and

$$\bar{X} = n^{-\frac{1}{2}} Y_n, \qquad S = \sum_1^{n-1} Y_i^2.$$

From (2.8.4), the joint density of Y_n^2 and S is

$$\frac{\pi^{(n-1)/2}}{\Gamma((n-1)/2)} s^{(n-1)/2-1} (y_n^2)^{-\frac{1}{2}} f(s + y_n^2).$$

Then the joint density of $\bar{X}$ and S is

$$\frac{n^{\frac{1}{2}} \pi^{(n-1)/2}}{\Gamma((n-1)/2)} s^{(n-1)/2-1} f(s + n\bar{x}^2).$$

By Theorem 2.8.1, $\bar{X}$ and s are independent if and only if $\boldsymbol{x} \sim N_n(\boldsymbol{0}, \sigma^2\boldsymbol{I}_n)$ for some $\sigma^2 > 0$.

2.8.2. Cochran's Theorem for the Normal Case

Assume that $\boldsymbol{x} \sim N_n(\mu, \boldsymbol{I}_n)$ and $\boldsymbol{C}$ is an $n \times n$ constant symmetric matrix. We shall discuss when $\boldsymbol{x}' \boldsymbol{C}\boldsymbol{x}$ has a chi-squared distribution in this subsection. Also we shall give the necessary and sufficient condition for two quadratic forms being independent.

Lemma 2.8.2. *Assume that* $\boldsymbol{x} \sim N_n(\mu, \boldsymbol{I}_n)$ *and* $\boldsymbol{C}$ *is an* $n \times n$ *symmetric matrix with eigenvalues* $\alpha_1, \ldots, \alpha_n$. *Then*

$$E(e^{t\boldsymbol{x}'\boldsymbol{C}\boldsymbol{x}}) = \prod_{j=1}^{n} (1-2t\alpha_j)^{-\frac{1}{2}} \exp\left\{\frac{t\alpha_j\lambda_j^2}{1-2t\alpha_j}\right\}, \tag{2.8.5}$$

where $\lambda = (\lambda_1, \ldots, \lambda_n)' = \Gamma'\mu$ *and* $\Gamma'\boldsymbol{C}\Gamma = \operatorname{diag}(\alpha_1, \ldots, \alpha_n)$.

Proof. Let $\boldsymbol{y} = \Gamma'\boldsymbol{x}$. Then $\boldsymbol{y} \sim N_n(\Gamma'\mu, \boldsymbol{I}_n) = N_n(\lambda, \boldsymbol{I}_n)$ and

$$E(e^{t\boldsymbol{x}'\boldsymbol{C}\boldsymbol{x}}) = E(e^{t\boldsymbol{y}'\Gamma'\boldsymbol{C}\Gamma\boldsymbol{y}}) = \prod_{j=1}^{n} E(e^{t\alpha_j y_j^2})$$
$$= \prod_{j=1}^{n}\left[(1-2t\alpha_j)^{-\frac{1}{2}}\exp\left\{\frac{t\alpha_j\lambda_j^2}{1-2t\alpha_j}\right\}\right]. \qquad \square$$

Theorem 2.8.2. (Cochran) *Assume that* $\boldsymbol{x} \sim N_n(\mu, \boldsymbol{I}_n)$ *and* $\boldsymbol{C}$ *is a symmetric matrix. Then* $\boldsymbol{x}'\boldsymbol{C}\boldsymbol{x} \sim X_k^2(\mu'\boldsymbol{C}\mu)$ *if and only if*

$$\boldsymbol{C}^2 = \boldsymbol{C} \qquad \text{and} \qquad \operatorname{rk}\boldsymbol{C} = k. \tag{2.8.6}$$

Proof. Assume that (2.8.6) holds. Then there is an orthogonal matrix Γ such that $\Gamma'\boldsymbol{C}\Gamma = \begin{bmatrix} \boldsymbol{I}_k & \boldsymbol{O} \\ \boldsymbol{O} & \boldsymbol{O} \end{bmatrix}$. Let $\boldsymbol{y} = \Gamma'\boldsymbol{x}$. Then $\boldsymbol{y} \sim N_n(\Gamma'\mu, \boldsymbol{I}_n)$ and $\boldsymbol{x}'\boldsymbol{C}\boldsymbol{x} = \boldsymbol{y}'\Gamma'\boldsymbol{C}\Gamma\boldsymbol{y} = \sum_1^p y_i^2$. From Definition 2.4.1 we have $\boldsymbol{x}'\boldsymbol{C}\boldsymbol{x} \sim \chi_k^2(\mu'\ \boldsymbol{C}\mu)$. Conversely, assume $\boldsymbol{x}'\boldsymbol{C}\boldsymbol{x} \sim \chi_k^2$ $(\mu'\ \boldsymbol{C}\mu)$. By Lemma 2.8.2 we have

$$(1-2t)^{-k/2}\exp\left\{\frac{t\lambda}{1-2t}\right\} = \prod_{j=1}^{n}(1-2\alpha_j t)^{-\frac{1}{2}}\exp\left\{\frac{\alpha_j\lambda_j^2}{1-2\alpha_j}\right\},$$

where $\lambda = \mu'\boldsymbol{C}\mu$ and $\{\alpha_j\}$, $\{\lambda_j\}$ have the meaning as in Lemma 2.8.2. Comparing the singular points on the two sides, we must have $\alpha_{i_1} = \cdots = \alpha_{i_k} = 1$ for some $1 \leqslant i_1 < i_2 < \cdots < i_k \leqslant n$ and $\alpha_j = 0$, $j \neq i_1, \cdots, i_k$. That means $\boldsymbol{C}^2 = \boldsymbol{C}$ and rk $\boldsymbol{C} = k$. $\square$

Corollary 1. *If the assumption of Theorem 2.8.2 holds, then* $\boldsymbol{x}'\boldsymbol{C}\boldsymbol{x} \sim \chi_k^2$ *if and only if* $\boldsymbol{C}^2 = \boldsymbol{C}$, rk $\boldsymbol{C} = k$ *and* $\boldsymbol{C}\mu = \boldsymbol{0}$.

Theorem 2.8.3. (Craig) *Assume that* $\boldsymbol{x} \sim N_n(\mu, \boldsymbol{I}_n)$ *and* $\boldsymbol{C}_i$ $(1 \leqslant i \leqslant m)$ *are* $n \times n$ *symmetric matrices. Then* $\boldsymbol{x}'\boldsymbol{C}_i\boldsymbol{x}$, $i = 1, \ldots, m$ *are independent if and only if* $\boldsymbol{C}_i\boldsymbol{C}_j = 0$ *for all* $i \neq j$.

Before proving this theorem, we need the following lemma which is from P.L. Hsu's lecture in 1962.

Lemma 2.8.3. *Assume that* $\boldsymbol{A}$ *and* $\boldsymbol{B}$ *are* $n \times n$ *symmetric matrices with nonzero eigenvalues* $\{\lambda_1, \ldots, \lambda_r\}$ *and* $\{\mu_1, \ldots, \mu_s\}$, *respectively. If the non-zero eigenvalues of* $\boldsymbol{A} + \boldsymbol{B}$ *are* $\{\lambda_1, \ldots, \lambda_r, \mu_1, \ldots, \mu_s\}$, *then* $\boldsymbol{A}\boldsymbol{B} = \boldsymbol{B}\boldsymbol{A} = \boldsymbol{O}$.

Proof. First we assume $r+s = n$. Without loss of generality, we can assume $\boldsymbol{A} = \begin{bmatrix} \boldsymbol{D}_\lambda & \boldsymbol{O} \\ \boldsymbol{O} & \boldsymbol{O} \end{bmatrix}$ with $\boldsymbol{D}_\lambda = \operatorname{diag}(\lambda_1, \ldots, \lambda_r)$. Take an orthogonal matrix Γ such that $\Gamma'\boldsymbol{B}\Gamma$

$= \begin{bmatrix} O & O \\ O & D_\mu \end{bmatrix}$ with $D_\mu = \mathrm{diag}(\mu_1, \ldots, \mu_s)$. According to the fashion of partition of $\Gamma' B \Gamma$, let $\Gamma = \begin{bmatrix} C & F \\ D & G \end{bmatrix}$. We have

$$B = \begin{bmatrix} C & F \\ D & G \end{bmatrix} \begin{bmatrix} O & O \\ O & D_\mu \end{bmatrix} \begin{bmatrix} C' & D' \\ F' & G' \end{bmatrix}$$

$$= \begin{bmatrix} I & F \\ O & G \end{bmatrix} \begin{bmatrix} O & O \\ O & D_\mu \end{bmatrix} \begin{bmatrix} I & O \\ F' & G' \end{bmatrix}$$

$$A = \begin{bmatrix} I & F \\ O & G \end{bmatrix} \begin{bmatrix} D_\lambda & O \\ O & O \end{bmatrix} \begin{bmatrix} I & O \\ F' & G' \end{bmatrix}$$

and

$$A + B = \begin{bmatrix} I & F \\ O & G \end{bmatrix} \begin{bmatrix} D_\lambda & O \\ O & D_\mu \end{bmatrix} \begin{bmatrix} I & O \\ F' & G' \end{bmatrix}.$$

As AB has the same nonzero eigenvalues as BA, therefore $A + B$ has the same nonzero eigenvalues as

$$E = \begin{bmatrix} D_\lambda & O \\ O & D_\mu \end{bmatrix} \begin{bmatrix} I & O \\ F' & G' \end{bmatrix} \begin{bmatrix} I & F \\ O & G \end{bmatrix} = \begin{bmatrix} D_\lambda & O \\ O & D_\mu \end{bmatrix} \begin{bmatrix} I & F \\ F' & F'F + G'G \end{bmatrix}$$

$$= \begin{bmatrix} D_\lambda & O \\ O & D_\mu \end{bmatrix} \begin{bmatrix} I & F \\ F' & I \end{bmatrix},$$

for $\Gamma'\Gamma = I$ and $F'F + G'G = I$. Hence

$$\prod_1^\gamma \lambda_i \prod_1^s \mu_j = |A + B| = |E| = \begin{vmatrix} D_\lambda & O \\ O & D_\mu \end{vmatrix} \begin{vmatrix} I & F \\ F' & I \end{vmatrix} = \prod_1^\gamma \lambda_i \prod_1^s \mu_j |I - F'F| .$$

Then the consequence is that $|I - F'F| = 1$ and $|G'G| = 1$. Thus the eigenvalues of $G'G$ are all equal to 1 and $F'F = O$, i.e., $F = O$. and

$$B = \begin{bmatrix} I & O \\ O & G \end{bmatrix} \begin{bmatrix} O & O \\ O & D_\mu \end{bmatrix} \begin{bmatrix} I & O \\ O & G' \end{bmatrix} = \begin{bmatrix} O & O \\ O & GD_\mu G' \end{bmatrix}.$$

The assertion $AB = BA = O$ follows.

Secondly, if $r + s < n$. *Without loss of generality we can assume* $A = \begin{bmatrix} D_\lambda & O \\ O & O \end{bmatrix}$. There exists an orthogonal matrix Q such that

$$Q'BQ = \begin{bmatrix} O & O & O \\ O & D_\mu & O \\ O & O & O \end{bmatrix} \begin{matrix} r \\ s \\ n-r-s \end{matrix}.$$

Applying the first part of the proof to the $(r+s)$-dimensional first principal minor of $Q'BQ$, $AB = BA = O$ follows. □

The proof of Theorem 2.8.3. Assume $C_iC_j = O$ for all $i \neq j$. This implies $C_iC_j = C_jC_i$ and there exists an orthogonal matrix Γ such that $\Gamma C_i \Gamma' = \Lambda_i = \mathrm{diag}(\lambda_1^{(i)}, \ldots, \lambda_n^{(i)})$, $i = 1, \ldots, m$ (cf. Section 1.2, (8)). Let $G_i = \{t\colon \lambda_t^{(i)} \neq 0,\ t = 1, \ldots, n\}$. As $\Lambda_i\Lambda_j = \Gamma C_i \Gamma' \Gamma C_j \Gamma' = \Gamma C_i C_j \Gamma' = O$, we have $G_i \cap G_j = \phi$ for $i \neq j$. Let $y = \Gamma x$. Then y

$\sim N_n(\Gamma\mu, \boldsymbol{I}_n)$ and $\boldsymbol{x}'\boldsymbol{C}_i\boldsymbol{x} = \boldsymbol{y}'\Lambda_i\boldsymbol{y} = \sum_{l\in G_i} y_l^2\lambda_l^{(i)}$. The independence of $\{\boldsymbol{x}'\boldsymbol{C}_1\boldsymbol{x}, \ldots, \boldsymbol{x}'\boldsymbol{C}_m\boldsymbol{x}\}$ follows from the fact that $G_i \bigcap G_j = \phi$ for $i \neq j$ and $y_1, \cdots, y_n$ are independent. Conversely we prove $\boldsymbol{C}_i\boldsymbol{C}_j = \boldsymbol{O}$. Let $\boldsymbol{A} = \boldsymbol{C}_i$, $\boldsymbol{B} = \boldsymbol{C}_j$, $\text{rk}(\boldsymbol{A}) = r$, $\text{rk}(\boldsymbol{B}) = s$ and $\text{rk}(\boldsymbol{A} + \boldsymbol{B}) = q$. Denote the eigenvalues of $\boldsymbol{A}$, $\boldsymbol{B}$, and $\boldsymbol{A}+\boldsymbol{B}$ by $\lambda_1, \ldots, \lambda_r$; $\mu_1, \ldots, \mu_s$; $\nu_1, \ldots, \nu_q$, respectively. By using (2.8.5),

$$E(e^{t\boldsymbol{x}'\boldsymbol{A}\boldsymbol{x}}) = \prod_{i=1}^{r} (1 - 2t\lambda_i)^{-\frac{1}{2}}\exp\left\{\frac{t\lambda_i a_i}{1 - 2\lambda_i t}\right\}, \quad a_i \geqslant 0, \tag{2.8.7}$$

$$E(e^{t\boldsymbol{x}'\boldsymbol{B}\boldsymbol{x}}) = \prod_{j=1}^{s} (1 - 2t\mu_j)^{-\frac{1}{2}}\exp\left\{\frac{t\mu_j b_j}{1 - 2\mu_j t}\right\}, \quad b_j \geqslant 0, \tag{2.8.8}$$

$$E(e^{t\boldsymbol{x}'(\boldsymbol{A}+\boldsymbol{B})\boldsymbol{x}}) = \prod_{k=1}^{q} (1 - 2t\nu_k)^{-\frac{1}{2}}\exp\left\{\frac{t\nu_k c_k}{1 - 2\nu_k t}\right\}, \quad c_k \geqslant 0. \tag{2.8.9}$$

As $\boldsymbol{x}'\boldsymbol{A}\boldsymbol{x}$ and $\boldsymbol{x}'\boldsymbol{B}\boldsymbol{x}$ are independent, (2.8.7) $\times$ (2.8.8) $=$ (2.8.9). Comparing the singular points we have $\{\nu_1, \ldots, \nu_q\} = \{\lambda_1, \ldots, \lambda_r, \mu_1, \ldots, \mu_s\}$. Then the assertion follows from Lemma 2.8.3. □

Lemma 2.8.4. *Assume that $\boldsymbol{C}_1, \ldots, \boldsymbol{C}_m$ are $n\times n$ symmetric matrices with rank $r_1, \ldots, r_m$ respectively. Let $\boldsymbol{C} = \sum_1^m \boldsymbol{C}_i$ and $\text{rk}(\boldsymbol{C}) = r$. Consider the following conditions:*

(a) $\boldsymbol{C}_i^2 = \boldsymbol{C}_i$, $i = 1, \ldots, m$;
(b) $\boldsymbol{C}_i\boldsymbol{C}_j = \boldsymbol{O}$, *for* $i \neq j$;
(c) $\boldsymbol{C}^2 = \boldsymbol{C}$;
(d) $r = r_1 + \ldots + r_m$.

Then (a)(b) $\Rightarrow$ (c)(d), (a)(c) $\Rightarrow$ (b)(d), (b)(c) $\Rightarrow$ (a)(d) and (c)(d) $\Rightarrow$ (a)(b).

The proof is left to the reader. (cf. Anderson and Styan (1983) or Zhang and Fang (1982).)

Summarizing the above results we obtain the following theorem.

Theorem 2.8.4. *Assume that $\boldsymbol{x} \sim N_n(\mu, \boldsymbol{I}_n)$, $\boldsymbol{C}_1, \ldots, \boldsymbol{C}_m$ are $n \times n$ symmetric matrices with rank $r_1, \ldots, r_m$ respectively. Let $\boldsymbol{C} = \sum_1^m \boldsymbol{C}_i$ and $r = \text{rk}(\boldsymbol{C})$. Consider the following statements:*

(A1) $\boldsymbol{C}_i^2 = \boldsymbol{C}_i$, $i = 1, \ldots, m$;
(A2) $\boldsymbol{C}_i\boldsymbol{C}_j = \boldsymbol{O}$, *for* $i \neq j$; $i,j = 1, \ldots, m$;
(A3) $\boldsymbol{C}^2 = \boldsymbol{C}$;
(B1) $\boldsymbol{x}'\boldsymbol{C}_i\boldsymbol{x} \sim \chi^2_{r_i}(\mu'\boldsymbol{C}_i\mu)$, $i = 1, \ldots, m$;
(B2) $\{\boldsymbol{x}'\boldsymbol{C}_i\boldsymbol{x}, i = 1, \ldots, m\}$ *are independent;*
(B3) $\boldsymbol{x}'\boldsymbol{C}\boldsymbol{x} \sim \chi^2_r(\mu'\boldsymbol{C}\mu)$;
(D) $\sum_1^m r_i = r$.

Then (a) (Ai) $\Longleftrightarrow$ (Bi), $i = 1, 2, 3$;
(b) *two of* (Ai) *and* (Bj), $i \neq j$ *imply others;*
(c) (A3) *and* (D) *or* (B3) *and* (D) *imply others.*

2.8.3. Cochran's Theorem for the Case of ECD

Suppose $\mathbf{x} \sim S_n(\phi)$ and $\boldsymbol{C}' = \boldsymbol{C}$. If $\boldsymbol{C}^2 = \boldsymbol{C}$ and $\text{rk}(\boldsymbol{C}) = k$, it is easily seen that $\boldsymbol{x}'\boldsymbol{C}\boldsymbol{x} = \boldsymbol{x}^{(1)\prime}\boldsymbol{x}^{(1)} \sim G_2(k/2; (n-k)/2; \phi)$ where $\boldsymbol{x}^{(1)}$ is a vector with the first k

components of $\boldsymbol{x}$. Conversely, if $\boldsymbol{x}'\boldsymbol{C}\boldsymbol{x} \sim G_2(k/2; (n-k)/2; \phi)$ we want to know whether $\boldsymbol{C}^2 = \boldsymbol{C}$ and $\mathrm{rk}(\boldsymbol{C}) = k$ or not. First we need the following lemma.

Lemma 2.8.5. *Assume that* $\boldsymbol{x} = (X_1, \ldots, X_n)' \sim N_n(\mathbf{0}, \mathbf{I}_n)$, $(Z_1, \ldots, Z_n) \sim D_n(\frac{1}{2}, \ldots, \frac{1}{2})$, *and* $\boldsymbol{C}_1, \ldots, \boldsymbol{C}_m$ $(1 \leqslant m \leqslant n)$ *are* $n \times n$ *symmetric matrices. Then*

$$\begin{bmatrix} \boldsymbol{x}'\boldsymbol{C}_1\boldsymbol{x} \\ \cdot \\ \cdot \\ \cdot \\ \boldsymbol{x}'\boldsymbol{C}_m\boldsymbol{x} \end{bmatrix} \stackrel{d}{=} \begin{bmatrix} \sum_1^n l_{1j}X_j^2 \\ \cdot \\ \cdot \\ \cdot \\ \sum_1^n l_{mj}X_j^2 \end{bmatrix} \Longleftrightarrow \begin{bmatrix} \boldsymbol{x}'\boldsymbol{C}_1\boldsymbol{x}/\|\boldsymbol{x}\|^2 \\ \cdot \\ \cdot \\ \cdot \\ \boldsymbol{x}'\boldsymbol{C}_m\boldsymbol{x}/\|\boldsymbol{x}\|^2 \end{bmatrix} \stackrel{d}{=} \begin{bmatrix} \sum_1^n l_{1j}Z_j \\ \cdot \\ \cdot \\ \cdot \\ \sum_1^n l_{mj}Z_j \end{bmatrix} \tag{2.8.10}$$

Proof. For ease of exposition we prove the lemma for $m=1$. If the left side of (2.8.10) holds, then $\|\boldsymbol{x}\|^2[\boldsymbol{x}'\boldsymbol{C}_1\boldsymbol{x}/\|\boldsymbol{x}\|^2] = \boldsymbol{x}'\boldsymbol{C}_1\boldsymbol{x} \stackrel{d}{=} \sum_{j=1}^n l_{1j}X_j^2 = \|\boldsymbol{x}\|^2[\sum_{j=1}^n l_{1j}X_j^2/\|\boldsymbol{x}\|^2]$. As $\|\boldsymbol{x}\|^2 \sim \chi_n^2$, the c.f. of $\log \|\boldsymbol{x}\|^2$ is different from zero almost everywhere and $\|\boldsymbol{x}\|^2$ and $\boldsymbol{x}/\|\boldsymbol{x}\|^2$ are independent. The right-side follows from (c) of Section 2.1.7. The proof in the other direction is easy. □

By Lemma 2.8.5, we can transform the theory of distributions from quadratic forms of $\boldsymbol{x}$ to that of $\boldsymbol{x}/\|\boldsymbol{x}\|$.

Corollary 1. *Assume that* $\boldsymbol{x} \sim \boldsymbol{N}_n(\mathbf{0}, \boldsymbol{I}_n)$ *and* $\boldsymbol{C}$ *is an* $n \times n$ *symmetric matrix. Then* $\boldsymbol{x}'\boldsymbol{C}\boldsymbol{x}/\|\boldsymbol{x}\|^2 \sim B(k/2, (n-k)/2)$ *if and only if* $\boldsymbol{C}^2 = \boldsymbol{C}$ *and* $rk(\boldsymbol{C}) = k$.

Corollary 2. *Assume that* $\boldsymbol{x} \sim \boldsymbol{N}_n(\mathbf{0}, \boldsymbol{I}_n)$ *and* $\boldsymbol{C}$ *and* $\boldsymbol{D}$ *are* $n \times n$ *symmetric matrices. Then* $(\boldsymbol{x}'\boldsymbol{C}\boldsymbol{x}/\|\boldsymbol{x}\|^2, \boldsymbol{x}'\boldsymbol{D}\boldsymbol{x}/\|\boldsymbol{x}\|^2) \sim D_3(k/2, m/2; (n-m-k)/2)$ *if and only if* $\boldsymbol{C}\boldsymbol{D} = \boldsymbol{O}$, $\boldsymbol{C}^2 = \boldsymbol{C}$, $\boldsymbol{D}^2 = \boldsymbol{D}$, $rk(\boldsymbol{C}) = k$ *and* $rk(\boldsymbol{D}) = m$.

Theorem 2.8.5. *Suppose that* $\boldsymbol{x} \stackrel{d}{=} R\boldsymbol{u}^{(n)} \sim S_n^+(\phi)$ *and* $\boldsymbol{C}$ *is an* $n \times n$ *symmetric matrix. Then* $\boldsymbol{x}'\boldsymbol{C}\boldsymbol{x} \sim G_2(k/2; (n-k)/2; \phi)$ *if and only if* $\boldsymbol{C}^2 = \boldsymbol{C}$ *and* $\mathrm{rk}(\boldsymbol{C}) = k$.

Proof. The "if" part is trivial. Suppose $\boldsymbol{x}'\boldsymbol{C}\boldsymbol{x} \sim G_2(k/2; (n-k)/2; \phi)$. Then $\|\boldsymbol{x}\|^2(\boldsymbol{x}'\boldsymbol{C}\boldsymbol{x}/\|\boldsymbol{x}\|^2) = \boldsymbol{x}'\boldsymbol{C}\boldsymbol{x} \stackrel{d}{=} R^2 Z$, where $Z \sim B(k/2, (n-k)/2)$ is independent of $\boldsymbol{R}$. Obviously, $P(Z > 0) = 1$. Thus, $P(\boldsymbol{x}'\boldsymbol{C}\boldsymbol{x}/\|\boldsymbol{x}\|^2 > 0) = P(R^2 > 0) = P(\boldsymbol{x} \neq \mathbf{0}) = 1$. As $P(0 < Z < 1) = 1$, $\phi_{\log Z} \in (\mathrm{U})$ (cf. Definition 2.1.4). By (c) of Section 2.1.7, we have $\boldsymbol{x}'\boldsymbol{C}\boldsymbol{x}/\|\boldsymbol{x}\|^2 \stackrel{d}{=} Z$. The assertion follows from **Corollary 1 of Lemma 2.8.5**. □

Corollary 1. *Assume that* $\boldsymbol{x} \sim EC_n(\mathbf{0}, \Sigma, \phi)$ *with* $\Sigma > 0$ *and* $P(\boldsymbol{x}=\mathbf{0}) = 0$, *and* $\boldsymbol{C}$ *is an* $n \times n$ *symmetric matrix. Then* $\boldsymbol{x}'\boldsymbol{C}\boldsymbol{x} \sim G_2(k/2; (n-k)/2; \phi)$ *if and only if* $\boldsymbol{C}\boldsymbol{\Sigma}\boldsymbol{C} = \boldsymbol{C}$ *and* $\mathrm{rk}(\boldsymbol{C}) = k$.

Theorem 2.8.6. *Suppose that* $\boldsymbol{x} \stackrel{d}{=} R\boldsymbol{u}^{(n)} \sim S_n^+(\phi)$ *and* $\boldsymbol{C}_1, \ldots, \boldsymbol{C}_m$ *are* $n \times n$ *symmetric matrices. Then* $(\boldsymbol{x}'\boldsymbol{C}_1\boldsymbol{x}, \ldots, \boldsymbol{x}'\boldsymbol{C}_m\boldsymbol{x}) \sim G_{m+1}(n_1/2, \ldots, n_m/2; n_{m+1}/2; \phi)$ *with* $\sum_1^{m+1} n_i = n$ *if and only if* $\boldsymbol{C}_i\boldsymbol{C}_j = \delta_{ij}\boldsymbol{C}_i$ *and* $\mathrm{rk}(\boldsymbol{C}_i) = n_i$, $i,j = 1, \ldots, m$, *where* $\delta_{ii} = 1$ *and* $\delta_{ij} = 0$, $i \neq j$.

Proof. For ease of exposition, we only prove the theorem for $m=2$. Assume $(\boldsymbol{x}'\boldsymbol{C}_1\boldsymbol{x},$

$x'C_2x) \sim G_3(n_1/2, n_2/2{:}\ n_3/2; \phi)$. We have

$$(x'C_1x,\ x'C_2x) \stackrel{d}{=} R^2(Z_1, Z_2),$$

where R is independent of (Z_1, Z_2) and $(Z_1, Z_2) \sim D_3(n_1/2, n_2/2{:}\ n_3/2)$. Thus $x'C_1x \sim G_2(n_1/2,\ (n_2+n_3)/2;\ \phi)$, $x'C_2x \sim G_2(n_2/2,\ (n_1+n_3)/2;\ \phi)$, and $x'(C_1 + C_2)x \sim G_2((n_1+n_2)/2;\ n_3/2;\ \phi)$. By Theorem 2.8.5, we have $C_1^2 = C_1$, $\mathrm{rk}(C_1) = n$, $C_2^2 = C_2$, $\mathrm{rk}(C_2) = n_2$ and $(C_1 + C_2)^2 = C_1 + C_2$ which implies $C_1C_2 = O$ and the assertion follows. The "if" part is trivial. □.

Corollary 1. *Suppose that* $x \sim EC_n(\mathbf{0}, \Sigma, \phi)$ *with* $\Sigma > 0$ *and* $P(x = \mathbf{0}) = 0$, $C_1, \ldots, C_m$ *are* $n\times n$ *symmetric matrices. Then* $(x'C_1x, \ldots, x'C_mx) \sim G_{m+1}(n_1/2, \ldots, n_m/2; n_{m+1}/2; \phi)$ *if and only if* $C_i\Sigma C_j = \delta_{ij}C_i$, $\mathrm{rk}(C_i) = n_i$, $i,j = 1, \ldots, m$.

We can extend Cochran's theorem to the more general case, for example, to the case of $C^3 = C$ (cf. Anderson and Fang (1982a), Fang and Wu (1984), and Exercise 2.13).

2.9. Some Non-Central Distributions

In this section, we will obtain the generalized non-central χ^2-distribution, the generalized non-central t-distribution and generalized non-central F-distribution.

2.9.1. Generalized Non-Central χ^2-Distribution

Let $x \sim EC_n(\mu, I_n, \phi)$. The distribution of $x'x$ is called the generalized non–central χ^2-distribution and we write $x'x \sim G\chi_n^2(\delta^2, \phi)$ with $\delta^2 = \mu'\mu$ or $x'x \sim G\chi_n^2(\delta^2, f)$ if x has a density $f(x'x)$. Let Γ be an orthogonal matrix such that $\Gamma\mu = (\|\mu\|, 0, \ldots, 0)' \stackrel{\Delta}{=} \nu$ say, and $y = \Gamma x$, then $y \sim EC_n(\nu, I_n, \phi)$ and $x'x = y'y$. Thus, the distribution of $x'x$ depends on μ only through $\delta = \|\mu\|$. It is easily seen that if $x \sim EC_n(\mu, \Sigma, \phi)$ with $\Sigma > 0$, then $x'\Sigma^{-1}x \sim G\chi_n^2(\delta^2, \phi)$ with $\delta^2 = \mu'\Sigma^{-1}\mu$.

Now we will give the density of $x'x$ for both cases of x having density or not.

Theorem 2.9.1. *Assume* $x \stackrel{d}{=} Ru^{(n)} + \mu \sim EC_n(\mu, I_n, \phi)$ *with* $\mu \neq \mathbf{0}$ *and* $P(x = \mu) = 0$. *Then the density of* $U = x'x$ *is*

$$\frac{1}{2\delta B(\frac{1}{2}, (n-1)/2)} \int_{|u^{\frac{1}{2}} - \delta|}^{u^{\frac{1}{2}} + \delta} r^{-1}\left\{1 - \left(\frac{u - \delta^2 - r^2}{2r\delta}\right)^2\right\}^{(n-3)/2} dF(r) \tag{2.9.1}$$

for $u > 0$, *where* $R \sim F(r)$.

Proof. From assumption we have

$$U = x'x \stackrel{d}{=} \mu'\mu + R^2 + 2R\mu'u^{(n)} = \delta^2 + R^2 + 2R\mu'u^{(n)},$$

where R is independent of $u^{(n)}$. Denote $u^{(n)} = (u_1, \ldots, u_n)'$. Clearly,

$$U \stackrel{d}{=} \delta^2 + R^2 + 2R\delta u_1,$$

where the density of u_1 is $[B(\frac{1}{2}, (n-1)/2)]^{-1}(1-u_1^2)^{(n-3)/2}$(cf. (2.5.13)). For each nonnegative Borel function $h(.)$, we have

$$Eh(U) = Eh(\delta^2 + R^2 + 2\delta R u_1)$$
$$= \frac{1}{B(\frac{1}{2}, (n-1)/2)} \int_0^\infty \int_{-1}^1 h(\delta^2 + 2\delta r u_1 + r^2)(1-u_1^2)^{(n-3)/2} du_1 dF(r).$$

Making transformation $\mu = \delta^2 + 2\delta r u_1 + r^2$, the above integral becomes

$$E(h(U)) = \frac{1}{B(\frac{1}{2}, (n-1)/2)} \int_0^\infty \int_{(\delta-r)^2}^{(\delta+r)^2} h(u) \frac{1}{2\delta r}\left[1 - \left(\frac{u-\delta^2-r^2}{2r\delta}\right)^2\right]^{(n-3)/2} du\, dF(r).$$

Exchanging the order of integration, (2.9.1) follows immediately. ∎

Corollary 1. *The density of* $G\chi_n^2(\delta^2, f)$ *is*

$$(2.9.2)\qquad \frac{\pi^{(n-1)/2}}{\Gamma((n-1)/2)\delta} \int_{|u^{\frac{1}{2}}-\delta|}^{u^{\frac{1}{2}}+\delta} r^{n-2}\left[1 - \left(\frac{u-\delta^2-r^2}{2r\delta}\right)^2\right]^{(n-3)/2} f(r^2)dr.$$

Proof. The density (2.9.2) is a consequence of (2.9.1) and (2.5.17). ∎

Sometimes, the integration of (2.9.2) is not easily integrated; we give another formula as follows.

Theorem 2.9.2. *The density of* $U \sim G\chi_n^2(\delta^2 f)$ *is given by*

$$(2.9.3)\qquad \frac{\pi^{(n-1)/2}}{\Gamma((n-1)/2)} u^{\frac{1}{2}n-1} \int_0^\pi f(u - 2\delta u^{\frac{1}{2}}\cos\phi + \delta^2)\sin^{n-2}\phi\, d\phi.$$

Proof. Let $h(.)$ be any nonnegative Borel function and $\boldsymbol{x} \sim EC_n(\nu, \boldsymbol{I}_n, f)$ with $\nu = (\delta, 0, \ldots, 0)'$ and a density $f((\boldsymbol{x}-\nu)'(\boldsymbol{x}-\nu))$. Then $U \overset{d}{=} \boldsymbol{x}'\boldsymbol{x}$ and

$$E(h(U)) = \int h(\boldsymbol{x}'\boldsymbol{x}) f[(x_1-\delta)^2 + x_2^2 + \cdots + x_n^2]\quad d\boldsymbol{x}.$$

Taking the generalized spherical coordinate transformation (cf. Example 1.6.8), we have

$$E(h(U)) = \frac{2\pi^{(n-1)/2}}{\Gamma((n-1)/2)} \int_0^\infty \int_0^\pi h(r^2) r^{n-1} f(r^2 - 2r\delta\cos\phi_1 + \delta^2)\sin^{n-2}\phi_1 d\phi_1 dr$$

$$(2.9.4)\qquad = \frac{\pi^{(n-1)/2}}{\Gamma((n-1)/2)} \int_0^\infty h(u) u^{\frac{1}{2}n-1}\, du \int_0^\pi f(u - 2u^{\frac{1}{2}}\cos\phi + \delta^2)\sin^{n-2}\phi\quad d\phi$$

which implies (2.9.3). ∎

Corollary 1. *If $U \sim G\chi_n^2(\delta^2, f)$ and $h(.)$ is an arbitrary function such that $E|h(U)| < \infty$, then*

(2.9.5)
$$E(h(U)) = \frac{2\pi^{(n-1)/2}}{\Gamma((n-1)/2)} \int_0^\infty M(\rho)\rho^{n-1} f(\rho^2)d\rho,$$

where

$$M(\rho) = \int_0^\pi h(\rho^2 + 2\rho\delta\cos\phi + \delta^2)\sin^{n-2}\phi \quad d\phi.$$

Let $I_v(.)$ denote the modified Bessel function of the first kind. Its integral representation is

(2.9.6)
$$I_m(z) = \frac{(z/2)^m}{\pi^{\frac{1}{2}}\Gamma(m+\frac{1}{2})} \int_0^\pi \exp(\pm z\cos\theta)\sin^{2m}\theta \ d\theta.$$

Let

(2.9.7)
$$J_m(z) = \frac{(z/2)^m}{\pi^{\frac{1}{2}}\Gamma(m+\frac{1}{2})} \int_0^\pi \exp(\pm iz\cos\theta)\sin^{2m}\theta \ d\theta.$$

Corollary 2. *The c.f. of $U \sim G\chi_n^2(\delta^2, f)$ is given by the integral*

(2.9.8)
$$\phi_U(t) = 2\pi^{m+1}\frac{\exp(it\delta^2)}{(\delta t)^m} \int_0^\infty \exp(it\rho^2)\rho^{m+1} \ f(\rho^2)J_m(2\delta t\rho) \ d\rho,$$

where $m = \frac{1}{2}n - 1$ and $i = (-1)^{\frac{1}{2}}$.

Proof. (2.9.8) follows from (2.9.5), and (2.9.7). □

Putting $h(U) = U^k$ in (2.9.5), we obtain the moments of U immediately.

Corollary 3. *Suppose $E(U^k) < \infty$. Then*

(2.9.9)
$$E(U^k) = 2\pi^{\frac{1}{2}n}\sum_{l=0}^{[k/2]}\sum_{m=0}^{k-2l}\frac{k!\delta^{2(k-l-m)}}{l!m!(k-2l-m)!\Gamma(1+\frac{1}{2}n)}\int_0^\infty \rho^{2l+2m+n-1} f(\rho^2)d\rho.$$

In particular, for the first two moments, we find

(2.9.10)
$$E(U) = \delta^2 + \frac{n}{2\pi c_{n+2}}, \qquad E(U^2) = \delta^4 + \frac{n+2}{\pi c_{n+2}}\delta^2 + \frac{n(n+2)}{4\pi^2 c_{n+4}}$$

(2.9.11)
$$\mathrm{Var}(U) = \frac{2\delta^2}{\pi c_{n+2}} + \frac{n(n+2)}{4\pi^2 c_{n+4}} - \frac{n^2}{4\pi^2 c_{n+2}^2},$$

where

(2.9.12) $$c_l \sim \frac{\Gamma(\frac{1}{2}l)}{2\pi^{\frac{1}{2}l}\int_0^\infty r^{l-1}f(r^2)\,dr} \qquad (\textit{noting } c_n = 1).$$

The readers may check these results by (2.9.5) of Corollary 1.

In view of what follows, we establish a representation of a $U \sim G\chi_n^2(\delta^2, g)$ in terms of two independent random variables: $\boldsymbol{R}$ and θ, which is based on the following fact.

Suppose that $\boldsymbol{x} = (X_1, ..., X_n)' \sim S_n(\phi)$, $n \geqslant 2$, and has a density $f(\boldsymbol{x}'\boldsymbol{x})$. Then there exists a unique set of random variables $R \geqslant 0$, $\theta_k \in [0, \pi]$, $k = 1, ..., n-2$, $\theta_{n-1} \in [0, 2\pi]$ for which

(2.9.13) $$\begin{cases} X_j = R(\prod_{k=1}^{j-1} \sin\theta_k)\cos\theta_j & 1 \leqslant j \leqslant n-1 \\ X_n = R(\prod_{k=1}^{n-2} \sin\theta_k)\sin\theta_{n-1}. \end{cases}$$

Furthermore, R, θ_1, ..., θ_{n-1} are independent and have respective densities

(2.9.14) $$\begin{cases} f_R(r) = \dfrac{2\pi^{\frac{1}{2}n}}{\Gamma(\frac{1}{2}n)}\, r^{n-1}\, f(r^2) & r \geqslant 0 \\ f_{\theta_k}(\theta) = \dfrac{1}{B(\frac{1}{2}, (n-k)/2)} \sin^{n-k-1}\theta & 0 \leqslant \theta \leqslant \pi, \quad k = 1, ..., n-2 \\ f_{\theta_{n-1}}(\theta) = 1/(2\pi) & 0 \leqslant \theta \leqslant 2\pi. \end{cases}$$

Conversely if R, θ_1, ..., θ_{n-1} are independent and have densities given in (2.9.14) and $\boldsymbol{x}$ is defined in (2.9.13), then $\boldsymbol{x}$ has the spherical density $f(\boldsymbol{x}'\boldsymbol{x})$.

By recalling the proof of Theorem 2.9.1, the above fact yields the next representation theorem.

Theorem 2.9.3. *Assume* $U \sim G\chi_n^2(\delta^2, f)$. *Then*

(2.9.15) $$U \stackrel{\mathrm{d}}{=} R^2 + 2R\delta\cos\theta + \delta^2 \stackrel{\mathrm{d}}{=} R^2 - 2R\delta\cos\theta + \delta^2,$$

where R, θ *are independent variates with densities* $f_R(r)$ (*cf.* (2.9.14)) *and*

(2.9.16) $$f_\theta(\theta) = \frac{1}{B(\frac{1}{2}, (n-1)/2)} \sin^{n-2}\theta \qquad 0 \leqslant \theta \leqslant \pi,$$

respectively.

Example 2.9.1. Let $N = 1$ and $s = 1$ in Example 2.6.1. The density of $\boldsymbol{x}$ is

(2.9.17) $$\pi^{-\frac{1}{2}n}\, r^{\frac{1}{2}n}\, \exp(-r\boldsymbol{x}'\boldsymbol{x}),$$

and the corresponding density of $G\chi_n^2(\delta^2, f)$ is

$$r(u/\delta^2)^{\frac{1}{2}m}\exp(-r(u+\delta^2))I_m(2r\delta u^{\frac{1}{2}}) \qquad u > 0$$

with $m = \frac{1}{2}n - 1$. When $r = \frac{1}{2}$ we obtain the usual density of $\chi_n^2(\delta^2)$.

The reader can find more examples of $G\chi_n^2(\delta^2, f)$ in Cacoullos and Koutras (1984). The main results in this section are due to Cacoullos and Koutras, too, but Theorem 2.9.1 is from Fan (1984).

2.9.2. Ceneralized Non-Central t-Distribution

Let $\boldsymbol{x} = \begin{bmatrix} \boldsymbol{x}_1 \\ \boldsymbol{x}^{(2)} \end{bmatrix} \sim EC_{n+1}(\mu, \boldsymbol{I}_{n+1}, \phi)$, where $\boldsymbol{x}^{(2)}$: $n \times 1$ and $\mu = (\delta, 0, \ldots, 0)'$. The distribution of

(2.9.18)
$$t = \frac{n^{\frac{1}{2}} X_1}{(\boldsymbol{x}^{(2)\prime}\, \boldsymbol{x}^{(2)})^{\frac{1}{2}}}$$

is called the generalized non-central t-distribution and we write $t \sim Gt_n(\delta, \phi)$ or $t \sim Gt_n(\delta, f)$ if $\boldsymbol{x}$ has a density $f((\boldsymbol{x}-\mu)'(\boldsymbol{x}-\mu))$.

Exercise 2.18 and Exercise 2.19 show us some applications of the generalized non–central t-distribution.

Theorem 2.9.4. *Suppose $t \sim Gt_n(\delta, f)$. Then the density of t is*

(2.9.19)

$$\frac{2(n\pi)^{\frac{1}{2}n}}{\Gamma(\frac{1}{2}n)} (n+t^2)^{-\frac{1}{2}(n+1)} \int_0^\infty f(y^2 - 2\delta_1 y + \delta^2) y^n dy, \qquad -\infty < t < +\infty,$$

where $\delta_1 = t\delta/(n+t^2)^{\frac{1}{2}}$.

Proof. Assume that $\boldsymbol{x} \sim \boldsymbol{EC}_{n+1}(\mu, \boldsymbol{I}_{n+1}, f)$ with $\mu = (\delta, 0, \ldots, 0)'$ and $h(.)$ is a Borel function such that $E|h(t)| < \infty$. By using (2.4.15) for $X_2, \ldots, X_{n+1}$, we have

(2.9.20)
$$\begin{aligned} E(h(t)) &= \frac{2\pi^{\frac{1}{2}n}}{\Gamma(\frac{1}{2}n)} \int_{-\infty}^\infty \int_0^\infty h(n^{\frac{1}{2}} x_1/r) f((x_1-\delta)^2 + r^2) r^{n-1} dr dx_1 \\ &= \frac{2\pi^{\frac{1}{2}n}}{\Gamma(\frac{1}{2}n) n^{\frac{1}{2}}} \int_{-\infty}^\infty \int_0^\infty h(t) f((tr/n^{\frac{1}{2}} - \delta)^2 + r^2) r^n dr dt. \end{aligned}$$

Thus the density of t is

$$\frac{2\pi^{\frac{1}{2}n}}{n^{\frac{1}{2}}\Gamma(\frac{1}{2}n)} \int_0^\infty f((t^2+n) r^2 n^{-1} - 2t\delta r n^{-\frac{1}{2}} + \delta^2) r^n \, dr,$$

which can be reduced to (2.9.19) immediately by making $y = ((t^2+n)/n)^{\frac{1}{2}} r$. □

When $\delta = 0$, (2.9.19) reduces to the density t, with which we are familiar.

Corollary 1. *Suppose $t \sim Gt_n(\delta, f)$, $E|h(t)| < \infty$. Then*

(2.9.21)
$$E(h(t)) = \frac{2\pi^{\frac{1}{2}n}}{\Gamma(\frac{1}{2}n)} \int_0^\infty M(\rho) \rho^n f(\rho^2) d\rho,$$

where

$$M(\rho) = \int_0^\pi h(n^{\frac{1}{2}}(\delta + \rho\cos\theta)/(\rho\sin\theta))\sin^{n-1}\theta \, d\theta. \tag{2.9.22}$$

Proof. The assertion follows from (2.9.20) by making the transformation $x_1 = \delta + \rho\cos\theta$, $r = \rho\sin\theta$. □

Corollary 2. *Assume* $E|t|^k < \infty$. *Then*

$$E(t^k) = \frac{n^{\frac{1}{2}k}\Gamma(\frac{1}{2}(n-k))k!}{\Gamma(\frac{1}{2}n)} \sum_{j=0}^{[k/2]} \frac{\delta^{k-2j}\pi^{\frac{1}{2}(k-2j)}}{2^{2j}j!(k-2j)!c_{n-k+2j+1}} \tag{2.9.23}$$

where $[x]$ *denotes the integral part of* x *and* $c.$ *is defined by* (2.9.12). *In particular, (noting* $c_{n+1} = 1$)

$$\begin{cases} E(t) = \dfrac{(n\pi)^{\frac{1}{2}}\delta\Gamma(\frac{1}{2}(n-1))}{\Gamma(\frac{1}{2}n)c_n} & n > 1 \\ E(t^2) = \dfrac{n}{n-2}\left[\dfrac{2\delta^2\pi}{c_{n-1}} + 1\right] & n > 2 \\ \mathrm{Var}(t) = \dfrac{n}{n-2}\left[\dfrac{2\delta^2\pi}{c_{n-1}} + 1\right] - n\pi\delta^2(\Gamma(\frac{1}{2}(n-1))/(\Gamma(\frac{1}{2}n)c_n))^2. \end{cases} \tag{2.9.24}$$

Corollary 2 follows from (2.9.21), (2.9.22) and Legendre's duplication formula

$$\Gamma(2a) = \frac{2^{2a-1}}{\pi^{\frac{1}{2}}}\Gamma(a)\Gamma(a + \tfrac{1}{2}).$$

2.9.3. Generalized Non-Central F-Distribution

Let $\boldsymbol{x} = \begin{bmatrix} \boldsymbol{x}^{(1)} \\ \boldsymbol{x}^{(2)} \end{bmatrix} \sim EC_{m+n}(\mu, \boldsymbol{I}_{m+n}, \phi)$, where $\boldsymbol{x}^{(1)}$: $m \times 1$ and $\mu = (v' \ \boldsymbol{0}')$ with v: $m \times 1$. We call the distribution of

$$F = \frac{n}{m}\frac{\boldsymbol{x}^{(1)\prime}\boldsymbol{x}^{(1)}}{\boldsymbol{x}^{(2)\prime}\boldsymbol{x}^{(2)}} \tag{2.9.25}$$

the generalized non-central F-distribution and write $F \sim GF_{m,n}(\delta^2, \phi)$ or $F \sim GF_{m,n}(\delta^2, f)$ if $\boldsymbol{x}$ has a density $f((\boldsymbol{x}-\mu)'(\boldsymbol{x}-\mu))$, where $\delta^2 = v'v$. Clearly, if $t \sim Gt_n(\delta, \phi)$, then $t^2 \sim GF_{1,n}(\delta^2, f)$. It is easy to show that the distribution of $GF_{m,n}(\delta^2, \phi)$ depends on v only through $\delta = \|v\|$ (cf. Section 2.9.1).

Theorem 2.9.5. *Assume* $F \sim GF_{m,n}(\delta^2, f)$. *Then the density of* F *is*

$$\frac{2\pi^{\frac{1}{2}(m+n-1)}}{\Gamma(\frac{1}{2}(m-1))\Gamma(\frac{1}{2}n)}\frac{m}{n}\left(\frac{m}{n}F\right)^{\frac{1}{2}(m-2)}\left(1 + \frac{m}{n}F\right)^{-\frac{1}{2}(m+n)} \tag{2.9.26}$$

$$\cdot \int_0^\pi \int_0^\infty \sin^{m-2}\theta \, y^{m+n-1} f(y^2 - 2\delta_1 y\cos\theta + \delta^2) \, d\theta \, dy, \qquad F > 0,$$

where $\delta_1 = (mF/(n + mF))^{\frac{1}{2}}\delta$.

Proof. By a technique similar to Theorem 2.9.2, for any Borel function $h(.)$ such that $E|h(F)| < \infty$, we have

$$E(h(F)) = \int h\left[(n/m)\frac{y_1^2 + \cdots + y_m^2}{y_{m+1}^2 + \cdots + y_{m+n}^2}\right] f((y_1 - \delta)^2$$

$$+ y_2^2 + \cdots + y_{m+n}^2) dy_1 \cdots dy_{m+n}$$

$$= \frac{4\pi^{\frac{1}{2}(m+n-1)}}{\Gamma(\frac{1}{2}(m-1))\,\Gamma(\frac{1}{2}n)} \int_0^\infty \int_0^\infty \int_0^\pi h(nr_1^2/(mr_2^2))\, f(r_1^2 - 2r_1\delta\cos\theta + \delta^2 + r_2^2)$$

$$\sin^{m-2}\theta\, r_1^{m-1} r_2^{n-1}\, dr_1\, dr_2\, d\theta$$

$$= \frac{2\pi^{\frac{1}{2}(m+n-1)}}{\Gamma(\frac{1}{2}(m-1))\,\Gamma(\frac{1}{2}n)} (m/n) \int_0^\infty \int_0^\infty \int_0^\pi h(F) f(r_1^2 - 2r_1\delta\cos\theta + \delta^2$$

$$+ (nr_1^2/(mF)))(mF/n)^{-\frac{1}{2}(n+1)} r_1^{m+n-1} \sin^{m-2}\theta\, d\theta\, dr_1\, dF.$$

Thus the density of F is

$$\frac{2\pi^{\frac{1}{2}(m+n-1)}}{\Gamma(\frac{1}{2}(m-1))\,\Gamma(\frac{1}{2}n)} (m/n) \int_0^\infty \int_0^\pi (mF/n)^{-\frac{1}{2}(n+1)} \sin^{m-2}\theta\, r^{m+n-1}$$

$$f(r_1^2 - 2r_1\delta_1\cos\theta + \delta^2 + (nr_1^2/(mF)))d\theta dr_1$$

which implies (2.9.26) if we make the transformation $y = (mF/(n + mF))^{\frac{1}{2}} r_1$. □

Corollary 1. *Assume $F \sim GF_{m,n}(\delta^2, f)$. Then for each Borel function $h(.)$ such that $E|h(F)| < \infty$ we have*

(2.9.27)

$$E(h(F)) = \frac{2\pi^{\frac{1}{2}(m+n)-1}}{\Gamma(\frac{1}{2}(m-1))\,\Gamma(\frac{1}{2}n)} \int_0^\infty \int_0^\infty M(R_1, R_2) R_1^{m-1} R_2^{n-1} f(R_1^2 + R_2^2) dR_1 dR_2$$

where

$$M(R_1, R_2) = \int_0^\pi h\left[(n/m)\frac{R_1^2 - 2R_1\delta\cos\theta + R_2^2}{R_2^2}\right] \sin^{m-2}\theta d\theta.$$

The proof is left to the reader. Let $h(u) = \exp(itu)$, we may obtain the c.f. of F.

Corollary 2. *The c.f. of $F \sim GF_{m,n}(\delta^2, f)$ is*

(2.9.28) $$\frac{2\pi^{\frac{1}{2}(m+n)}}{\Gamma(\frac{1}{2}n)} (n\delta t/m)^{-\frac{1}{2}(m-2)} \int_0^\infty \int_0^\infty \exp(itn(R_1^2 + \delta^2)/$$

$$(mR_2^2))J_{\frac{1}{2}(m-2)}\left(\frac{2nR_1\delta t}{mR_2^2}\right)$$

$$\cdot(R_1/R_2^2)^{-\frac{1}{2}(m-2)}R_1^{m-1}R_2^{n-1}f(R_1^2+R_2^2)\,dR_1\,dR_2.$$

Corollary 3. *Assume $F \sim GF_{m,n}(\delta^2, f)$ and $E(F^2) < \infty$. Then*

(2.9.29) $$E(F) = \frac{1}{n-2}(n/m)\left(\frac{2\pi\delta^2}{c_{m+n-2}} + m\right), \qquad \text{for } n > 2,$$

(2.9.30)

$$E(F^2) = \frac{1}{(n-2)(n-4)}(n/m)^2\left[\frac{\pi^2\delta^4}{c_{m+n-4}} + \frac{(m+2)\pi\delta^2}{c_{m+n-2}} + \frac{(m+2)m}{4}\right],$$

$$\text{for } n > 4,$$

where c_1 is defined by (2.9.12) and $c_{m+n} = 1$.

Proof. Let $h(u) = u^k$, $k = 1, 2$. The formulae (2.9.29) and (2.9.30) follow. □

Letting $h(u) = u^k$, we can obtain the k–th moment of F for $k = 1, 2, \ldots$ if $E(F^k) < \infty$. Analogously to Theorem 2.9.3, we have the following Stochastic representation.

Corollary 4. *Assume $F \sim GF_{m,n}(\delta^2, f)$. Then*

(2.9.31) $$F \overset{d}{=} (n/m)\frac{R_1^2 + 2R_1\delta\cos\theta + \delta^2}{R_2^2} \overset{d}{=} (n/m)\frac{R_1^2 - 2R_1\delta\cos\theta + \delta^2}{R_2^2},$$

where θ is independent of (R_1, R_2) and their densities are

$$g(\theta) = (1/B(\tfrac{1}{2}, \tfrac{1}{2}(m-1)))\sin^{m-2}\theta \qquad 0 \leqslant \theta \leqslant \pi$$

$$h(v_1, v_2) = \frac{2\pi^{\frac{1}{2}(m+n)}}{\Gamma(\frac{1}{2}m)\Gamma(\frac{1}{2}n)} r_1^{m-1} r_2^{n-1} f(r_1^2 + r_2^2) \qquad r_1 > 0,\ r_2 > 0,$$

respectively.

The results in Sections 2.9.2 and 2.9.3 are due to Fan (1984).

References

Anderson (1984), Anderson and Fang (1982a), (1984), Anderson and Styan (1982), Bei, Su, Fang and Chen (1980), Cacoullos and Kourtras (1984), Cambanis, Huang and Simons (1981), Esseen (1945), Fan (1984), Fang and Wu (1984), Feller (1971), Hsu (1954, 1983), Johnson and Kotz (1982), Kelker (1970), Kingman (1972), Kotz (1975), Laurent (1974), Li (1984), Loéve (1960), Lukacs (1956), Marcinkiewicz (1938), Schoenberg (1938), Zhang and Fang (1982), Zygmund (1951).

Exercises 2

2.1. Let $F(x)$ be a nonsingular cdf and integers $2 \leqslant r < n$. Prove that there exist random variables $X_1, \ldots, X_n$ such that

a) $X_i \sim F(x)$, $i = 1, 2, \ldots, n$;

b) $X_{i_1}, \ldots, X_{i_r}$ are independent for every r-subset $\{i_1, \ldots, i_r\}$ of $\{1, 2, \ldots, n\}$;

c) $X_{i_1}, \ldots, X_{i_{r+1}}$ are dependent for any $(r+1)$-subset $\{i_1, \ldots, i_{r+1}\}$ of $\{1, 2, \ldots, n\}$.

(cf. Bei, Su, Fang and Chen (1980)).

2.2. Let $X \sim B(a, b)$. Show $\log X \in (U)$.

2.3. Make a counterexample that Z is independent of X and of Y respectively, $Z + X \overset{d}{=} Z + Y$, but $X \overset{d}{\neq} Y$.

2.4. Prove the formula (2.2.19).

2.5. Suppose that $\boldsymbol{x}$ has the $2k$-th moment and $\boldsymbol{A}_1, \ldots, \boldsymbol{A}_m$ are symmetric matrices, Prove the following formula:

$$E[(\boldsymbol{x}'\boldsymbol{A}_1 x)^{k_1}(\boldsymbol{x}'\boldsymbol{A}_2\boldsymbol{x})^{k_2} \ldots (\boldsymbol{x}'\boldsymbol{A}_m\boldsymbol{x})^{k_m}] = \operatorname{tr}[(\underbrace{\boldsymbol{A}_1 \otimes \ldots \otimes \boldsymbol{A}_1}_{k_1}) \otimes \cdots \otimes (\underbrace{\boldsymbol{A}_m \otimes \cdots \otimes \boldsymbol{A}_m}_{k_m}) \Gamma_{2k}],$$

where $k = k_1 + k_2 + \ldots + k_m$.

2.6. Let $(Z_1, \ldots, Z_m) \sim D_m(\alpha_1, \ldots, \alpha_m)$. show that

a) $(Z_1, \ldots, Z_k) \sim D_{k+1}(\alpha_1, \ldots, \alpha_k; \alpha_{k+1} + \cdots + \alpha_m)$, $1 \leqslant k < m$;

b) $Z_1 + \cdots + Z_k \sim B(\alpha_1 + \cdots + \alpha_k, \alpha_{k+1} + \cdots + \alpha_m)$, $1 \leqslant k < m$.

2.7. Let $\boldsymbol{x} = \begin{bmatrix} \boldsymbol{x}^{(1)} \\ \boldsymbol{x}^{(2)} \end{bmatrix} \sim N_n\left(\begin{bmatrix} \mu^{(1)} \\ \mu^{(2)} \end{bmatrix}, \begin{bmatrix} \Sigma_{11} & \Sigma_{12} \\ \Sigma_{21} & \Sigma_{22} \end{bmatrix}\right)$ with $\boldsymbol{x}^{(1)}$: $m \times 1$. The random vector $\boldsymbol{x}_{1.2} = \boldsymbol{x}^{(1)} - \mu^{(1)} - \Sigma_{12}\Sigma_{22}^{-1}(\boldsymbol{x}^{(2)} - \mu^{(2)})$ is called the set of residual variates. Show that

(a) $E(\boldsymbol{x}^{(1)} - \mu^{(1)})\boldsymbol{x}'_{1.2} = \Sigma_{11} - \Sigma_{12}\Sigma_{22}^{-1}\Sigma_{21}$;

(b) $E(\boldsymbol{x}^{(2)} - \mu^{(2)})\boldsymbol{x}'_{1.2} = \boldsymbol{0}$.

2.8. (The multivariate log-normal distribution). Let $\boldsymbol{x} \sim N_n(\mu, \Sigma), \Sigma > 0$, and $\log \boldsymbol{y} \hat{=} (\log Y_1, \ldots, \log Y_n)' \overset{d}{=} \boldsymbol{x}$. Then $\boldsymbol{y}$ is said to have a p- variate log-normal distribution with density

$$(2\pi)^{-\frac{1}{2}n}|\Sigma|^{-\frac{1}{2}} \prod_{i=1}^{n} y_i^{-1} \exp\left\{-\frac{1}{2}(\log \boldsymbol{y} - \mu)'\Sigma^{-1}(\log \boldsymbol{y} - \mu)\right\}$$

when $y_i > 0$, $i = 1, \ldots, p$, and is zero otherwise.

(a) Show that, for any positive integer r,

$E(y_i^r) = \exp(r\mu_i + \frac{1}{2}r^2\sigma_{ii})$

$\operatorname{Var}(y_i) = \exp(2\mu_i + 2\sigma_{ii}) - \exp(2\mu_i + \sigma_{ii})$

$\operatorname{Cov}(y_i, y_j) = \exp\left\{\mu_i + \mu_j + \frac{1}{2}(\sigma_{ii} + \sigma_{jj}) + \sigma_{ij}\right\} - \exp\left\{\mu_i + \mu_j + \frac{1}{2}(\sigma_{ii} + \sigma_{jj})\right\}$.

(b) Find the marginal density of $(Y_1, \ldots, Y_m)$, $m < n$.

2.9. With the help of (2.4.16) prove that the surface area of a sphere of radius r in R^n is

$$S_n = \frac{2\pi^{\frac{1}{2}n} r^{n-1}}{\Gamma(\frac{1}{2}n)}.$$

2.10 Let $\Omega_n(\|\boldsymbol{t}\|^2)$ denote the c.f. of $\boldsymbol{u}^{(n)}$. Prove that

$$\Omega_n(u) = \Gamma(\tfrac{1}{2}n)(2/u)^{\frac{1}{2}(n-2)} J_{\frac{1}{2}(n-2)}(u)$$

$$= 1 - \frac{u^2}{2n} + \frac{u^4}{2 \cdot 4n(n+2)} - \frac{u^6}{2 \cdot 4 \cdot 6 \cdot n(n+2)(n+4)} + \cdots,$$

where $J_\nu(u)$ is the Bessel function of the first kind and

$$J_\gamma(u) = \sum_{k=0}^{\infty} \frac{(-1)^k}{k!\Gamma(\nu+k+1)} (u/2)^{\nu+2k}.$$

2.11. Prove that $\Omega_n(\|\boldsymbol{t}\|^2)$ with $\boldsymbol{t} \in R^{n+1}$ is not a c.f. in R^{n+1}. Therefore Φ_{n+1} is a proper subset of Φ_n for $n \geqslant 1$.

2.12. Show that $\phi(t) = \exp(-\frac{1}{2}t)$ belongs to Φ_∞.

2.13. Without using Theorem 2.5.4 if $\boldsymbol{x} \overset{d}{=} R\boldsymbol{u}^{(n)} = R^*\boldsymbol{u}^{(n)}$ where R and R^* are independent of $\boldsymbol{u}^{(n)}$, show that $R \overset{d}{=} R^*$.

(Hint: From assumption we have $X_1 \overset{d}{=} Ru_1 \overset{d}{=} R^*u_1$.)

2.14. Assume that $\boldsymbol{x} \sim S_n(\phi)$ has a density $f(x_1^2 + \cdots + x_n^2)$ and its marginal densities are $f_j(x_1^2 + \cdots + x_j^2)$, $1 \leqslant j < n$. If $f(u)$ is bounded in a neighborhood of a point z, then z is a point of continuity of $f_j(u)$, $j = 1, \ldots, (n-1)$.

2.15. Let $\boldsymbol{x} = (\boldsymbol{x}^{(1)\prime}, \boldsymbol{x}^{(2)\prime}, \boldsymbol{x}^{(3)\prime})' \sim S_n^+(\phi)$ with $\boldsymbol{x}^{(1)} \in R^p$, $\boldsymbol{x}^{(2)} \in R^q$ and $\boldsymbol{x}^{(3)} \in R^r$, $p+q+r = n$. Denote the distribution of $\boldsymbol{x}^{(1)\prime}\boldsymbol{x}^{(1)} - \boldsymbol{x}^{(2)\prime}\boldsymbol{x}^{(2)}$ by $H(\frac{1}{2}p, \frac{1}{2}q, \frac{1}{2}r; \phi)$. $\boldsymbol{A}$ is an $n \times n$ symmetric matrix. Prove that $\boldsymbol{x}'\boldsymbol{A}\boldsymbol{x} \sim H(\frac{1}{2}p, \frac{1}{2}q, \frac{1}{2}r; \phi)$ if and only if $\boldsymbol{A}^3 = \boldsymbol{A}$ with p 1's, q (-1)'s and r 0's as its eigenvalues.

2.16. Let $U \sim G\chi_n^2(\delta^2, f)$ with $f(t) = c(1 + t/s)^{-N}$, $N > n/2$, where $c = (\pi s)^{-n/2}\Gamma(N)/\Gamma(N - \frac{1}{2}n)$. Prove that the density of U is

$$\frac{\Gamma(N)\, s^{N-\nu-1}}{\Gamma(\nu+1)\,\Gamma(n-\nu-1)} u^\nu (s+u+\delta^2)^{-N} H(\tfrac{1}{2}N, \tfrac{1}{2}(N+1); \nu+1; z).$$

where $\nu = \frac{1}{2}n - 1$, $z = (2\delta u^{\frac{1}{2}}/(s+u+\delta^2))^2$, and H denotes the hypergeometric function which is defined as

$$H(\alpha, \beta; r; z) = \sum_{k=0}^{\infty} \frac{[\alpha]_k [\beta]_k}{[r]_k} \frac{z^k}{k!}, \ |z| < 1,\ r \neq 0, -1, \ldots$$

with $[\lambda]_k = \lambda(\lambda+1)\cdots(\lambda+k-1)$.

2.17. Let $F \sim GF_{m,n}(\delta^2, f)$ where $f(.)$ is the same with Exercise 2.16. Prove that the density of F is

$$\frac{m\,\Gamma(\frac{1}{2}(m+n))}{n\Gamma(\frac{1}{2}m)\,\Gamma(\frac{1}{2}n)} \left(\frac{s}{s+\delta^2}\right)^{N-\frac{1}{2}n} (mu/n)^{\frac{1}{2}m-1}(1+mu/n)^{-\frac{1}{2}(m+n)}$$

$$\cdot F\left(\tfrac{1}{2}(m+n), N - \tfrac{1}{2}(m+n); \tfrac{1}{2}m; \frac{mu}{n+mu}\frac{\delta^2}{s+\delta^2}\right).$$

2.18. If $\boldsymbol{x} \sim EC_n(\mu, \boldsymbol{I}_n, f)$ with $\mu = (\delta, \ldots, \delta)' = \delta\boldsymbol{1}_n$, then

$$\frac{n^{\frac{1}{2}}(\bar{x} - \delta_1)}{\sqrt{\frac{1}{n-1}\sum_1^n (X_i - \bar{X})^2}} \sim Gt_{n-1}(n^{\frac{1}{2}}(\delta - \delta_1), f),$$

where $\bar{X} = (1/n)(X_1 + \cdots + X_n)$.

2.19. Suppose that $(X_1, \ldots, X_m, Y_1, \ldots, Y_n) \sim EC_{m+n}(\mu, \boldsymbol{I}_{m+n}, f)$, where $\mu = \begin{bmatrix} \delta_1 \boldsymbol{1}_m \\ \delta_2 \boldsymbol{1}_n \end{bmatrix}$. Show

$$\sqrt{\frac{mn(m+n-2)}{mn}} \frac{\bar{X} - \bar{Y} - \delta_3}{\sqrt{\sum_1^m (X_i - \bar{X})^2 + \sum_1^n (Y_i - \bar{Y})^2}} \sim Gt_{m+n-2}(\delta, g)$$

where $\delta = \sqrt{\dfrac{mn}{m+n}}(\delta_1 - \delta_2 - \delta_3)$ and $g(x) = 2\displaystyle\int_0^\infty f(x+u^2)\,du$.

2.20. Assume that $\boldsymbol{x} \sim S_n(\phi)$ with $P(\boldsymbol{x} = \boldsymbol{0}) < 1$ and $\boldsymbol{x} = \begin{pmatrix} \boldsymbol{x}^{(1)} \\ \boldsymbol{x}^{(2)} \end{pmatrix}$ where $\boldsymbol{x}^{(1)}$: $m \times 1$ and $1 \leqslant m < n$. Prove that $\boldsymbol{x}^{(1)}$ and $\boldsymbol{x}^{(2)}$ are independent if and only if $\boldsymbol{x}^{(1)\prime}\boldsymbol{x}^{(1)}$ and $\boldsymbol{x}^{(2)\prime}\boldsymbol{x}^{(2)}$ are independent.

2.21. By using the proposition in the preceding exercise, prove Theorem 2.8.1. under the same assumption with the preceding exercise.

CHAPTER III
SPHERICAL MATRIX DISTRIBUTIONS

In the preceding chapter, we have defined and studied the classes of ECD. It is required to develop the thcory of sample based on ECD. In order to do this, we need to extend the concepts of spherical symmetry from the case of vector to the case of matrix. In this chapter several classes of spherical matrix distribution and of elliptically contoured matrix distributions are defined and studied. Then the relationships among them are discussed in Section 3.2. Some related distributions to spherical matrix distributions (such as quadratic forms, the matrix variate Beta, the matrix variate Dirichlet, the matrix variate t, the matrix variate F, the eigenvalues of matrix variate) are derived in Sections 3.4 and 3.5. Some further problems will be treated in the last section.

3.1. Introduction

In the preceding chapter, we have studied the spherical distributions. How to extend the concept of spherical symmetry from the case of vector to the case of matrix is an interesting and important problem in multivariate analysis. Let $\boldsymbol{X}$ be an $n \times p$ matrix. We express it in terms of elements, columns, and rows as

(3.1.1.) $$\boldsymbol{X} = (x_{ij}) = (\boldsymbol{x}_1, \dots, \boldsymbol{x}_p) = (\boldsymbol{x}_{(1)}, \dots, \boldsymbol{x}_{(n)})'.$$

Here $\boldsymbol{x}_{(1)}, \dots, \boldsymbol{x}_{(n)}$ can be thought of as a sample of size n from a p-dimensional population, but it is not necessary that $\boldsymbol{x}_{(1)}, \dots, \boldsymbol{x}_{(n)}$ are independent. Hence the study of spherical matrix distribution is the base of sampling theory in multivariate analysis. Recalling Theorem 2.5.3, we can define spherical matrix distribution by many ways. In the case of a vector, all of those ways introduce the same class of distributions. But it is a little complicated in the case of matrix that the different ways lead to different classes of distributions. So there are many definitions for spherical matrix distributions. Now we start to define several classes of them.

3.1.1. Left-Spherical Distributions

Definition 3.1.1. Let $\boldsymbol{X}$ be an $n \times p$ random matrix. If $\boldsymbol{X} \overset{d}{=} \Gamma \boldsymbol{X}$ for every $\Gamma \in O(n)$, we call $\boldsymbol{X}$ *left-spherical.*

The left-spherical distributions were defined by Dawid (1977) first. The following facts are from Dawid (1977), (1978), and Fang and Chen (1984).

Lemma 3.1.1. *Let $\boldsymbol{X}$ be left-spherical. Then the c.f. of $\boldsymbol{X}$ has the form $\phi(\boldsymbol{T}'\boldsymbol{T})$, where $\boldsymbol{T}: n \times p$.*

Proof. Denote the c.f. of $\boldsymbol{X}$ by $\psi(\boldsymbol{T})$. For each $\Gamma \in O(n)$ we have

$$\begin{aligned}\psi(\boldsymbol{T}) &= E[\exp(i\boldsymbol{T}'\boldsymbol{X})] = E[\exp(i(\Gamma\, \boldsymbol{T})'\Gamma \boldsymbol{X})] = E[\exp(i(\Gamma \boldsymbol{T})'\boldsymbol{X})] \\ &= \psi(\Gamma \boldsymbol{T}).\end{aligned}$$

So $\psi(\boldsymbol{T})$ is an invariant under $O(n)$ and is a function of a maximal invariant under $O(n)$. Hence $\psi(\boldsymbol{T}) = \phi(\boldsymbol{T}'\boldsymbol{T})$ for some ϕ(cf. Example 1.7.1). □

By the above lemma, if $\boldsymbol{X}$ is left-spherical we can write $\boldsymbol{X} \sim LS_{n\times p}(\phi)$ or $\boldsymbol{X} \sim LS(\phi)$ for simplicity.

Lemma 3.1.2. *Assume* $\boldsymbol{X} \sim LS_{n\times p}(\phi)$.

(i) *If* $\boldsymbol{Q}$ *is a* $p \times q$ *constant matrix, then* $\boldsymbol{XQ} \sim LS_{n\times p}(\psi)$ with

$$\psi(\boldsymbol{T}'\boldsymbol{T}) = \phi(\boldsymbol{QT}'\boldsymbol{TQ}'),\ \boldsymbol{T}: n \times q. \tag{3.1.2}$$

(ii) *Partition* $\boldsymbol{X}$ *into* $\boldsymbol{X} = \begin{bmatrix} \boldsymbol{X}_1 \\ \boldsymbol{X}_2 \end{bmatrix}$ *where* $\boldsymbol{X}_1: m \times p$. *Then* $\boldsymbol{X}_1 \sim LS_{m\times p}(\phi)$.

The proof is straightforward and is left to the reader.

It follows that, if $\boldsymbol{X}$ $(n \times p)$ is left-spherical conditional on a random variable ζ and $\boldsymbol{Q}(p \times q)$ is a function of ζ, then $\boldsymbol{XQ}$ is left-spherical. This is useful in particular if $\boldsymbol{X}$ is left-spherical and independent of $\boldsymbol{Q}$.

There is an important distribution in the class of left-spherical distributions. It is uniform distribution.

Definition 3.1.2. If $\boldsymbol{U}$ is left-spherical and $\boldsymbol{U}'\boldsymbol{U} = \boldsymbol{I}_p$, we say that $\boldsymbol{U}$ is *distributed according to uniform distribution* and write $\boldsymbol{U} \sim \mathcal{U}_{n,p}$.

Let $O(n,p)$ $(n \geqslant p)$ denote the set, the Stiefel manifold, of $n \times p$ matrices $\boldsymbol{U}$ having orthonormal columns, so that $\boldsymbol{U}'\boldsymbol{U} = \boldsymbol{I}_p$. If $\boldsymbol{U} \sim \mathcal{U}_{n,p}$, then $\boldsymbol{U}$ is uniformly distributed over $O(n,p)$.

Theorem 3.1.1. *The distribution of the left-spherical matrix* $\boldsymbol{X}$ *is fully determined by that of* $\boldsymbol{X}'\boldsymbol{X}$.

Proof. It amounts to proving that if $\boldsymbol{Y}(n \times p)$ is left-spherical and $\boldsymbol{Y}'\boldsymbol{Y} \overset{d}{=} \boldsymbol{X}'\boldsymbol{X}$, then $\boldsymbol{X} \overset{d}{=} \boldsymbol{Y}$. Let F and $\Omega(\boldsymbol{T}'\boldsymbol{T})$ be the mcdf and the c.f. of $\mathcal{U}_{n,n}$, respectively. From the assumption $\boldsymbol{X} \sim LS(\phi)$ and $\boldsymbol{Y} \sim LS(\psi)$ we want to point out $\phi = \psi$. As $O(n,n) = O(n)$ and $\boldsymbol{X}'\boldsymbol{X} \overset{d}{=} \boldsymbol{Y}'\boldsymbol{Y}$, we have

$$\begin{aligned}
\phi(\boldsymbol{T}'\boldsymbol{T}) &= \phi(\boldsymbol{T}'\boldsymbol{T}) \int_{O(n)} dF = \int_{O(n)} \phi((\boldsymbol{UT})'(\boldsymbol{UT}))\, dF(\boldsymbol{U}) \\
&= E\left[\int_{O(n)} \exp(i\boldsymbol{T}'\boldsymbol{U}'\boldsymbol{X})\, dF(\boldsymbol{U})\right] = E\left[\int_{O(n)} \exp(i\boldsymbol{TX}'\boldsymbol{U})\, dF(\boldsymbol{U})\right] \\
&= E[\Omega(\ \boldsymbol{TX}'\boldsymbol{XT}')] = E(\Omega(\boldsymbol{TY}'\boldsymbol{YT}')) \\
&= \psi(\boldsymbol{T}'\boldsymbol{T})
\end{aligned}$$

which completes the proof. □

Corollary 1. *The uniform distribution is a unique left-spherical distribution over* $O(n,p)$.

Proof. The assertion follows from Theorem 3.1.1. and $\boldsymbol{U}'\boldsymbol{U} = \boldsymbol{I}_p$ where $\boldsymbol{U} \sim \mathcal{U}_{n,p}$. □

Corollary 2. *If $X \sim LS_{n\times p}(\phi)$ and $K(m \times n)$ is fixed, then the distribution of KX depends on K only through KK'.*

Proof. If H is an $m \times n$ constant matrix such that $HH' = KK'$, then $H = K\Gamma$ for some $\Gamma \in O(n)$ (cf. Section 1.2). So

$$HX = K\Gamma X \overset{d}{=} KX. \qquad \square$$

In general, it is not necessary that KX is left-spherical. If $KK' = I_m$, then KX is left-spherical. (Why? the proof is left to the reader.)

Definition 3.1.3. If X' is left-spherical, we call X *right-spherical* and write $X \sim RS_{n\times p}(\phi)$. We say X is *spherical* if it is both left- and right-spherical, and write $X \sim SS_{n\times p}(\phi)$.

Lemma 3.1.3. *Assume $U = (U_1, U_2) \sim \mathcal{U}_{n,p}$ where $U_1: n \times q$, $0 < q < p$.*
(i) *If $V \in O(p, q)(p \geqslant q)$ is fixed, then $UV \sim \mathcal{U}_{n,q}$.*
(ii) $U_1 \sim \mathcal{U}_{n,q}$.
(iii) *$U \sim SS_{n\times p}(\phi)$ for some ϕ.*
(iv) *If $n = p$, then $U' = U^{-1} \sim \mathcal{U}_{n,n}$.*

Proof. (i) and (ii) are directly from Definition 3.1.2. Taking $q = p$ in (i), we find that U is right-spherical, hence U is spherical. When $n = p$, U' is left-spherical and $(U')'U' = I_n$, i.e., $U' \sim \mathcal{U}_{n,n}$. $\square$

As we know, the stochastic representation of spherical distribution in the case of vector plays an important role. The similar result can be found for left-spherical distribution.

Theorem 3.1.2. *Assume $X \sim LS_{n\times p}(\phi)$. There exists a $p \times p$ random matrix A such that*

$$X \overset{d}{=} UA. \tag{3.1.3}$$

where $U \sim \mathcal{U}_{n,p}$ is independent of A.

Proof. There exists a $p \times p$ random matrix A such that $A'A \overset{d}{=} X'X$ and A is independent of $U \sim \mathcal{U}_{n,p}$. Let $Y = UA$. Then Y is left-spherical and $Y'Y = A'U'UA = A'A \overset{d}{=} X'X$. By Theorem 3.1.1, $Y \overset{d}{=} X$. $\square$

In the decomposition (3.1.3), A is not quite unique. For most of the cases, we take A to be an upper triangular matrix with nonnegative diagonal elements or take A to be left-spherical with $A \geqslant 0$. The existence of A in the first case is from the well-known Cholesky decomposition and the existence of A in the second case is from the following statement. Let $B = (X'X)^{\frac{1}{2}}$ be independent of $V \sim \mathcal{U}_{p,p}$. Take $A = VB$. Then A is left-spherical and $A'A = B'B = X'X$.

If we add the condition $P(|X'X| = 0) = 0$, the above A for the two cases has a unique distribution. We only prove the first case. In this section let $UT(p)$ denote the set of upper triangular matrices with positive diagonal elements.

Lemma 3.1.4. *Suppose $X \sim LS_{n\times p}(\phi)$* and $P(|X'X| = 0) = 0$. *Then*
(i) *$X \overset{d}{=} UA \overset{d}{=} UB$ implies $A \overset{d}{=} B$, where $A \in UT(p)$ and $B \in UT(p)$;*

(ii) $X = QT$ *where* $Q'Q = I_p$ *and* $T \in UT(p)$ *implies that* $Q \stackrel{d}{=} U$ *and* Q *is independent of* T.

Proof. We consider a mapping f: $f(A) = A'A$, where $A \in UT(p)$. Clearly, f is a one-to-one mapping. As $A'A \stackrel{d}{=} B'B$, (i) follows from

$$E[h(A)] = E[h(f^{-1}(A'A))] = E[h(f^{-1}(B'B))] = E[h(B)]$$

for each Borel function $h \geqslant 0$. Note that if $|X'X| \neq 0$ there is a unique decomposition $X = QT$, where $Q'Q = I_p$ and $T \in UT(p)$. Let the function $g(X) = (Q, T)$. We have $(Q, T) = g(X) \stackrel{d}{=} g(UA) = (U, A)$ as $X \stackrel{d}{=} UA$, which completes the proof. □

From now on when we write $X \stackrel{d}{=} UA$ for $X \sim LS(\phi)$ we always mean that A is an upper triangular matrix with nonnegative diagonal elements.

Example 3.1.1. Let $x_{(1)}, \dots, x_{(n)}$ be i.i.d, $x_{(1)} \sim N_p(0, \Sigma)$, and $X = (x_{(1)}, \dots, x_{(n)})'$. The c.f. of $x_{(j)}$ is (cf. Theorem 2.3.1) $\exp(-\frac{1}{2}t'_{(j)}\Sigma t_{(j)})$, $j = 1, \dots, n$. Hence the c.f. of X is

$$\prod_{j=1}^{n} \exp(-\tfrac{1}{2}t'_{(j)}\Sigma t_{(j)}) = \prod_{j=1}^{n} \exp(-\tfrac{1}{2}\Sigma t_{(j)}t'_{(j)})$$

$$= \exp(-\frac{1}{2}\Sigma \sum_{j=1}^{n} t_{(j)}t'_{(j)}) = \exp(-\tfrac{1}{2}\Sigma T'T),$$

where $T = (t_{(1)}, \cdots, t_{(n)})'$. That means $X \sim LS_{n \times p}(\phi)$ for some ϕ.

3.1.2. Spherical Distributions

Spherical distributions were defined in the preceding section. Here we want to discuss some of their properties. Let $A \geqslant 0$ be a $p \times p$ matrix and $\lambda_1 \geqslant \cdots \geqslant \lambda_p \geqslant 0$ be the eigenvalues of A. We write $\lambda(A) = \operatorname{diag}(\lambda_1, \dots, \lambda_p)$.

Theorem 3.1.3. X *is spherical if and only if*

$$X \stackrel{d}{=} U\Lambda V, \tag{3.1.4}$$

where U, Λ and V are independent, $\Lambda = \lambda((X'X)^{\frac{1}{2}})$, $U \sim \mathcal{U}_{n,p}$ and $V \sim \mathcal{U}_{p,p}$.

Proof. Let $X = G\Lambda H$ be a measurable singular decomposition (cf. Section 1.2) and let $U^* \sim \mathcal{U}_{n,n}$ and $V^* \sim \mathcal{U}_{p,p}$ be independent of each other and of (G, Λ, H). Write $U = U^*G$, $V = HV^*$, $X^* = U^*XV^*$. Given (G, Λ, H), we have $U \sim \mathcal{U}_{n,p}$ and $V \sim \mathcal{U}_{p,p}$ are independent by Lemma 3.1.3. So U, V, Λ are all independent, having given distributions. Also, $X^* = U\Lambda V$. We must therefore show that $X^* \stackrel{d}{=} X$. Clearly, X^* is spherical from Lemma 3.1.2. Hence its distribution depends on U^* and V^* only through $U^{*\prime}U^* = I_p$ and $V^{*\prime}V^* = I_p$ (Corollary 2 of Theorem 3.1.1). So does X. We have $X = X^*$. □

Theorem 3.1.4. *If* X *is spherical, then the c.f. of* X *has the form* $\phi(\lambda(T'T))$.

Proof. From Definition 3.1.3 the c.f. of X, $\psi(T)$, satisfies $\psi(T) = \psi(PTQ)$ for each $P \in O(n)$ and each $Q \in O(p)$. The maximum invariant of $T(n \times p)$ under the transformation $\boldsymbol{PTQ}$ for each $\boldsymbol{P} \in O(n)$ and each $\boldsymbol{Q} \in O(p)$ is $\lambda(\boldsymbol{T}'\boldsymbol{T})$ (cf. Example 1.7.3) which completes the proof. □

3.1.3. Multivariate Spherical Distributions

Definition 3.1.4. An $n \times p$ random matrix $\boldsymbol{X}$ is said to be *distributed according to multivariate spherical distribution* if its c.f. has the form $\phi(\boldsymbol{t}_1'\boldsymbol{t}_1, \dots, \boldsymbol{t}'_p\boldsymbol{t}_p)$, where $\boldsymbol{T} = (\boldsymbol{t}_1, \dots, \boldsymbol{t}_p): n \times p$. We write $\boldsymbol{X} \sim MS_{n\times p}(\phi)$.

Multivariate spherical distributions were studied by Anderson and Fang (1982b) first. The following are some of their elementary properties. Hence the proofs are omitted and left to the reader.

Lemma 3.1.5. *$\boldsymbol{X} \sim MS_{n\times p}(\phi)$ if and only if*

$$\boldsymbol{X} = (\boldsymbol{x}_1, \dots, \boldsymbol{x}_p) \stackrel{d}{=} (R_1\boldsymbol{u}_1, \dots, R_p\boldsymbol{u}_p) = \boldsymbol{U}_2\boldsymbol{R} \tag{3.1.5}$$

where $\boldsymbol{R} = \operatorname{diag}(R_1, \dots, R_p)$ and $\boldsymbol{U}_2 = (\boldsymbol{u}_1, \dots, \boldsymbol{u}_p)$ are independent, $\boldsymbol{R} \geqslant 0$, and $\boldsymbol{u}_1, \dots, \boldsymbol{u}_p$ i.i.d. $\boldsymbol{u}_1 \stackrel{d}{=} \boldsymbol{u}^{(n)}$.

Lemma 3.1.6. *$\boldsymbol{X} \sim MS_{n\times p}(\phi)$ if and only if*

$$(\boldsymbol{P}_1\boldsymbol{x}_1, \dots, \boldsymbol{P}_p\boldsymbol{x}_p) \stackrel{d}{=} (\boldsymbol{x}_1, \dots, \boldsymbol{x}_p) \tag{3.1.6}$$

for every $\boldsymbol{P}_j \in O(n)$, $j = 1, \dots, p$.

Lemma 3.1.7. *If $\boldsymbol{X} \sim MS_{n\times p}(\phi)$, then its distribution is fully determined by $(\boldsymbol{x}_1'\boldsymbol{x}_1, \dots, \boldsymbol{x}_p'\boldsymbol{x}_p) \stackrel{d}{=} (R_1^2, \dots R_p^2)$, where $(R_1, \dots, R_p)$ is defined by* Lemma 3.1.5.

Example 3.1.2. Let $\boldsymbol{x}_1, \dots, \boldsymbol{x}_p$ be i.i.d. $\boldsymbol{x}_1 \sim S_n(\psi)$, and let $\boldsymbol{X} = (\boldsymbol{x}_1, \dots, \boldsymbol{x}_p)$. Then $\boldsymbol{X}$ is multivariate spherical, because the c.f. of $\boldsymbol{X}$ is

$$\prod_{j=1}^{p} \psi(\boldsymbol{t}_j'\boldsymbol{t}_j) = \phi(\boldsymbol{t}'_1\boldsymbol{t}_1, \dots, \boldsymbol{t}'_p\boldsymbol{t}_p),$$

say. Later on we shall see $\boldsymbol{X}$ is not left-spherical except that $\boldsymbol{x}_1$ is normal.

3.1.4. Vector-Spherical Distributions

Let $\boldsymbol{X} = (\boldsymbol{x}_1, \dots, \boldsymbol{x}_p)$ be an $n \times p$ matrix, and define $\operatorname{vec}(\boldsymbol{X}) = (\boldsymbol{x}_1', \dots, \boldsymbol{x}'_p)'$. Therefore $\operatorname{vec}(\boldsymbol{X}) \in R^{np}$.

Definition 3.1.5. Let $\boldsymbol{X}$ be an $n \times p$ random matrix. If $\operatorname{vec}(\boldsymbol{X}) \sim S_{np}(\phi)$ we call $\boldsymbol{X}$ vector-spherical distribution and write $\boldsymbol{X} \sim VS_{n\times p}(\phi)$.

The following results are directly from Theorem 2.5.3.

Lemma 3.1.8. *Assume $\boldsymbol{X}$ is an $n \times p$ random matrix. Then the following statements are equivalent:*

(i) $\boldsymbol{X} \sim VS_{n\times p}(\phi)$;

(ii) *the c.f. of $\boldsymbol{X}$ has the form $\phi(\operatorname{tr}(\boldsymbol{T}'\boldsymbol{T}))$, where $\phi \in \Phi_{np}$ (cf. Section 2.5);*

(iii) *$\boldsymbol{X}$ has a stochastic representation*

$$\boldsymbol{X} \stackrel{d}{=} R\boldsymbol{U}_3, \tag{3.1.7}$$

where $R \geqslant 0 \leftrightarrow \phi \in \Phi_{np}$ *is independent of* $\boldsymbol{U}_3$ *and* $vec(\boldsymbol{U}_3) \overset{d}{=} \boldsymbol{u}^{(np)}$.

(iv) $vec(\boldsymbol{X}) \overset{d}{=} \Gamma(vec(\boldsymbol{X}))$ *for every* $\Gamma \in O(np)$.

3.2. Relationships among Classes of Spherical Matrix Distributions

We defined four kinds of spherical matrix distributions in the preceding section. To study the relationships among them, define

$$\mathcal{F}_1 = \{\mathcal{L}(\boldsymbol{X})\colon \boldsymbol{X} \text{ is left-spherical}\},$$
$$\mathcal{F}_2 = \{\mathcal{L}(\boldsymbol{X})\colon \boldsymbol{X} \text{ is multivariate spherical}\},$$
$$\mathcal{F}_3 = \{\mathcal{L}(\boldsymbol{X})\colon \boldsymbol{X} \text{ is vector-spherical}\},$$
$$\mathcal{F}_s = \{\mathcal{L}(\boldsymbol{X})\colon \boldsymbol{X} \text{ is spherical}\},$$

where $\boldsymbol{X}: n \times p$ and $\mathcal{L}(\boldsymbol{X})$ is the distribution function of $\boldsymbol{X}$. For convenience, throughout this book $\boldsymbol{X} \in \mathcal{F}_i$ means $\mathcal{L}(\boldsymbol{X}) \in \mathcal{F}_i$, $i = 1, 2, 3, \text{s}$. In this section, we will discuss these classes in the following aspects: inclusion relation, the coordinate system and the coordinate transformations, marginal distributions, marginal densities and sphericity.

3.2.1. Inclusion Relation

Lemma 3.2.1. $\mathcal{F}_1 \supset \mathcal{F}_2 \supset \mathcal{F}_3$ *and* $\mathcal{F}_1 \supset \mathcal{F}_s \supset \mathcal{F}_3$.

Proof. The assertions can be seen in two aspects. One is from the group of transformations:

$$\boldsymbol{X} \in \mathcal{F}_1 \Longleftrightarrow \text{diag}\,(\boldsymbol{P}, \dots, \boldsymbol{P})\,\text{vec}(\boldsymbol{X}) \overset{d}{=} \text{vec}(\boldsymbol{X}) \text{ for every } \boldsymbol{P} \in O(n);$$

$$\boldsymbol{X} \in \mathcal{F}_2 \Longleftrightarrow \text{diag}\,(\boldsymbol{P}_1, \dots, \boldsymbol{P}_p)\,\text{vec}(\boldsymbol{X}) \overset{d}{=} \text{vec}(\boldsymbol{X}) \text{ for every } \boldsymbol{P}_i \in O(n),\quad i = 1, \dots, p;$$

$$\boldsymbol{X} \in \mathcal{F}_3 \Longleftrightarrow \boldsymbol{P}\,\text{vec}(\boldsymbol{X}) \overset{d}{=} \text{vec}(\boldsymbol{X}) \text{ for every } \boldsymbol{P} \in O(np).$$

So we have $\mathcal{F}_1 \supset \mathcal{F}_2 \supset \mathcal{F}_3$. Another is from the form of c.f.'s:

$\boldsymbol{X} \in \mathcal{F}_1 \Longleftrightarrow$ the c.f. of $\boldsymbol{X}$ has the form $\phi(\boldsymbol{T}'\boldsymbol{T})$;

$\boldsymbol{X} \in \mathcal{F}_2 \Longleftrightarrow$ the c.f. of $\boldsymbol{X}$ has the form $\phi(\boldsymbol{t}'_1\boldsymbol{t}_1, \dots, \boldsymbol{t}'_p\boldsymbol{t}_p) = \phi(\text{diag}(\boldsymbol{t}'_1\boldsymbol{t}_1, \dots, \boldsymbol{t}'_p\boldsymbol{t}_p))$, where $\boldsymbol{t}'_1\boldsymbol{t}_1, \dots, \boldsymbol{t}'_p\boldsymbol{t}_p$ are the diagonal elements of $\boldsymbol{T}'\boldsymbol{T}$;

$\boldsymbol{X} \in \mathcal{F}_s \Longleftrightarrow$ the c.f. of $\boldsymbol{X}$ has the form $\phi(\lambda(\boldsymbol{T}'\boldsymbol{T}))$;

$\boldsymbol{X} \in \mathcal{F}_3 \Longleftrightarrow$ the c.f. of $\boldsymbol{X}$ has the form $\phi(\text{tr}(\boldsymbol{T}'\boldsymbol{T})) = \phi\left(\sum_{j=1}^{p} \boldsymbol{t}'_j\boldsymbol{t}_j\right)$.

Obviously, $\mathcal{F}_1 \supset \mathcal{F}_2 \supset \mathcal{F}_3$ and $\mathcal{F}_1 \supset \mathcal{F}_s \supset \mathcal{F}_3$. □

Further we want to show the above inclusion relations are proper ones.

Lemma 3.2.2. *Suppose* $X \in \mathcal{F}_1$ *and* $F(x_i = 0) = 0, i = 1, \dots, p$. *Then* $X \in \mathcal{F}_2$ *if and only if* X *satisfies the following conditions:*

(i) $x_1 / \|x_1\|, \dots, x_p / \|x_p\|$ *are independent; and*

(ii) $(\|x_1\|, \dots, \|x_p\|)$ *and* $(x_1 / \|x_1\|, \dots, x_p / \|x_p\|)$ *are independent.*

Proof. The "if" part is a consequence of Lemma 3.1.5. If X satisfies (i) and (ii), we have

$$X = (x_1, \dots, x_p) = \left(\frac{x_1}{\|x_1\|}, \cdots, \frac{x_p}{\|x_p\|} \right) \operatorname{diag} (\|x_1\|, \dots, \|x_p\|).$$

Let $U_2 = (x_1 / \|x_1\|, \dots, x_p / \|x_p\|)$ and $R = \operatorname{diag}(\|x_1\|, \dots, \|x_p\|)$. As $X \in \mathcal{F}_1, x_j \sim S_n(\phi)$ for some ϕ, $j = 1, \dots, p$. It is easy to check that U_2 and R satisfy all conditions in Lemma 3.1.5. i.e., $X \in \mathcal{F}_2$. □

Corollary 1. *The uniform distribution* $\mathcal{U}_{n,p} \notin \mathcal{F}_2$.

Proof. As $U'U = I_p$ and $U \in \mathcal{F}_1$, $U = (u_1, \dots, u_p)$ where $u_1 = u_1 / \|u_1\|, \dots, u_p = u_p / \|u_p\|$ are not independent, the corollary follows from Lemma 3.2.2. □

Corollary 2. $U_2 \notin \mathcal{F}_3$.

Proof. Suppose $U_2 = (u_1, \dots, u_p) = (u_{(1)}, \dots, u_{(n)})' \in \mathcal{F}_3$. Then $U_2 \overset{d}{=} RU_3$ for some $R \geqslant 0$ (cf. (3.1.7)) and R is independent of U_3. As $u_1, \dots, u_p$ are independent and $u_{(i)}$ $(i = 1, \dots, n)$ has a spherical distribution, the distribution of $u_{(i)}$ must be normal (cf. Theorem 2.7.2). obviausly $u_{(i)}$ can not be normal. The contradiction proves the corollary. □

Exercise 3.4 shows that the condition (ii) in Lemma 3.2.2 is necessary.

3.2.2. Classes of Marginal Distributions

Let $\mathcal{F}_i^c$ $(i = 1, 2, 3, \mathrm{s})$ denote the set of first columns of X's in $\mathcal{F}_i$ $(i = 1, 2, 3, \mathrm{s})$, i.e., $x \in \mathcal{F}_i^c$ if and only if there exist $x_2, \dots, x_p$ such that $X = (x, x_2, \dots, x_p) \in \mathcal{F}_i$, $i = 1, 2, 3, \mathrm{s}$. Similarly, $\mathcal{F}_i^r$ indicates a set of first row vector of X's in $\mathcal{F}_i$, $i = 1, 2, 3, \mathrm{s}$. Clearly $\mathcal{F}_1^c \supset \mathcal{F}_2^c \supset \mathcal{F}_3^c$, $\mathcal{F}_1^c \supset \mathcal{F}_s^c \supset \mathcal{F}_3^c$, $\mathcal{F}_1^r \supset \mathcal{F}_2^r \supset \mathcal{F}_3^r$, and $\mathcal{F}_1^r \supset \mathcal{F}_s^r \supset \mathcal{F}_3^r$ by Lemma 3.2.1. Further, we want to know whether or not the above inclusion relations are proper.

Lemma 3.2.3. $\mathcal{F}_2^c = \mathcal{F}_1^c = \mathcal{F}_s^c$.

Proof. If $x \in \mathcal{F}_1^c$, then $x \sim S_n(\phi)$ for some $\phi \in \Phi_n$. Let $x_2, \dots, x_p$ be $p - 1$ $n \times 1$ random vectors such that $x, x_2, \dots, x_p$ are i.i.d. Thus $X = (x, x_2, \dots, x_p) \in \mathcal{F}_2$ by Example 3.1.2, i.e., $\mathcal{F}_2^c = \mathcal{F}_1^c$. Let U, V, x be independent and let U and V be defined by Theorem 3.1.3. Take $X = U \operatorname{diag}(\|x\|, \dots, \|x\|) V = \|x\| UV$. Clearly $X \in \mathcal{F}_s$ by Theorem 3.1.3. We must show that the first column x_1 of X has the same distribution as x. This is true, because $x_1 \sim S_n(\psi)$ for some ψ, $X'X = \|x\|^2 V'U'UV \overset{d}{=} \|x\|^2 I_p$ and $x_1' x_1 = \|x\|^2$ which implies $x_1 \overset{d}{=} x$ (Exercise 2.13 or Theorem 3.1.1 with $p = 1$). □

One may ask whether or not $\mathcal{F}_3^c = \mathcal{F}_2^c$ holds. But that is not true in fact. First, we need the following lemma.

Lemma 3.2.4. *Let* $\Omega_n(t't)$, $t \in R^n$, *be the c.f. of* $u^{(n)}$. *Then* $\Omega_n(t't)$ *with* $t \in R^{n+1}$ *is not an* $(n + 1)$*-dimensional c.f.*

Proof. Suppose that $\Omega_n(\boldsymbol{t}'\boldsymbol{t})$ with $\boldsymbol{t} \in R^{n+1}$ is an $(n+1)$-dimensional c.f. Then there exists a cdf F such that (Theorem 2.5.2)

$$\Omega_n(u) = \int_0^\infty \Omega_{n+1}(ur^2)\, dF(r),\ u \geqslant 0$$

i.e., $\boldsymbol{u}^{(n)} \overset{d}{=} R\boldsymbol{u}_n$, where $R \geqslant 0$ is independent of $\boldsymbol{u}_n$ and $\boldsymbol{u}_n$ is the subvector of $\boldsymbol{u}^{(n+1)}$ with the first n components. Since $\boldsymbol{u}_n$ has a density and $P(R=0) = P(R\boldsymbol{u}_n = \boldsymbol{0}) = P(\boldsymbol{u}^{(n)} = 0) = 0$, $\boldsymbol{u}^{(n)}$ has a density. This contradiction completes the proof. □

Theorem 3.2.1. *The set $\mathscr{F}_3^c$ is a proper subset of $\mathscr{F}_2^c$ if $p > 1$.*

Proof. Let $\boldsymbol{u} \overset{d}{=} \boldsymbol{u}^{(n)}$. Clearly, $\boldsymbol{u} \in \mathscr{F}_2^c$. We want to point out $\boldsymbol{u} \notin \mathscr{F}_3^c$. If $\boldsymbol{u} \in \mathscr{F}_3^c$, then there exist $\boldsymbol{u}_2, \dots, \boldsymbol{u}_p$ such that $\boldsymbol{U}^* = (\boldsymbol{u}, \boldsymbol{u}_2, \dots, \boldsymbol{u}_p) \in \mathscr{F}_3$. Let $\phi(\mathrm{tr}(\boldsymbol{T}'\boldsymbol{T}))$ denote the c.f. or $\boldsymbol{U}^*$. Then the c.f. of $\boldsymbol{u}$ is $\phi(\boldsymbol{t}'_1\boldsymbol{t}_1) = \Omega_n(\boldsymbol{t}'_1\boldsymbol{t}_1)$, $\boldsymbol{t}_1 \in R^n$, i.e., $\phi(.) = \Omega_n(.)$. This means that $\Omega_n(\boldsymbol{t}'\boldsymbol{t}) = \phi(\boldsymbol{t}'\boldsymbol{t})$, $\boldsymbol{t} \in R^{np}$, is a c.f., but it is impossible by Lemma 3.2.4. □

Let us consider the row marginal distributions. First, we want to point out that $\mathscr{F}_2^r$ is a proper subset of $\mathscr{F}_1^r$.

Lemma 3.2.5. *Suppose $\boldsymbol{X} = (\boldsymbol{x}_1, \dots, \boldsymbol{x}_p) = (\boldsymbol{x}_{(1)}, \dots, \boldsymbol{x}_{(n)})' \in \mathscr{F}_2$ and the covariance of $\boldsymbol{x}_{(1)}$ exists. Then*

(i) $\mathrm{cov}(\boldsymbol{x}_{(i)}, \boldsymbol{x}_{(j)}) = \delta_{ij}\Lambda_j$, *where Λ_i is a diagonal matrix and $\delta_{ij} = 0$, and $\delta_{ii} = 1$, $i \neq j$, $i, j = 1, \dots, n$, and*

(ii) $\mathrm{cov}(\boldsymbol{x}_i, \boldsymbol{x}_j) = \delta_{ij}\sigma_{ii}^2 \boldsymbol{I}_n$, *where $\sigma_{ii}^2 = ER_i^2/n$ and R_i is defined by* Lemma 3.1.5, *$i, j = 1, \dots, p$.*

Proof. Clearly, $\boldsymbol{x}_{(1)}, \dots, \boldsymbol{x}_{(n)}$ are identically distributed and $\boldsymbol{X}$ has the stochastic decompostion (3.1.5). As $\mathcal{E}\boldsymbol{U}_2 = \boldsymbol{0}$, we have $\mathcal{E}\boldsymbol{x}_{(k)} = \boldsymbol{0}$ and $\mathcal{E}\boldsymbol{x}_j = \boldsymbol{0}$ for $k = 1, \dots, n$; $j = 1, \dots, p$. By (3.1.5) and $\boldsymbol{U}_2 = (u_{ij})$,

$$\mathrm{cov}(\boldsymbol{x}_{(i)}, \boldsymbol{x}_{(j)}) = \mathcal{E}\boldsymbol{x}_{(i)}\boldsymbol{x}'_{(j)} = \mathrm{diag}(E[R_1^2 u_{i1} u_{j1}], \dots, E[R_p^2 u_{ip} u_{jp}]).$$

The first assertion follows from $Eu_{ik}u_{jk} = 0$ and $Eu_{ik}^2 = 1/n$ for $i \neq j$; $k = 1, \dots, p$. Similarly, $\mathcal{E}\boldsymbol{x}_i\boldsymbol{x}'_j = E(R_iR_j)\mathcal{E}(\boldsymbol{u}_i\boldsymbol{u}'_j) = \delta_{ij}(ER_i^2/n)\boldsymbol{I}_n = \delta_{ij}\sigma_{ii}^2\boldsymbol{I}_n$. □

Corollary 1. *The set $\mathscr{F}_2^r$ is a proper subset of $\mathscr{F}_1^r$.*

Proof. Let $\boldsymbol{X} \sim N_{n\times p}(\boldsymbol{O}, \boldsymbol{I}_n \otimes \Sigma)$ and Σ be not a diagonal matrix. By Lemma 3.2.5 $\boldsymbol{x}_{(1)} \notin \mathscr{F}_2^r$, but $\boldsymbol{x}_{(1)} \in \mathscr{F}_1^r$ (Example 3.1.1). Thus $\mathscr{F}_2^r$ is a proper subset of $\mathscr{F}_1^r$. □

Theorem 3.2.2. *Assume $\boldsymbol{X} \in \mathscr{F}_2$. Then $\boldsymbol{X} \in \mathscr{F}_3$ if and only if $\boldsymbol{x}_{(1)} \in \mathscr{F}_3^r$.*

Proof. The "only if" part is obvious. Suppose $\boldsymbol{x}_{(1)} \in \mathscr{F}_3^r$. Then $\boldsymbol{x}_{(1)}$ has a c.f. $\phi(\boldsymbol{t}'_{(1)}\boldsymbol{t}_{(1)})$, where $\phi(\boldsymbol{t}'_{(1)}\boldsymbol{t}_{(1)} + \cdots + \boldsymbol{t}'_{(n)}\boldsymbol{t}_{(n)})$ is a c.f. in R^{np}. On the other hand, since $\boldsymbol{X} \in \mathscr{F}_2$, $\boldsymbol{X}$ has a c.f. $\psi(\boldsymbol{t}'_1\boldsymbol{t}_1, \dots, \boldsymbol{t}'_p\boldsymbol{t}_p)$. Hence we must have

$$\phi(r_1^2 + \cdots + r_p^2) = \psi(r_1^2, \dots, r_p^2),\ \text{for } r_i^2 \geqslant 0,\ i = 1, \dots, p, \tag{3.2.1}$$

because they are the same c.f. of $\boldsymbol{x}_{(1)}$. By (3.2.1), we have $\phi(\boldsymbol{t}'_1\boldsymbol{t}_1 + \cdots + \boldsymbol{t}'_p\boldsymbol{t}_p) = \psi(\boldsymbol{t}'_1\boldsymbol{t}_1, \dots, \boldsymbol{t}'_p\boldsymbol{t}_p)$ for all $\boldsymbol{t}_i \in R^n$, $i = 1, \dots, p$, i.e., $\boldsymbol{X} \in \mathscr{F}_3$. The theorem follows. □

Corollary 1. *Suppose $\boldsymbol{X} \in \mathscr{F}_2$. Then $x_{(1)} \in \mathscr{F}_3^r$ if and only if $\boldsymbol{x}_{(1)} \sim S_p(\phi)$.*
The proof is easy and is left to the reader.

Corollary 2. *The first row* $\boldsymbol{u}_{(1)}$ *of* $\boldsymbol{U}$ *is not in* $\mathcal{F}_2'$.

Proof. Assume $\boldsymbol{u}_{(1)} \in \mathcal{F}_2'$. Then there exists a $\boldsymbol{Y}$ such that $\boldsymbol{X} = (\boldsymbol{u}_{(1)}, \boldsymbol{Y})' \in \mathcal{F}_2$. As $\boldsymbol{U}$ is right-spherical (Lemma 3.1.3), therefore $\boldsymbol{u}_{(1)} \sim S_p(\phi)$ and $\boldsymbol{u}_{(1)} \in \mathcal{F}_3'$ by $\boldsymbol{X} \in \mathcal{F}_2$ and Corollary 1 of Theorem 3.2.2. However, this is imposible (cf. the following example 3.2.1). Hence, $\boldsymbol{u}_{(1)} \notin \mathcal{F}_2'$. □

Corollary 3. $\mathcal{F}_3 = \mathcal{F}_s \bigcap \mathcal{F}_2$.

Proof. Clearly $\mathcal{F}_3 \subset \mathcal{F}_s \bigcap \mathcal{F}_2$. Conversely, if $\boldsymbol{X} \in \mathcal{F}_s \bigcap \mathcal{F}_2$, the fact $\boldsymbol{X} = (\boldsymbol{x}_{(1)}, \dots, \boldsymbol{x}_{(n)})' \in \mathcal{F}_s$ implies that $\boldsymbol{x}_{(1)}$ is spherical and $\boldsymbol{X} \in \mathcal{F}_3$ by the theorem. □

Before coming to the next corollary, the concept of matrix variate normal distribution is needed.

Definition 3.2.1. Let $\boldsymbol{X}$ be an $n \times p$ random matrix. If $\text{vec}(\boldsymbol{X}') \sim N_{np}(\mu, \boldsymbol{C} \otimes \boldsymbol{D})$ with $\mu = \text{vec}(\boldsymbol{M}')$, $\boldsymbol{M}$: $n \times p$, $\boldsymbol{C}$: $n \times n$, $\boldsymbol{D}$: $p \times p$, $\boldsymbol{C} \geqslant 0$ and $\boldsymbol{D} \geqslant 0$, $\boldsymbol{X}$ is said to have a *matrix variate normal distribution* and we write $\boldsymbol{X} \sim N_{n \times p}(\boldsymbol{M}, \boldsymbol{C} \otimes \boldsymbol{D})$.

If $\boldsymbol{Y} = (y_{ij}) \sim N_{n \times p}(\boldsymbol{O}, \boldsymbol{I}_n \otimes \boldsymbol{I}_p)$, then all y_{ij}'s are i.i.d and $y_{ij} \sim N(0, 1)$. By the above definition, we obtain directly the following facts whose proofs are left to the reader.

(1) If $\boldsymbol{X} \sim N_{n \times p}(\boldsymbol{M}, \boldsymbol{C} \otimes \boldsymbol{D})$ with $\boldsymbol{C} > 0$ and $\boldsymbol{D} > 0$, then $\boldsymbol{X}$ has a density which is

$$(2\pi)^{-\frac{1}{2}np} |\boldsymbol{C}|^{-\frac{1}{2}p} |\boldsymbol{D}|^{-\frac{1}{2}n} \operatorname{etr}[-\tfrac{1}{2}\boldsymbol{C}^{-1}(\boldsymbol{X} - \boldsymbol{M})\boldsymbol{D}^{-1}(\boldsymbol{X} - \boldsymbol{M})'].$$

(2) $\boldsymbol{X} \sim N_{n \times p}(\boldsymbol{M}, \boldsymbol{C} \otimes \boldsymbol{D})$ if and only if

$$\boldsymbol{X} \overset{\mathrm{d}}{=} \boldsymbol{M} + \boldsymbol{C}^{\frac{1}{2}} \boldsymbol{Y} \boldsymbol{D}^{\frac{1}{2}},$$

where $\boldsymbol{Y} \sim N_{n \times p}(\boldsymbol{O}, \boldsymbol{I}_n \otimes \boldsymbol{I}_p)$.

Corollary 4. *Assume* $\boldsymbol{X} \in \mathcal{F}_s$ *with independent columns (or rows). Then* $\boldsymbol{X}$ *must be normal.*

Proof. From the assumption, the c.f. of $\boldsymbol{X}$ is $\prod_1^p \phi(\boldsymbol{t}_i'\boldsymbol{t}_i)$, i.e., $\boldsymbol{X} \in \mathcal{F}_2$. By Corollary 3, $\boldsymbol{X} \in \mathcal{F}_3$. The assertion follows from Theorem 2.7.2. □

Example 3.2.1. Suppose $\boldsymbol{X} \overset{\mathrm{d}}{=} \boldsymbol{U}\boldsymbol{A}$, where $\boldsymbol{U}$ and $\boldsymbol{A}$ are independent, and $\boldsymbol{A} = \text{diag}(a_1, \dots, a_p)$, $0 < p_i = P(a_i = 1) = 1 - P(a_i = 0) < 1$, $i = 1, \dots, p$, $a_1 + \cdots + a_p = 1$. Clearly $\boldsymbol{X} \in \mathcal{F}_1$, but $\boldsymbol{X} \in \mathcal{F}_2$, because $(\boldsymbol{x}_1 / \|\boldsymbol{x}_1\|, \cdots, \boldsymbol{x}_p / \|\boldsymbol{x}_p\|) \overset{\mathrm{d}}{=} (\boldsymbol{u}_1, \dots, \boldsymbol{u}_p) = \boldsymbol{U}$ whose columns are dependent (Lemma 3.2.2). However, we want to show $\boldsymbol{x}_{(1)} \overset{\mathrm{d}}{=} \boldsymbol{A}\boldsymbol{u}_{(1)} \in \mathcal{F}_2'$, where $\boldsymbol{u}_{(1)}$ is the first row of $\boldsymbol{U}$. It can be shown that the c.f. of $\boldsymbol{u}_{(1)}$ is $\Omega_n(t_1^2 + \cdots + t_p^2)$ by the sphericity of $\boldsymbol{U}$. And $\boldsymbol{x}_{(1)}$ has a c.f.

$$\psi(t_1^2, \dots, t_p^2) = \int \Omega_n(a_1^2 t_1^2 + \cdots + a_p^2 t_p^2)\, dF(a_1, \dots, a_p) \tag{3.2.2}$$

$$= \sum_{i=1}^{p} p_i \Omega_n(t_i^2).$$

Thus, we have

$$\psi(t_1't_1, \dots, t_p't_p) = \sum_{i=1}^{p} p_i \Omega_n(t_i't_i), \qquad t_i \in R^n,\ i = 1, \dots, p.$$

As $\Omega_n(t_i't_i)$ is a c.f. in R^n and $\sum_{i=1}^{p} p_i = 1, p_i > 0, i = 1, \dots, p$, hence $\psi(t'_1t_1, \dots, t_p't_p)$ is the c.f. of some Y in $\mathcal{F}_2$ and $y_{(1)} \stackrel{d}{=} x_{(1)}$, where $y_{(1)}$ is the first row of Y; that means $x_{(1)} \in \mathcal{F}_2^r$.

Theorem 3.2.2 and Example 3.2.1 show that $\mathcal{F}_2$, related to $\mathcal{F}_1$, cannot be characterized by its row marginal distributions. But for $\mathcal{F}_3$, related to $\mathcal{F}_2$, it is feasible. Some other relationships can be found in Exercises 3.5 and 3.6.

3.2.3. Coordinate Systems

In Section 3.1, we obtained the stochastic decompositions for $\mathcal{F}_i$, $i = 1, 2, 3, s$. They are

(3.1.3) $$X \stackrel{d}{=} UA \qquad \text{for } \mathcal{F}_1,$$

(3.1.4) $$X \stackrel{d}{=} U\Lambda V \qquad \text{for } \mathcal{F}_s,$$

(3.1.5) $$X \stackrel{d}{=} U_2R \qquad \text{for } \mathcal{F}_2,$$

and

(3.1.6) $$X \stackrel{d}{=} RU_3 \qquad \text{for } \mathcal{F}_3.$$

Here U, U_2, U_3 and (U, V) play the role of "coordinate system" separately.

Lemma 3.2.6. *Let* $X \sim N_{n \times p}(O, I_n \otimes I_p)$. *Then*

(3.2.3) $$U \stackrel{d}{=} X(X'X)^{-\frac{1}{2}},$$

(3.2.4) $$U_2 \stackrel{d}{=} (x_1/\|x_1\|, \dots, x_p/\|x_p\|),$$

and

(3.2.5) $$U_3 \stackrel{d}{=} X/(\mathrm{tr}(X'X))^{\frac{1}{2}}.$$

Proof. Let $Y \stackrel{d}{=} X(X'X)^{-\frac{1}{2}}$. As $Y'Y = I_p$ and

$$PY = PX(X'X)^{-\frac{1}{2}} = PX[(PX)'(PX)]^{-\frac{1}{2}} \stackrel{d}{=} X(X'X)^{-\frac{1}{2}} = Y$$

for every $P \in O(n)$, (3.2.3) follows from Definition 3.1.2. Similarly, we can get (3.2.4) and (3.2.5). ∎

In fact, X in Lemma 3.2.6 can be replaced by $X \in \mathcal{F}_1$ with $P(|X'X| = 0) = 0$ in (3.2.3); $X \in \mathcal{F}_2$ with $P(x_j = 0) = 0$, $j = 1, \dots, p$, in (3.2.4); and $X \in \mathcal{F}_3$ with $P(X = O) = 0$ in (3.2.5), individually.

A natural problem is how to give stochastic representations for $\mathcal{F}_i$ ($i = 1, 2, 3, s$) under the same coordinate system. A detailed discussion will be given in Section 3.6.

3.2.4. Densities

Let $X \in \mathcal{F}_i$, $i = 1,2,3$, or s. In general, it is not necessary that X has a density. Suppose X has a density. It is easily shown that

(i) $X \in \mathcal{F}_1$ if and only if the density of X has the form $f(X'X)$;
(ii) $X \in \mathcal{F}_2$ if and only if the density of X has the form $f(x'_1x_1, \dots, x'_px_p)$;
(iii) $X \in \mathcal{F}_3$ if and only if the density of X has the form $f(\mathrm{tr}(X'X))$;
(iv) $X \in \mathcal{F}_s$ if and only if the density of X has the form $f(\lambda(X'X))$.

When X has a density $f(.)$, we write $X \sim LS_{n\times p}(f)$ instead of $X \sim LS_{n\times p}(\phi)$ (the same way for MS, VS and SS). By Theorem 2.5.5, if $X \overset{d}{=} RU_3 \in \mathcal{F}_3$, then X has a density $f(.)$ if and only if R has a density $g(.)$. And there is a relationship between $f(.)$ and $g(.)$ as follows:

$$g(r) = \frac{2\pi^{\frac{1}{2}np}}{\Gamma(\frac{1}{2}np)} r^{np-1} f(r^2), \; r > 0. \tag{3.2.6}$$

Similarly, if $X \overset{d}{=} U_2 R \in \mathcal{F}_2$, then X has a density $f(.)$ if and only if R has a density $g(.)$, and there is a relationship between them:

$$g(r_1, \dots, r_p) = \frac{2^p \pi^{\frac{1}{2}np}}{(\Gamma(\frac{1}{2}n))^p} \prod_{i=1}^p r_i^{n-1} f(r_1^2, \dots, r_p^2), \; r_1, \dots, r_p > 0. \tag{3.2.7}$$

Further, suppose $X \overset{d}{=} UA \in \mathcal{F}_1$ with $A \in UT(p)$. Then X has a density $f(.)$ if and only if A has a density $g(.)$ and g is related to f as follows:

$$g(A) = \frac{2^p \pi^{\frac{1}{2}np - \frac{1}{4}p(p-1)}}{\prod_{i=1}^p \Gamma\left(\frac{1}{2}(n-i+1)\right)} \prod_{i=1}^p a_{ii}^{n-i} f(A'A), \tag{3.2.8}$$

where $A = (a_{ij})$. The proof is left to the next section. There is a similar result for $\mathcal{F}_s$. Let $X \overset{d}{=} U\Lambda V \in \mathcal{F}_s$, Then X has a density $f(.)$ if and only if Λ has a density $g(.)$, and they have the following relationship:

$$g(\lambda_1, \dots, \lambda_p) = \frac{\pi^{\frac{1}{2}np + \frac{1}{2}p}}{\prod_{i=1}^p \Gamma(\frac{1}{2}(n-i+1))} \prod_{i<j} (\lambda_i - \lambda_j)(\lambda_1 \cdots \lambda_p)^{\frac{1}{2}(n-p-1)} \cdot f(\mathrm{diag}(\lambda_1, \cdots, \lambda_p)) \qquad \lambda_1 > \lambda_2 > \cdots > \lambda_p > 0. \tag{3.2.9}$$

The proof will be given in section 3.4.

3.3. Elliptically Contoured Matrix Distributions

Let X be an $n \times p$ random matrix and $X \in \mathcal{F}_i$, $i = 1,2,3$, or s. Let M and B be $m \times p$ and $n \times m$ constant matrices respectively. If

$$Y \overset{d}{=} M + B'X, \qquad \Sigma = B'P, \tag{3.3.1}$$

we say that Y is distributed according to an elliptically contoured matrix distribution, and write $Y \sim ELS_{m\times p}(M, \Sigma, \phi)$ for $X \sim LS(\phi)$; $P \sim EMS_{m\times p}(M, \Sigma, \phi)$ for X

$\sim MS(\phi)$; $Y \sim EMS_{m\times p}(M, \Sigma, \phi)$ for $X \sim VS(\phi)$; and $M \sim ESS_{m\times p}(M, \Sigma, \phi)$ for $X \sim SS(\phi)$. In order to show the distribution of Y depends on B only through $\Sigma = B'B$, we need the following lemma:

Lemma 3.3.1. *The c.f. of* $Y \sim ELS_{m\times p}(M, \Sigma, \phi)$ *is*

$$\exp(iT'M)\,\phi\,(T'\Sigma T); \tag{3.3.2}$$

the c. f. of $Y \sim EMS_{m\times p}(M, \Sigma, \phi)$ *is*

$$\exp(iT'M)\,\phi\,(t_1'\Sigma t_1, \cdots, t_p'\Sigma t_p); \tag{3.3.3}$$

the c.f. of $Y \sim EVS_{m\times p}(M, \Sigma, \phi)$ *is*

$$\exp(iT'M)\,\phi\,(\mathrm{tr}(T'\Sigma T)); \tag{3.3.4}$$

and the c.f. of $Y \sim ESS_{m\times p}(M, \Sigma, \phi)$ *is*

$$\exp(iT'M)\,\phi\,(\lambda(T'\Sigma T)), \tag{3.3.5}$$

where $T = (t_1, \ldots, t_p)$: $m \times p$.

The proof is straightforward and is omitted (cf. Section 3.1). Similarly, we have the following results.

Lemma 3.3.2. *Assume* $Y(m \times p)$ *has a density. Then the density of* Y *has the following form*

$$|\Sigma|^{-\frac{1}{2}p} f((Y - M)'\Sigma^{-1}(Y - M)), \quad \textit{for ELS}; \tag{3.3.6}$$

$$|\Sigma|^{-\frac{1}{2}p} f(y_1'\Sigma^{-1}y_1, \ldots, y_p'\Sigma^{-1}y_p), \quad \textit{for EMS}; \tag{3.3.7}$$

$$|\Sigma|^{-\frac{1}{2}p} f(\mathrm{tr}(\Sigma^{-1}(Y - M)(Y - M)')), \quad \textit{for EVS}; \tag{3.3.8}$$

$$|\Sigma|^{-\frac{1}{2}p} f(\lambda((Y - M)'\Sigma^{-1}(Y - M))), \quad \textit{for ESS}, \tag{3.3.9}$$

where $Y = (y_1, \ldots, y_p)$.

By definition (3.3.1), if $Y \sim ELS_{m\times p}(M, \Sigma, \phi)$ and

$$Z = C'Y + D, \tag{3.3.10}$$

where $C(m \times l)$ and $D(l \times p)$ are constant matrices, then $Z \sim ELS_{l\times p}(C'M + D, C'\Sigma C, \phi)$. Partition

$$Y = \begin{bmatrix} Y_1 \\ Y_2 \end{bmatrix}, \quad M = \begin{bmatrix} M_1 \\ M_2 \end{bmatrix} \quad \Sigma = \begin{bmatrix} \Sigma_{11} & \Sigma_{12} \\ \Sigma_{21} & \Sigma_{22} \end{bmatrix}, \tag{3.3.11}$$

where Y_1: $q \times p$, M_1: $q \times p$ and Σ_{11}: $q \times q$. Taking $C' = (I_q, O)$ and $D = O$ into (3.3.10), we have $Y_1 \sim ELS_{q\times p}(M_1, \Sigma_{11}, \phi)$. Similar result holds for other classes of elliptically contoured matrix distributions.

Lemma 3.3.3. *Assume* $Y \sim ELS_{m\times p}(M, \Sigma, \phi)$ *and*

$$Y = \begin{bmatrix} Y_{11} & Y_{12} \\ Y_{21} & Y_{22} \end{bmatrix}, \qquad M = \begin{bmatrix} M_{11} & M_{12} \\ M_{21} & M_{22} \end{bmatrix}, \qquad \Sigma = \begin{bmatrix} \Sigma_{11} & \Sigma_{12} \\ \Sigma_{21} & \Sigma_{22} \end{bmatrix} \tag{3.3.12}$$

where Y_{11}: $n_1 \times p_1$, M_{11}: $n_1 \times p_1$ *and* Σ_{11}: $n_1 \times n_1$. *Then* $Y_{11} \sim ELS_{n_1\times p_1}(M_{11}, \Sigma_{11}, \phi^*)$ *with*

$$\phi^*(A_{11}) = \phi\begin{pmatrix} A_{11} & O \\ O & O \end{pmatrix} \textit{ and } A_{11}: p_1 \times p_1. \tag{3.3.13}$$

Proof. The c.f. of $\boldsymbol{Y}$ is (3.3.2). Hence the c.f. of $\boldsymbol{Y}_{11}$ is $\exp(i\boldsymbol{T}_{11}'\boldsymbol{M}_{11})\phi^*(\boldsymbol{T}_{11}'\Sigma_{11}\boldsymbol{T}_{11})$ which completes the proof. □

Now we consider moments of elliptically contoured distributions.

Theorem 3.3.1. *Assume $\boldsymbol{X} \sim LS_{n\times p}(\phi)$ has the second moments. Then*

$$\mathcal{E}\boldsymbol{X} = \boldsymbol{O}, \qquad \mathcal{D}(\mathrm{vec}(\boldsymbol{X})) = \boldsymbol{V} \otimes \boldsymbol{I}_n, \tag{3.3.14}$$

where $\boldsymbol{V} = E(\boldsymbol{x}_{(1)}\boldsymbol{x}'_{(1)})$.

Proof. Let $\boldsymbol{U} = (\boldsymbol{u}_1, \dots, \boldsymbol{u}_p) \sim \mathcal{U}_{n,p}$. It is easy to see $\boldsymbol{u}_1 \overset{d}{=} \cdots \overset{d}{=} \boldsymbol{u}_p \overset{d}{=} \boldsymbol{u}^{(n)}$. By Corollary 2 of Theorem 2.5.3, we have $\mathcal{E}\boldsymbol{U} = (\mathcal{E}\boldsymbol{u}_1, \dots, \mathcal{E}\boldsymbol{u}_p) = \boldsymbol{O}$ and $\mathcal{E}\boldsymbol{u}_1\boldsymbol{u}_1' = (1/n)\boldsymbol{I}_n$. As $\boldsymbol{U} \in \mathcal{F}_s$, we must have $(\boldsymbol{u}_i, \boldsymbol{u}_j) \overset{d}{=} (\boldsymbol{u}_i, -\boldsymbol{u}_j)$, $i,j = 1, \dots, p$ which implies $\mathcal{E}[(\mathrm{vec}(\boldsymbol{U}))(\mathrm{vec}(\boldsymbol{U}))'] = \boldsymbol{I}_p \otimes \mathcal{D}(\boldsymbol{u}_1) = (1/n)\boldsymbol{I}_p \otimes \boldsymbol{I}_n = (1/n)\boldsymbol{I}_{np}$.

Now we return to $\boldsymbol{X}$. As $\boldsymbol{X} \overset{d}{=} \boldsymbol{U}\boldsymbol{A}$ has the second moment, there exists the second moment of $\boldsymbol{A}$. Hence

$$\begin{aligned}
\mathcal{E}(\boldsymbol{X}) &= \mathcal{E}(\boldsymbol{U})\mathcal{E}(\boldsymbol{A}) = \boldsymbol{O}\\
\mathcal{D}(\mathrm{vec}(\boldsymbol{X})) &= \mathcal{E}[\mathrm{vec}(\boldsymbol{U}\boldsymbol{A})(\mathrm{vec}(\boldsymbol{U}\boldsymbol{A}))']\\
&= \mathcal{E}[(\boldsymbol{A}' \otimes \boldsymbol{I}_n)\,\mathrm{vec}(\boldsymbol{U})][(\mathrm{vec}(\boldsymbol{U}))'(\boldsymbol{A} \otimes \boldsymbol{I}_n)]\\
&= \mathcal{E}[(\boldsymbol{A}' \otimes \boldsymbol{I}_n)\mathcal{E}[(\mathrm{vec}(\boldsymbol{U}))(\mathrm{vec}(\boldsymbol{U}))'] \mid \boldsymbol{A}](\boldsymbol{A} \otimes \boldsymbol{I}_n)]\\
&= (1/n)\,\mathcal{E}[(\boldsymbol{A}' \otimes \boldsymbol{I}_n)(\boldsymbol{A} \otimes \boldsymbol{I}_n)] = (1/n)\mathcal{E}(\boldsymbol{A}'\boldsymbol{A}) \otimes \boldsymbol{I}_n\\
&= (1/n)\mathcal{E}(\boldsymbol{X}'\boldsymbol{X}) \otimes \boldsymbol{I}_n.
\end{aligned}$$

Noting that $\boldsymbol{x}_{(1)} \overset{d}{=} \cdots \overset{d}{=} \boldsymbol{x}_{(n)}$ and $\boldsymbol{X}'\boldsymbol{X} = \boldsymbol{x}_{(1)}\boldsymbol{x}'_{(1)} + \cdots + \boldsymbol{x}_{(n)}\boldsymbol{x}'_{(n)}$, we find $\mathcal{E}(\boldsymbol{X}'\boldsymbol{X}) = n\mathcal{E}(\boldsymbol{x}_{(1)}\boldsymbol{x}'_{(1)}) = n\boldsymbol{V}$. □

Corollary 1. *If $\boldsymbol{Y} = \boldsymbol{B}'\boldsymbol{X} + \boldsymbol{M} \sim ELS_{m\times p}(\boldsymbol{M}, \Sigma, \phi)$, where $\boldsymbol{X}$ satisfies the condition of Theorem 3.3.1, then*

$$\mathcal{E}(\boldsymbol{Y}) = \boldsymbol{M}, \qquad \mathcal{D}(\mathrm{vec}(\boldsymbol{Y})) = \boldsymbol{V} \otimes (\boldsymbol{B}'\boldsymbol{B}). \tag{3.3.15}$$

Proof. Clearly, $\mathcal{E}(\boldsymbol{Y}) = \boldsymbol{M}$. And

$$\begin{aligned}
\mathcal{D}(\mathrm{vec}(\boldsymbol{Y})) = \mathcal{D}(\mathrm{vec}(\boldsymbol{B}'\boldsymbol{X})) &= \mathcal{E}[(\mathrm{vec}(\boldsymbol{B}'\boldsymbol{X}))(\mathrm{vec}(\boldsymbol{B}'\boldsymbol{X}))']\\
&= (\boldsymbol{I}_p \otimes \boldsymbol{B}')\mathcal{E}[(\mathrm{vec}(\boldsymbol{X}))(\mathrm{vec}(\boldsymbol{X}))'](\boldsymbol{I}_p \otimes \boldsymbol{B})\\
&= (\boldsymbol{I}_p \otimes \boldsymbol{B}')(\boldsymbol{V} \otimes \boldsymbol{I}_n)(\boldsymbol{I}_p \otimes \boldsymbol{B}) = \boldsymbol{V} \otimes (\boldsymbol{B}'\boldsymbol{B}).
\end{aligned}$$ □

Naturally, we can define another kind of elliptically contoured matrix distributions as

$$\boldsymbol{Y} \overset{d}{=} \boldsymbol{M} + \boldsymbol{X}\boldsymbol{B}. \tag{3.3.16}$$

Some results about them can be found in Exercise 3.8.

3.4. Distributions of Quadratic Forms

Let $\boldsymbol{X}$ be an $n \times p$ random matrix. In this section, the distributions of $\boldsymbol{W} = \boldsymbol{X}'\boldsymbol{X}$

and their properties will be studied for $\boldsymbol{X} \in \mathfrak{F}_1$. As $\boldsymbol{W}$ is a symmetric random matrix, the distribution of $\boldsymbol{W}$ means that of

$$(w_{11}, \ldots, w_{1p}, w_{22}, \ldots, w_{2p}, \ldots, w_{p-1,p-1}, w_{p-1,p}, w_{pp}).$$

3.4.1. Densities of $\boldsymbol{W}$

First of all, we need the following lemma.

Lemma 3.4.1. *For every nonnegative Borel function $f(.)$ we have*

$$\int_{R^n} f(\boldsymbol{a}'\boldsymbol{x}, \boldsymbol{x}'\boldsymbol{x})\, d\boldsymbol{x} = \frac{\pi^{\frac{1}{2}(n-1)}}{\Gamma(\frac{1}{2}(n-1))} \int_{-\infty}^{\infty} \mathrm{d}y \int_0^{\infty} u^{\frac{1}{2}(n-1)-1} f(y \| \boldsymbol{a} \|, y^2 + u)\, du, \tag{3.4.1}$$

where $\boldsymbol{0} \neq \boldsymbol{a} \in R^n$.

Proof. Let $\boldsymbol{T} \in O(n)$ with the first row $\boldsymbol{a}' / \|\boldsymbol{a}\|$ and $\boldsymbol{y} = \boldsymbol{T}\boldsymbol{x}$. We get

$$\int_{R^n} f(\boldsymbol{a}'\boldsymbol{x}, \boldsymbol{x}'\boldsymbol{x})\, d\boldsymbol{x} = \int_{R^n} f(\|\boldsymbol{a}\| y_1, y_1^2 + \sum_2^n y_i^2)\, d\boldsymbol{y}.$$

Applying (2.4.15) $(n-1)$-times to $y_2, \ldots, y_n$ completes the proof. □

The following theorem is very useful.

Theorem 3.4.1. *Let $\mathbf{X}$ be an $n \times p$ matrix and $\boldsymbol{T} \in UT(p)$. Then*

$$\int f(\mathbf{X}'\mathbf{X})d\mathbf{X} = \frac{2^p \pi^{\frac{1}{2}np}}{\Gamma_p(\frac{1}{2}n)} \int_{D_p} \Big(\prod_{i=1}^p t_{ii}^{n-i}\Big) f(\boldsymbol{T}'\boldsymbol{T})\, d\boldsymbol{T}, \tag{3.4.2}$$

where $D_p = \{\boldsymbol{T} \mid \boldsymbol{T} \in UT(p)$ with positive diagonal elements$\}$, $\Gamma_p(.)$ is the multivariate Gamma function and $\Gamma_p(\frac{1}{2}n) = \pi^{\frac{1}{4}p(p-1)} \prod_{i=1}^p \Gamma(\frac{1}{2}(n-i+1))$.

Proof. We prove (3.4.2) by induction. When $p = 1$, (3.4.2) holds. *Assume* (3.4.2) holds for $p-1$. Then

$$\int f(\mathbf{X}'\mathbf{X})d\mathbf{X} = \int d\boldsymbol{x}_1 \int f(\mathbf{X}'\mathbf{X})\, d\boldsymbol{x}_2 \cdots d\boldsymbol{x}_p. \tag{3.4.3}$$

For a given $\boldsymbol{x}_1 \neq \boldsymbol{0}$, let $\boldsymbol{T} \in O(n)$ with the first row $\boldsymbol{x}_1' / \|\boldsymbol{x}_1\|$ and let $\boldsymbol{y}_i = \boldsymbol{T}\boldsymbol{x}_i$, $i = 2, \ldots, p$. We obtain

$$\int f\left[\begin{bmatrix} \boldsymbol{x}_1'\boldsymbol{x}_1 & \cdot & \cdot & \cdot & \boldsymbol{x}_1'\boldsymbol{x}_p \\ & \cdot & & & \\ & \cdot & & & \\ & \cdot & & & \\ & \cdot & & & \\ & \boldsymbol{x}_p'\boldsymbol{x}_p & & & \end{bmatrix}\right] d\boldsymbol{x}_2 \cdots d\boldsymbol{x}_p$$

$$= \int f \left[\begin{bmatrix} \|\boldsymbol{x}_1\|^2 & \|\boldsymbol{x}_1\| y_{12} & \cdots & \|\boldsymbol{x}_1\| y_{1p} \\ & \boldsymbol{y}_2'\boldsymbol{y}_2 & \cdots & \boldsymbol{y}_2'\boldsymbol{y}_p \\ & & \ddots & \vdots \\ & & & \boldsymbol{y}_p'\boldsymbol{y}_p \end{bmatrix} \right] d\boldsymbol{y}_2 \cdots d\boldsymbol{y}_p.$$

Let $t_{1i} = y_{1i}, i = 2, \cdots, p$; $\boldsymbol{u}_i = (y_{2i}, \cdots, y_{ni})', i = 2, \cdots, p$. Therefore

$$\int f(\boldsymbol{X}'\boldsymbol{X}) d\boldsymbol{x}_2 \cdots d\boldsymbol{x}_p = \int f \left[\begin{bmatrix} \|\boldsymbol{x}_1\|^2 & \|\boldsymbol{x}_1\| t_{12} & \cdots & \|\boldsymbol{x}_1\| t_{1p} \\ & t_{12}^2 + \boldsymbol{u}_2'\boldsymbol{u}_2 & \cdots & t_{12}t_{1p} + \boldsymbol{u}_2'\boldsymbol{u}_p \\ & & & \vdots \\ & & & t_{1p}^2 + \boldsymbol{u}_p'\boldsymbol{u}_p \end{bmatrix} \right] \prod_{j=2}^{p} (dt_{1j} d\boldsymbol{u}_j)$$

$$= \int f \left[\boldsymbol{t}_1^* \boldsymbol{t}_1^{*\prime} + \begin{bmatrix} 0 & 0 & \cdots & 0 \\ 0 & \boldsymbol{u}_2'\boldsymbol{u}_2 & \cdots & \boldsymbol{u}_2'\boldsymbol{u}_p \\ \vdots & \vdots & & \vdots \\ 0 & \boldsymbol{u}_p'\boldsymbol{u}_2 & \cdots & \boldsymbol{u}_p'\boldsymbol{u}_p \end{bmatrix} \right] \prod_{j=2}^{p} (dt_{1j} d\boldsymbol{u}_j),$$

where $\boldsymbol{t}_1^* = (\|\boldsymbol{x}_1\|, t_{12}, \ldots, t_{1p})'$. Fixing $t_{12}, \ldots, t_{1p}$ and applying the assumption of the induction to

$$\begin{bmatrix} \boldsymbol{u}_2'\boldsymbol{u}_2 & \cdots & \boldsymbol{u}_2'\boldsymbol{u}_p \\ \vdots & & \vdots \\ \boldsymbol{u}_p'\boldsymbol{u}_2 & \cdots & \boldsymbol{u}_p'\boldsymbol{u}_p \end{bmatrix},$$

we have

$$\int f(\boldsymbol{X}'\boldsymbol{X})\, d\boldsymbol{x}_2 \cdots d\boldsymbol{x}_p = 2^{p-1} \prod_{j=1}^{p-1} \frac{\pi^{\frac{1}{2}(n-j)}}{\Gamma(\frac{1}{2}(n-j))} \int_{D_{p-1}} t_{22}^{n-2} \cdots t_{pp}^{n-p}.$$

$$f\left[\boldsymbol{t}_1^*\boldsymbol{t}_1^{*\prime} + \begin{pmatrix} \boldsymbol{O} & \boldsymbol{O} \\ \boldsymbol{O} & \boldsymbol{T}_1'\boldsymbol{T}_1 \end{pmatrix} \right] d\boldsymbol{T}_1 dt_{12} \cdots dt_{1p},$$

where $\boldsymbol{T}_1 \in UT(p-1)$. Putting the above integral into (3.4.3) and applying (2.4.15) to $\boldsymbol{x}_1$, we find

$$\int f(\boldsymbol{X}'\boldsymbol{X})\, d\boldsymbol{X} = \frac{2^{p-1}\pi^{\frac{1}{2}np}}{\Gamma_p(\frac{1}{2}n)} \int_0^\infty \int_{D_{p-1}} \left(\prod_{j=2}^{p} t_{jj}^{n-j} \right) y^{\frac{1}{2}n-1}.$$

$$\cdot f\left[\begin{bmatrix} y^{\frac{1}{2}} \\ t_{12} \\ \vdots \\ t_{1p} \end{bmatrix}(y^{\frac{1}{2}}, t_{12}, \cdots, t_{1p}) + \begin{pmatrix} O & O \\ O & T_1'T_1 \end{pmatrix}\right] dy dt_{12} \cdots dt_{1p} dT_1.$$

Formula (3.4.2) follows by making the transformation $t_{11} = y^{\frac{1}{2}}$. □

Corollary 1. *Let* $X \sim N_{n \times p}(O, I_n \otimes I_p)$ *and* $X'X = T'T$, *where* $T = (t_{ij}) \in UT(p)$ *with positive diagonal elements. Then*

(i) $t_{ii}^2 \sim \chi^2_{n-i+1}$, $i = 1, \ldots, p$;

(ii) $t_{ij} \sim N(0, 1)$, $i > j$, $i, j = 1, \ldots, p$;

(iii) $\{t_{ij}, i \geqslant j\}$ *are independent.*

Proof. From assumption,

$$f(X'X) = (2\pi)^{-\frac{1}{2}np} \exp(-\tfrac{1}{2}X'X).$$

By using the Jacobian of the transformation $X = UT$ given in Exercise 1.22, we can find the joint density of U and T as follows:

$$(2\pi)^{-\frac{1}{2}np} g_{n,p}(U)\left(\prod_{i=1}^{p} t_{ii}^{n-i}\right)\exp(-\tfrac{1}{2}T'T).$$

Integrating the above function with respect to U we obtain that the marginal density of T is

(3.4.4) $$c(2\pi)^{-\frac{1}{2}np}\prod_{i=1}^{p} t_{ii}^{n-i}\exp(-\tfrac{1}{2}T'T) = c(2\pi)^{-\frac{1}{2}np}\prod_{i=1}^{p} t_{ii}^{n-i}\exp\left(-\tfrac{1}{2}\sum_{i=1}^{p}\sum_{j=1}^{p} t_{ij}^2\right)$$

where

(3.4.5) $$c = \int_{U'U=I_p} g_{n,p}(U)dU.$$

As (3.4.4) is a density, the constant c can be calculated as follows:

$$c^{-1} = (2\pi)^{-\frac{1}{2}np}\int_{\substack{t_{ii}>0 \\ i=1,\ldots,p.}}\int_{\substack{t_{ij}\in R1 \\ i<j}} \prod_{i=1}^{p} t_{ii}^{n-i}\left(\prod_{i=1}^{p}\prod_{j=1}^{p}\exp(-\tfrac{1}{2}t_{ij}^2)dt_{ij}\right)$$

$$= (2\pi)^{-\frac{1}{2}np+\frac{1}{4}p(p-1)}\prod_{i=1}^{p}\left[\int_0^\infty t_{ii}^{n-i}\exp(-\tfrac{1}{2}t_{ii}^2)dt_{ii}\right] \quad (\text{let } y_i = \tfrac{1}{2}t_{ii}^2)$$

$$= 2^{-p}\pi^{-\frac{1}{2}np+\frac{1}{4}p(p-1)}\prod_{i=1}^{p}\left[\int_0^\infty y_i^{\frac{1}{2}(n-i+1)-1}\exp(-y_i)dy_i\right]$$

$$= 2^{-p}\pi^{\frac{1}{4}p(p-1)-\frac{1}{2}np}\prod_{i=1}^{p}\Gamma(\tfrac{1}{2}(n-i+1))$$

$$= 2^{-p}\pi^{-\frac{1}{2}np}\Gamma_p(\tfrac{1}{2}n).$$

Combining (3.4.5) and (3.4.4) we find out (i), (ii), and (iii) all together. □

As a series of results, we have

Corollary 2.

$$\int_{U'U=I_p} g_{n,p}(U)\,dU = \frac{2^p\pi^{\frac{1}{2}np}}{\Gamma_p(\frac{1}{2}n)}. \tag{3.4.6}$$

Corollary 3. *The density of* U (i.e., *the joint density of* u_{ij}, $j = 1, \dots, p$; $i = 1, \dots, n - j$) *is*

$$\frac{\Gamma_p(\frac{1}{2}n)}{2^p\pi^{\frac{1}{2}np}} g_{n,p}(U). \tag{3.4.7}$$

The decomposition $X'X = T'T$ in Corollary 1 is called the Bartlett decomposition which is a well-known fact.

Corollary 4. *Assume* $X \sim N_{n\times p}(O, I_n \otimes I_p)$ *and* $W = X'X$. *Then*

$$|W| \stackrel{d}{=} \prod_{j=1}^{p} y_j \tag{3.4.8}$$

where $y_1, \dots, y_p$ *are independent and* $y_j \sim \chi^2_{n-j+1}$, $j = 1, \dots, p$.

Proof. By Corollary 1, we have

$$|W| = |X'X| = |T'T| = |T|^2 = \prod_{j=1}^{p} t_{jj}^2.$$

Let $y_j = t_{jj}^2$, $j = 1, \dots, p$. So the corollary follows. □

The determinant of $W = X'X$ is called the generalized variance of X.

Theorem 3.4.2. *Assume* $X \sim LS_{n\times p}(f)$, $n \geqslant p$, i.e., X *has a density* $f(X'X)$. *Then the density of* $W = X'X$ *is*

$$\frac{\pi^{\frac{1}{2}np}}{\Gamma_p(\frac{1}{2}n)} |W|^{\frac{1}{2}(n-p-1)} f(W), \text{ for } W > 0. \tag{3.4.9}$$

Proof. For every nonnegative Borel function $h(.)$ we have

$$E(h(W)) = E(h(X'X)) = \int h(X'X) f(X'X)\,dX$$

$$= \frac{2^p\pi^{\frac{1}{2}np}}{\Gamma_p(\frac{1}{2}n)} \int_{D_p} h(T'T) f(T'T) \left(\prod_{j=1}^{p} t_{jj}^{n-j} \right) dT$$

by (3.4.2). Making the transformation $T \to W$, $W = T'T$ whose Jacobian is $2^{-p} \prod_{j=1}^{p} t_{jj}^{-(p-j+1)}$ and noting $|W| = |T'T| = |T|^2 = \prod_{j=1}^{p} t_{jj}^2$, we have

$$E(h(W)) = \frac{\pi^{\frac{1}{2}np}}{\Gamma_p(\frac{1}{2}n)} \int |W|^{\frac{1}{2}(n-p-1)} f(W)\, h(W)\, dW,$$

which completes the proof. □

Corollary 1. *Assume* $X \sim N_{n\times p}(O, I_n \otimes \Sigma)$ *with* $\Sigma > 0$. *Then the density of* $W = X'X$ *is*

$$(3.4.10)\qquad \frac{1}{2^{\frac{1}{2}np}\Gamma_p(\frac{1}{2}n)}|\Sigma|^{-\frac{1}{2}n}|\boldsymbol{W}|^{\frac{1}{2}(n-p-1)}\exp(-\tfrac{1}{2}\Sigma^{-1}\boldsymbol{W})\ \textit{for}\ \boldsymbol{W}>0.$$

Proof. When $\boldsymbol{Y} \sim N_{n\times p}(\boldsymbol{O}, \boldsymbol{I}_n \otimes \boldsymbol{I}_p)$, we can directly obtain the density of $\boldsymbol{V} = \boldsymbol{Y}'\boldsymbol{Y}$ as

$$(3.4.11)\qquad \frac{1}{2^{\frac{1}{2}np}\Gamma_p(\frac{1}{2}n)}|\boldsymbol{V}|^{\frac{1}{2}(n-p-1)}\exp(-\tfrac{1}{2}\boldsymbol{V})$$

by (3.4.6). Taking the transformation $\boldsymbol{W} = \Sigma^{\frac{1}{2}}\boldsymbol{V}\Sigma^{\frac{1}{2}} = \Sigma^{\frac{1}{2}}\boldsymbol{Y}'\boldsymbol{Y}\Sigma^{\frac{1}{2}} \stackrel{d}{=} \boldsymbol{X}'\boldsymbol{X}$, the density (3.4.10) follows. □

The distribution of (3.4.10) is called the Wishart distribution, which was obtained by Wishart in 1928, and we write $W_p(n, \Sigma)$. This is a very important distribution, so there is a rich bibliography about it and there are many ways to derive it.

Return to the case of $\boldsymbol{X} \sim LS_{n\times p}(f)$. Partition $\boldsymbol{X}$ into $\boldsymbol{X} = (\boldsymbol{X}_1', \dots, \boldsymbol{X}_m')'$, where $\boldsymbol{X}_i$: $n_i \times p$, $n_i \geqslant p$, $i = 1, \dots, m$. By the same technique the joint density of $\boldsymbol{W}_i = \boldsymbol{X}_i'\boldsymbol{X}_i$, $i = 1, \dots, m$, is

$$(3.4.12)\qquad \prod_1^m \left[c_{n_i,p}|\boldsymbol{W}_i|^{\frac{1}{2}(n_i-p-1)}\right] f\left(\sum_{i=1}^m \boldsymbol{W}_i\right)$$

where

$$(3.4.13)\qquad c_{k,p} = \frac{\pi^{\frac{1}{2}kp-\frac{1}{4}p(p-1)}}{\prod_{j=1}^p \Gamma(\frac{1}{2}(k-j+1))} = \frac{\pi^{\frac{1}{2}kp}}{\Gamma_p(\frac{1}{2}k)}.$$

Thus, the marginal density of $\boldsymbol{W}_1, \cdots, \boldsymbol{W}_k$ $(1 \leqslant k < m)$ is

$$(3.4.14)\qquad c_{n^*,p}\prod_{i=1}^k \left[c_{n_i,p}|\boldsymbol{W}_i|^{\frac{1}{2}(n-p-1)}\right]\cdot \int_{\boldsymbol{V}>0} f\left(\sum_1^k \boldsymbol{W}_i + \boldsymbol{V}\right)|\boldsymbol{V}|^{\frac{1}{2}(n^*-p-1)}d\boldsymbol{V}$$

with $n^* = n_{k+1} + \cdots + n_m$. We write $(\boldsymbol{W}_1, \dots, \boldsymbol{W}_m) \sim MG_{m,p}(\frac{1}{2}n_1, \dots, \frac{1}{2}n_m; f)$ and $(\boldsymbol{W}_1, \dots, \boldsymbol{W}_k) \sim MG_{k+1,p}(\frac{1}{2}n_1, \cdots, \frac{1}{2}n_k; \frac{1}{2}n^*; f)$ (cf. Section 2.8.1).

3.4.2. A Multivariate Analogue to Cochran's Theorem

In Section 2.8.3, we discuss Cochran's theorem for the case of ECD. Now we wish to extend Cochran's to the multivariate case.

Theorem 3.4.3. *Assume that* $\boldsymbol{X} \sim LS_{n\times p}(\phi)$, $P(\boldsymbol{X} = \boldsymbol{O}) = 0$, *and* $\boldsymbol{A}$ *is an* $n \times n$ *symmetric matrix. Then* $\boldsymbol{X}'\boldsymbol{A}\boldsymbol{X} \sim MG_{2,p}(\frac{1}{2}k; \frac{1}{2}(n-k); \phi)$ *if and only if* $\boldsymbol{A}^2 = \boldsymbol{A}$ *and* $\mathrm{rk}(\boldsymbol{A}) = k$.

Proof. The "if" part is obvious. Assume $\boldsymbol{X}'\boldsymbol{A}\boldsymbol{X} \sim MG_{2,p}(\frac{1}{2}k; \frac{1}{2}(n-k); \phi)$. As $P(\boldsymbol{X} = \boldsymbol{O}) = 0$, there exists $\boldsymbol{l} \in R^1$ such that $P(\boldsymbol{y} = \boldsymbol{0}) = 0$, where $\boldsymbol{y} = \boldsymbol{X}\boldsymbol{l} \sim S_n(\phi)$ with $\psi(\boldsymbol{t}'\boldsymbol{t}) = \phi(\boldsymbol{t}'\boldsymbol{t}\boldsymbol{l}\boldsymbol{l}')$. From assumption, $\boldsymbol{X}'\boldsymbol{A}\boldsymbol{X} \stackrel{d}{=} \boldsymbol{X}_1'\boldsymbol{X}_1$, where $\boldsymbol{X}_1$ is the first k rows of $\boldsymbol{X}$. Thus

$$\boldsymbol{y}'\boldsymbol{A}\boldsymbol{y} = \boldsymbol{l}'\boldsymbol{X}'\boldsymbol{X}\boldsymbol{l} \stackrel{d}{=} \boldsymbol{l}'\boldsymbol{X}_1'\boldsymbol{X}_1\boldsymbol{l} \sim G_2(\tfrac{1}{2}k; \tfrac{1}{2}(n-k); \psi)$$

because

$$\begin{bmatrix} y_1 \\ y_2 \end{bmatrix} = y = \begin{bmatrix} X_1 l \\ X_2 l \end{bmatrix}.$$

By Theorem 2.8.5, $A^2 = A$ and $\text{rk}(A) = k$. □

Similarly, by Theorem 2.8.6 we can obtain the following theorem.

Theorem 3.4.4. *Assume that* $X \sim LS_{n\times p}(\varphi)$, $P(X = O) = 0$ *and* $A_1, \dots, A_k$ *are symmetric matrices. Then* $(X'A_1X, \dots, X'A_kX) \sim \text{MG}_{k+1,p}(\frac{1}{2}n_1, \dots, \frac{1}{2}n_k; \frac{1}{2}n^*; \varphi)$ *with* $n^* = n - n_1 - \cdots - n_k$ *if and only if* $A_iA_j = \delta_{ij}A_i$ *and* $\text{rk}(A_i) = n_i$, $i, j = 1, \dots, k$.

3.5. Some Related Distributions with Spherical Matrix Distributions

In this section, we discuss some related distributions with spherical matrix distributions such as the matrix beta distribution, the matrix t-distribution, the matrix F-distribution, and the distribution of characteristic roots of quadratic form.

3.5.1. The matrix Variate Beta Distributions

Definition 3.5.1. Assume $(W_1, W_2) \sim \text{MG}_{2,p}(\frac{1}{2}n_1, \frac{1}{2}n_2; f, n_1 \geqslant p, n_2 \geqslant p$. The distribution of $B = (W_1 + W_2)^{-\frac{1}{2}} W_1 (W_1 + W_2)^{-\frac{1}{2}}$ is called the *matrix variate beta distribution*. We write $B \sim B_p(\frac{1}{2}n_1, \frac{1}{2}n_2)$. When $p = 1$, it is reduced to $B(\frac{1}{2}n_1, \frac{1}{2}n_2)$.

Theorem 3.5.1. *The density of* $B \sim B_p(\frac{1}{2}n_1, \frac{1}{2}n_2)$ *is*

$$\text{(3.5.1)} \qquad \frac{\Gamma_p(\frac{1}{2}(n_1+n_2))}{\Gamma_p(\frac{1}{2}n_1)\,\Gamma_p(\frac{1}{2}n_2)} |B|^{\frac{1}{2}(n_1-p-1)} |I - B|^{\frac{1}{2}(n_2-p-1)} \qquad 0 < B < I_p.$$

Proof. For each nonnegative Borel function $h(.)$ we have

$$\begin{aligned} E(h(B)) &= E(h((W_1+W_2)^{-\frac{1}{2}} W_1 (W_1+W_2)^{-\frac{1}{2}})) \\ &= \int h((W_1+W_2)^{-\frac{1}{2}} W_1 (W_1+W_2)^{-\frac{1}{2}}) \\ &\quad \cdot c_{n_1,p} c_{n_2,p} |W_1|^{\frac{1}{2}(n_1-p-1)} |W_2|^{\frac{1}{2}(n_2-p-1)} \cdot f(W_1+W_2)\,dW_1\,dW_2. \end{aligned}$$

Let $B = (W_1 + W_2)^{-\frac{1}{2}} W_1 (W_1 + W_2)^{-\frac{1}{2}}$ and $W = W_1 + W_2$. Then $W_1 = W^{-\frac{1}{2}} B W^{-\frac{1}{2}}$ and $W_2 = W^{\frac{1}{2}}(I - B)W^{\frac{1}{2}}$, and the Jacobian is

$$\begin{aligned} &|J((W_1, W_2) \to (W, B))| \\ &\quad = |J((W_1, W_2) \to (W_1, W))| |J((W_1, W) \to (B, W))| = |W|^{\frac{1}{2}(p+1)}. \end{aligned}$$

Noting $c_{k,p} = \pi^{\frac{1}{2}kp} / \Gamma_p(\frac{1}{2}k)$, *we have*

$$E(h(B)) = \frac{\Gamma_p(\frac{1}{2}(n_1+n_2))}{\Gamma_p(\frac{1}{2}n_1)\,\Gamma_p(\frac{1}{2}n_2)} \int_{0<B<I} h(B)\,|B|^{\frac{1}{2}(n_1-p-1)} |I - B|^{\frac{1}{2}(n_2-p-1)} dB$$

$$\cdot \frac{\pi^{\frac{1}{2}p(n_1+n_2)}}{\Gamma_p\left(\frac{1}{2}(n_1+n_2)\right)} \int\limits_{W>0} |W|^{\frac{1}{2}(n_1+n_2)-\frac{1}{2}(p+1)} f(W)\, dW.$$

The second factor on the rigth-hand side is equal to 1 (cf. (3.4.7)). □

Note that the distribution of B is independent of f, which is an interesting fact. In this section, we will meet with many statistics whose distributions are independent of f. The general theory of invariant statistics will be discussed in Chapter V.

Corollary 1. *Using the notations of Definition 3.5.1, we have* $W_1 + W_2$ *and* B are independent.

Corollary 2. *If* $B \sim B_p(\frac{1}{2}n_1, \frac{1}{2}n_2)$, *then* $I_p \sim B \sim B_p(\frac{1}{2}n_2, \frac{1}{2}n_1)$.

Recall the Bartlett decomposition of the Wishart distribution (Corollary 1 of Theorem 3.4.1). Are there any similar results for the matrix variate beta distribution? The following theorem is from Kshirsagar's (1961, 1972).

Theorem 3.5.2. *If* $B \sim B_p(\frac{1}{2}n_1, \frac{1}{2}n_2)$ *and* $B = T'T$, *where* $T \in UT(p)$, *then* $t_{11}, \ldots, t_{pp}$ *are independent and* $t_{ii}^2 \sim B(\frac{1}{2}(n_1 - i + 1), \frac{1}{2}n_2)$, $i = 1, \ldots, p$.

Proof. By means of a canonical technique, we can find the density of T as follows:

$$g(T;\, p, n_1, n_2) = \frac{2^p\Gamma_p(\frac{1}{2}(n_1+n_2))}{\Gamma_p(\frac{1}{2}n_1)\,\Gamma_p(\frac{1}{2}n_2)} \prod_{j=1}^{p} t_{jj}^{n_1-j} |I - T'T|^{\frac{1}{2}(n_2-p-1)}. \tag{3.5.2}$$

Partition T as

$$T = \begin{bmatrix} t_{11} & t' \\ O & T_{22} \end{bmatrix}$$

where T_{22}:$(p-1) \times (p-1)$ and $T_{22} \in UT(p-1)$. Note that

$$|I - T'T| = \begin{vmatrix} 1 - t_{11}^2 & -t_{11}t' \\ -t_{11}t & I - tt' - T_{22}'T_{22} \end{vmatrix}$$

$$= (1 - t_{11}^2)|I - T_{22}'T_{22}| \left[1 - \frac{1}{1 - t_{11}^2} t(I - T_{22}'T_{22})^{-1} t' \right].$$

By making a transformation from t_{11}, T_{22}, t to t_{11}, T_{22}, v, where

$$v = \frac{1}{(1 - t_{11}^2)^{\frac{1}{2}}} (I - T_{22}'T_{22})^{-\frac{1}{2}} t,$$

whose Jacobian is $(1 - t_{11}^2)^{\frac{1}{2}(p+1)} |I - T_{22}'T_{22}|^{\frac{1}{2}}$, the joint density of t_{11}, T_{22} and v will be

$$\frac{2^p\Gamma_p(\frac{1}{2}(n_1+n_2))}{\Gamma_p(\frac{1}{2}n_1)\,\Gamma_p(\frac{1}{2}n_2)} t_{11}^{n_1-1} (1 - t_{11}^2)^{\frac{1}{2}n_2-1} \prod_{i=2}^{p} t_{ii}^{n_1-i} |I - T_{22}'T_{22}|^{\frac{1}{2}(n_2-p)} (1 - v'v)^{\frac{1}{2}(n_2-p-1)}.$$

This shows immediately that t_{11}, T_{22}, and v are independent and that $t_{11}^2 \sim B(\frac{1}{2}n_1, \frac{1}{2}n_2)$. The density of T_{22} is proportional to

$$\prod_{i=2}^{p} t_{ii}^{n_1-i} |\boldsymbol{I}-\boldsymbol{T}_{22}\boldsymbol{T}_{22}'|^{\frac{1}{2}(n_2-p)}$$

which has the same form as the density (3.5.2) for $\boldsymbol{T}$, with p replaced by $p-1$ and $\boldsymbol{n}_1$ replaced by $\boldsymbol{n}_1-1$. Hence the density of $\boldsymbol{T}_{22}$ is $g(\boldsymbol{T}_{22}; p-1, n_1-1, n_2)$. Repeating the above argument for this density shows that $t_{22}^2 \sim B(\frac{1}{2}(n_1-1), \frac{1}{2}n_2)$, and is independent of $t_{33}, ..., t_{pp}$. Repeating this argument successively completes the proof. □

3.5.2. The Matrix Variate Dirichlet Distributions

As an extension of the matrix variate beta distribution, the matrix Dirichlet distribution is defined as follows:

Definition 3.5.2. Assume $(\boldsymbol{W}_1, ..., \boldsymbol{W}_{k+1}) \sim MG_{k+1,p}(\frac{1}{2}n_1, ..., \frac{1}{2}n_{k+1}, f)$, $n_j \geqslant p$, $j = 1, ..., k+1$, $n_{k+1} \geqslant 1$, and $\boldsymbol{W} = \boldsymbol{W}_1 + ... + \boldsymbol{W}_{k+1}$. The distribution of $(\boldsymbol{D}_1, ..., \boldsymbol{D}_k) = (\boldsymbol{W}^{-\frac{1}{2}}\boldsymbol{W}_1\boldsymbol{W}^{-\frac{1}{2}}, ..., \boldsymbol{W}^{-\frac{1}{2}}\boldsymbol{W}_k\boldsymbol{W}^{-\frac{1}{2}})$ is called the *matrix variate Dirichlet distribution* and write $(\boldsymbol{D}_1, ..., \boldsymbol{D}_k) \sim MD_{k+1,p}(\frac{1}{2}n_1, ..., \frac{1}{2}n_k; \frac{1}{2}n_{k+1})$.

By means of an argument similar to that in the proof of Theorem 3.5.1 we obtain the following theorem.

Theorem 3.5.3. *The density of* $(\boldsymbol{D}_1, ..., \boldsymbol{D}_k)$ *is*

$$\frac{\Gamma_p(\frac{1}{2}(n_1+, ..., n_{k+1}))}{\prod_{i=1}^{k+1}\Gamma_p(\frac{1}{2}n_i)}\left(\prod_{i=1}^{k}|\boldsymbol{D}_i|^{\frac{1}{2}(n_i-p-1)}\right)\left|\boldsymbol{I}-\sum_{1}^{k}\boldsymbol{D}_i\right|^{\frac{1}{2}(n_{k+1}-p-1)} \tag{3.5.3}$$

$$\boldsymbol{D}_i > 0,\ i = 1, ..., k,\ \boldsymbol{D}_1 + \cdots + \boldsymbol{D}_k < \boldsymbol{I}_p.$$

Thus, the density is independent of f.

The following lemma is an application of the matrix variate Dirichlet distribution.

Lemma 3.5.1. *Assume* $\boldsymbol{U} = \begin{bmatrix}\boldsymbol{U}_1\\ \vdots \\ \boldsymbol{U}_r\end{bmatrix} \sim \mathcal{U}_{n,p}$, *where* $\boldsymbol{U}_i$: $n_i \times p$, $n_i \geqslant p$, $i = 1, ..., r$. *Then*

$$\mathrm{U} \overset{d}{=} \begin{bmatrix}\boldsymbol{V}_1\boldsymbol{B}_1\\ \vdots \\ \boldsymbol{V}_r\boldsymbol{B}_r\end{bmatrix}, \tag{3.5.4}$$

where $\boldsymbol{V}_1, ..., \boldsymbol{V}_r, (\boldsymbol{B}_1, ..., \boldsymbol{B}_r)$ *are independent,* $\boldsymbol{V}_i \sim \mathcal{U}_{n_i,p}$, $i = 1, ..., r$, $(\boldsymbol{B}_1'\boldsymbol{B}_1, ..., \boldsymbol{B}_r'\boldsymbol{B}_r) \sim MD_{r,p}(\frac{1}{2}n_1, ..., \frac{1}{2}n_{r-1}, \frac{1}{2}n_r)$, *and* $\boldsymbol{B}_1'\boldsymbol{B}_1 + \cdots + \boldsymbol{B}_r'\boldsymbol{B}_r = \boldsymbol{I}_p$.

Proof. Let $\boldsymbol{X} = (\boldsymbol{X}_1', ..., \boldsymbol{X}_r')' \sim N_{n\times p}(\boldsymbol{O}, \boldsymbol{I}_n \otimes \boldsymbol{I}_p)$, $\boldsymbol{X}_i$: $n_i \times p$, $i = 1, ..., r$. By (3.2.3)

$$\boldsymbol{U} \overset{d}{=} \boldsymbol{X}(\boldsymbol{X}'\boldsymbol{X})^{-\frac{1}{2}} = \begin{bmatrix}\boldsymbol{X}_1(\boldsymbol{X}'\boldsymbol{X})^{-\frac{1}{2}}\\ \vdots \\ \boldsymbol{X}_r(\boldsymbol{X}'\boldsymbol{X})^{-\frac{1}{2}}\end{bmatrix} = \begin{bmatrix}\boldsymbol{X}_1(\boldsymbol{X}_1'\boldsymbol{X}_1)^{-\frac{1}{2}}[(\boldsymbol{X}_1'\boldsymbol{X}_1)^{\frac{1}{2}}(\boldsymbol{X}'\boldsymbol{X})^{-\frac{1}{2}}]\\ \vdots \\ \boldsymbol{X}_r(\boldsymbol{X}_r'\boldsymbol{X}_r)^{-\frac{1}{2}}[(\boldsymbol{X}_r'\boldsymbol{X}_r)^{\frac{1}{2}}(\boldsymbol{X}'\boldsymbol{X})^{-\frac{1}{2}}]\end{bmatrix}.$$

Let $\boldsymbol{V}_i = \boldsymbol{X}_i(\boldsymbol{X}_i'\boldsymbol{X}_i)^{-\frac{1}{2}}$, $\boldsymbol{B}_i = (\boldsymbol{X}_i'\boldsymbol{X}_i)^{\frac{1}{2}}(\boldsymbol{X}'\boldsymbol{X})^{-\frac{1}{2}}$, $i = 1, ..., r$. It is easy to check that $\{\boldsymbol{V}_i, \boldsymbol{B}_i, i = 1, ..., r\}$ satisfies all conditions in the statement of the theorem. □

3.5.3. The Matrix Variate *t*-Distributions

The matrix variate t-distribution is an extension of the usual t-distribution in the case of a matrix.

Definition 3.5.3. Assume $X = \begin{bmatrix} X_1 \\ X_2 \end{bmatrix} \sim SS_{n\times p}(f)$, X_i: $n_i \times p, i = 1, 2, n_2 \geqslant p$. Let $T = (n_2)^{\frac{1}{2}} X_1 (X_2' X_2)^{-\frac{1}{2}}$. We say that T is distributed according to a matrix variate t-distribution and write $T \sim MT(p, n_1, n_2)$.

Theorem 3.5.4. *The density of* $T \sim MT(p, n_1, n_2)$ *is*

$$(n_2\pi)^{-\frac{1}{2}n_1 p} \prod_{i=1}^{p} \frac{\Gamma(\frac{1}{2}(n_1 + n_2 - i + 1))}{\Gamma(\frac{1}{2}(n_2 - i + 1))} |I_p + \frac{1}{n_2} T'T|^{-\frac{1}{2}(n_1+n_2)} \tag{3.5.5}$$

which is independent of f.

Proof. The density of X is $f(\lambda(X'X)) = f(\lambda(X_1'X_1 + X_2'X_2))$. By Theorem 3.4.2, the joint density of X_1 and $B = X_2'X_2$ is

$$c_{n_2,p} |B|^{\frac{1}{2}(n_2 - p - 1)} f(\lambda(X_1'X_1 + B)), \tag{3.5.6}$$

where $c_{n_2,p}$ is defined by (3.4.13). For each nonnegative Borel function $h(.)$ we have

$$E(h(T)) = c_{n_2,p} \int h(n_2^{\frac{1}{2}} X_1 B^{-\frac{1}{2}}) |B|^{\frac{1}{2}(n_2-p-1)} f(\lambda(X_1'X_1 + B)) dX_1 dB.$$

Making the transformation $T = n_2^{\frac{1}{2}} X_1 B^{-\frac{1}{2}}$, we find the corresponding Jacobian $J(T \to X_1) = n_2^{\frac{1}{2}n_1 p} |B|^{-\frac{1}{2}n_1}$, $X_1 = n_2^{-\frac{1}{2}} T B^{\frac{1}{2}}$, $X_1'X_1 = n_2^{-1} B^{\frac{1}{2}} T'T B^{\frac{1}{2}}$ and $X_1'X_1 + B = B^{\frac{1}{2}}(I_p + n_2^{-1} T'T) B^{\frac{1}{2}}$. Hence

$$E(h(T)) = c_{n_2,p} \cdot n_2^{-\frac{1}{2}n_1 p} \int h(T) |B|^{\frac{1}{2}(n_1+n_2-p-1)} f[\lambda(B^{\frac{1}{2}}(I_p + n_2^{-1} T'T) B^{\frac{1}{2}})] dT dB.$$

Note

$$\lambda[B^{\frac{1}{2}}(I_p + n_2^{-1}T'T)B^{\frac{1}{2}}] = \lambda[(I_p + n_2^{-1}T'T)B] = \lambda[(I_p + n_2^{-1}T'T)^{\frac{1}{2}} \cdot B(I_p + n_2^{-1}T'T)^{\frac{1}{2}}],$$

and let

$$W = (I_p + n_2^{-1}T'T)^{\frac{1}{2}} B (I_p + n_2^{-1}T'T)^{\frac{1}{2}}.$$

By Example 1.6.2

$$J(W \to B) = |I_p + n_2^{-1}T'T|^{-\frac{1}{2}(p+1)},$$

then we have

$$E(h(T)) = c_{n_2,p} \cdot n_2^{-\frac{1}{2}n_1 p} \int h(T) |I_p + n_2^{-1}T'T|^{-\frac{1}{2}(n_1+n_2)} dT$$

$$\cdot \int_{W>0} f(\lambda(W)) |W|^{(n_1+n_2-p-1)/2} dW.$$

By Theorem 3.4.2, the second factor on the right side of the above formula is equal to $c_{n_1+n_2 p}^{-1}$ and (3.5.5) follows by

$$c_{n_2,p} c_{n_1+n_2,p}^{-1} n_2^{\frac{1}{2}n_1 p} = (n_2\pi)^{-\frac{1}{2}n_1 p} \prod_{i=1}^{p} [\Gamma(\frac{1}{2}(n_1+n_2-i+1)) / \Gamma(\frac{1}{2}(n_2 - i + 1))]. \qquad \square$$

When $n_1 = 1$, (3.5.5) is reduced to

$$(n_2\pi)^{-\frac{1}{2}p} \prod_{i=1}^{p} \frac{\Gamma(\frac{1}{2}(n_2 - i + 2))}{\Gamma(\frac{1}{2}(n_2 - i + 1))} (1 + n_2^{-1}\boldsymbol{t}'\boldsymbol{t})^{-\frac{1}{2}(n_2+1)} \tag{3.5.7}$$

which is the multivariate t-distribution defined in Example 2.6.2. When $n_1 = p = 1$, (3.5.5) becomes the usual t-distribution.

3.5.4. The Matrix Variate F-Distributions

Definition 3.5.4. Assume $\boldsymbol{X} = \begin{bmatrix} \boldsymbol{X}_1 \\ \boldsymbol{X}_2 \end{bmatrix} \sim SS_{n \times p}(f)$ with $\boldsymbol{X}_i$: $n_i \times p$, $n_2 \geqslant p$. Let $\boldsymbol{F} = (n_2/n_1)(\boldsymbol{X}_2'\boldsymbol{X}_2)^{-\frac{1}{2}}(\boldsymbol{X}_1'\boldsymbol{X}_1)(\boldsymbol{X}_2'\boldsymbol{X}_2)^{-\frac{1}{2}}$. We say that $\boldsymbol{F}$ is distributed according to matrix F-distribution and write $\boldsymbol{F} \sim MF(p, n_1, n_2)$.

Theorem 3.5.5. *The density of* $\boldsymbol{F} \sim MF(p, n_1, n_2)$ *is*

$$\frac{\Gamma_p(\frac{1}{2}(n_1 + n_2))}{\Gamma_p(\frac{1}{2}n_1)\,\Gamma_p(\frac{1}{2}n_2)} (n_1/n_2)^{\frac{1}{2}n_1 p} |\boldsymbol{F}|^{\frac{1}{2}(n_1 - p - 1)} |\boldsymbol{I} + (n_1/n_2)\boldsymbol{F}|^{-\frac{1}{2}(n_1+n_2)}, \quad \boldsymbol{F} > 0, \tag{3.5.8}$$

which is independent of f.

Proof. Let $\boldsymbol{W}_1 = \boldsymbol{X}_1'\boldsymbol{X}_1$ and $\boldsymbol{W}_2 = \boldsymbol{X}_2'\boldsymbol{X}_2$. The joint density of $\boldsymbol{W}_1$ *and* $\boldsymbol{W}_2$ is

$$c_{n_1,p}\, c_{n_2,p} |\boldsymbol{W}_1|^{\frac{1}{2}(n_1 - p - 1)} |\boldsymbol{W}_2|^{\frac{1}{2}(n_2 - p - 1)} f(\lambda(\boldsymbol{W}_1 + \boldsymbol{W}_2))$$

by (3.4.8). For each nonnegative Borel function $h(.)$

$$E(h(\boldsymbol{F})) = c_{n_1,p}\, c_{n_2,p} \int |\boldsymbol{W}_1|^{\frac{1}{2}(n_1 - p - 1)} |\boldsymbol{W}_2|^{\frac{1}{2}(n_2 - p - 1)} f(\lambda(\boldsymbol{W}_1 + \boldsymbol{W}_2))$$
$$\cdot\, h((n_2/n_1)\boldsymbol{W}_2^{-\frac{1}{2}}\boldsymbol{W}_1\boldsymbol{W}_2^{-\frac{1}{2}})\,d\boldsymbol{W}_1 d\boldsymbol{W}_2.$$

Taking the transformation

$$\boldsymbol{F} = (n_2/n_1)\boldsymbol{W}_2^{-\frac{1}{2}}\boldsymbol{W}_1\boldsymbol{W}_2^{-\frac{1}{2}}$$

whose Jacobian is $J(\boldsymbol{F} \to \boldsymbol{W}_1) = (n_2/n_1)^{\frac{1}{2}p(p+1)} |\boldsymbol{W}_2|^{-\frac{1}{2}(p+1)}$, we have

$$E(h(\boldsymbol{F})) = c_{n_1,p}\, c_{n_2,p} (n_1/n_2)^{\frac{1}{2}n_1 p} \int |\boldsymbol{F}_1|^{\frac{1}{2}(n_1 - p - 1)} |\boldsymbol{W}_2|^{\frac{1}{2}(n_1 + n_2 - p - 1)} h(\boldsymbol{F}).$$
$$\cdot f[\lambda(\boldsymbol{W}_2^{\frac{1}{2}}(\boldsymbol{I} + (n_1/n_2)\boldsymbol{F})\boldsymbol{W}_2^{\frac{1}{2}})]\ d\boldsymbol{F}d\boldsymbol{W}_2$$
$$= c_{n_1,p}\, c_{n_2,p} (n_1/n_2)^{\frac{1}{2}n_1 p} \int |\boldsymbol{F}_1|^{\frac{1}{2}(n_1 - p - 1)} |\boldsymbol{W}_2|^{\frac{1}{2}(n_1 + n_2 - p - 1)} h(\boldsymbol{F})$$
$$\cdot\, f(\boldsymbol{I} + (n_1/n_2)\boldsymbol{F})^{\frac{1}{2}} \boldsymbol{W}_2(\boldsymbol{I} + (n_1/n_2)\boldsymbol{F})^{\frac{1}{2}})]\ d\boldsymbol{F}d\boldsymbol{W}_2.$$

Let $\boldsymbol{W} = (\boldsymbol{I} + (n_1/n_2)\boldsymbol{F})^{\frac{1}{2}} \boldsymbol{W}_2(\boldsymbol{I} + (n_1/n_2)\boldsymbol{F})^{\frac{1}{2}}$. By making the same argument as that in the proof of Theorem 3.5.4, (3.5.8) follows. □

When $p = 1$, (3.5.8) is reduced to the usual F-distribution.

3.5.5. Some Inverted Matrix Variate Distributions

In the Bayes inference, one usually meets with some inverted matrix distributions. In this paragraph, we discuss some of them.

Theorem 3.5.6. *Assume* $X \sim LS_{n\times p}(f)$, $n \geqslant p$. *Then the density of* $V = W^{-1} = (X'X)^{-1}$ *is*

$$\text{(3.5.9)} \qquad \frac{\pi^{\frac{1}{2}np - \frac{1}{4}p(p-1)}}{\prod_{j=1}^{p} \Gamma(\frac{1}{2}(n-j+1))} |V|^{-\frac{1}{2}(n+p+1)} f(V^{-1}) \qquad \textit{for } V > 0.$$

Proof. From (3.4.8) and the fact that the Jacobian of the transformation $W = V^{-1}$ is $|V|^{-(p+1)}$ the theorem follows. □

When $X \sim N_{n\times p}(O, I_n \otimes I_p)$, the corresponding distribution of V is called the *inverted Wishart distribution.*

Similarly, we can find the density of $C = B^{-1}$ with $B \sim B_p(\frac{1}{2}n_1, \frac{1}{2}n_2)$ as follows:

$$\text{(3.5.10)} \qquad \frac{\Gamma_p(\frac{1}{2}(n_1+n_2))}{\Gamma_p(\frac{1}{2}n_1)\,\Gamma_p(\frac{1}{2}n_2)} |C|^{-\frac{1}{2}(n_1+n_2)} |C - I|^{\frac{1}{2}(n_2-p-1)} \qquad \text{for } C > I_p.$$

And the joint density of $(C_1, \ldots, C_k) = (D_1^{-1}, \ldots, D_k^{-1})$ with $(D_1, \ldots, D_k) \sim MD_{k+1,p}(\frac{1}{2}n_1, \ldots, \frac{1}{2}n_k; \frac{1}{2}n_{k+1})$ is

$$\frac{\Gamma_p(\frac{1}{2}\sum_1^{k+1} n_i)}{\prod_{i=1}^{k+1}\Gamma_p(\frac{1}{2}n_i)} \left| I - \sum_1^k C_i^{-1} \right|^{\frac{1}{2}(n_{k+1}-p-1)} \prod_1^k |C_i|^{-\frac{1}{2}(n_i+p+1)},$$

$$\text{for } C_i > I, \qquad i = 1, \ldots, k.$$

We write $(C_1, \ldots, C_k) \sim D_{k+1,p}^{-1}(\frac{1}{2}n_1, \ldots, \frac{1}{2}n_k; \frac{1}{2}n_{k+1})$ and call that $(C_1, \ldots, C_k)$ is distributed according to *inverted matrix variate Dirichlet distribution* which has the following properties:

(1) If $(C_1, \ldots, C_k) \sim D_{k+1,p}^{-1}(\frac{1}{2}n_1, \ldots, \frac{1}{2}n_k; \frac{1}{2}n_{k+1})$, then $(C_1, \ldots, C_{j-1}, (I - \sum_1^k C_i^{-1})^{-1}, C_{j+1}, \ldots, C_k) \sim D_{k+1,p}^{-1}(\frac{1}{2}n_1, \ldots, \frac{1}{2}n_{j-1}, \frac{1}{2}n_{k+1}, \frac{1}{2}n_{j+1}, \ldots, \frac{1}{2}n_k; \frac{1}{2}n_j)$.

(2) If $(C_1, \ldots, C_k) \sim D_{k+1,p}^{-1}(\frac{1}{2}n_1, \ldots, \frac{1}{2}n_k; \frac{1}{2}n_{k+1})$, then $(C_1, \ldots, C_j) \sim D_{j+1,p}^{-1}(\frac{1}{2}n_1, \ldots, \frac{1}{2}n_j; \frac{1}{2}(n_{j+1} + \cdots + n_{k+1}))$ for $1 \leqslant j \leqslant k$, and $C_j \sim D_{2,p}^{-1}(\frac{1}{2}n_j; \sum_{i\neq j}^{k+1} \frac{1}{2}n_i)$ for $1 \leqslant j \leqslant k+1$.

(3) If $(C_1, \ldots, C_k) \sim D_{k+1,p}^{-1}(\frac{1}{2}n_1, \ldots, \frac{1}{2}n_k; \frac{1}{2}n_{k+1})$ and

$$Z_j = (I - \sum_1^j C_i^{-1})^{\frac{1}{2}} C_j (I - \sum_1^j C_j^{-1})^{\frac{1}{2}}, \qquad j = 1, \cdots, k,$$

then $(Z_{j+1}, \ldots, Z_k) \sim D_{k+1-j,p}^{-1}(\frac{1}{2}n_{j+1}, \ldots, \frac{1}{2}n_k; \frac{1}{2}n_{k+1})$ is independent of $(C_1, \ldots, C_j)$.

(4) If $(C_1, \ldots, C_k) \sim D_{k+1,p}^{-1}(\frac{1}{2}n_1, \ldots, \frac{1}{2}n_k; \frac{1}{2}n_{k+1})$ and

$$C_i = \begin{bmatrix} C_{11}^i & C_{12}^i \\ C_{21}^i & C_{22}^i \end{bmatrix}, \qquad C_{11}^i : l \times l, \qquad i = 1, \ldots, k,$$

then $(C_{22}^1, \ldots, C_{22}^k) \sim D_{k+1,p-1}^{-1}(\frac{1}{2}(n_1 - 1), \cdots, \frac{1}{2}(n_{k-1}); \frac{1}{2}(n_{k+1} + (k-1)l))$ is independent of $(C_{11,2}^1, \ldots, C_{11,2}^k)$, where

$$C_{11,2}^j = C_{11}^j - C_{12}^j (C_{22}^j)^{-1} C_{21}^j, \qquad j = 1, \ldots, k.$$

The above results are due to Xu (1984). The reader can find their proofs and more interesting results in Xu's paper.

3.5.6. Some Distributions of the Characteristic Roots of Matrix Variate

Let $\boldsymbol{X} \sim SS_{n\times p}(f)(n \geqslant p)$ and $\boldsymbol{W} = \boldsymbol{X}'\boldsymbol{X}$. The density of $\boldsymbol{W}$ was obtained in (3.4.8). The spectral decomposition of $\boldsymbol{W}$ is $\boldsymbol{W} = \boldsymbol{V}'\Lambda\boldsymbol{V}$, where $\Lambda = \operatorname{diag}(\lambda_1, \ldots, \lambda_p)$, $\lambda_1 > \lambda_2 > \ldots > \lambda_p > 0$ are the eigenvalues of $\boldsymbol{W}$, $\boldsymbol{V} \sim \mathcal{U}_{p,p}$, and $\boldsymbol{V}$ is independent of Λ. The Jacobian $J(\boldsymbol{W} \to \Lambda, \boldsymbol{V}) = \prod_{1\leqslant i<j\leqslant p} (\lambda_i - \lambda_j)\, g_p(\boldsymbol{V})$ (cf. Example 1.6). Hence the joint density of $\boldsymbol{V}$ and Λ is

$$c_{n,p}\,|\Lambda|^{\frac{1}{2}(n-p-1)}f(\Lambda) \prod_{1\leqslant i<j\leqslant p} (\lambda_i - \lambda_j) g_p(\boldsymbol{V}).$$

Integrate the above function with respect to $\boldsymbol{V}$ over $\boldsymbol{V}'\boldsymbol{V} = \boldsymbol{I}_p$ and note that $\boldsymbol{V} \sim \mathcal{U}_{p,p}$ has a density (3.4.6). We find that the density of Λ is

$$\text{(3.5.11)} \qquad \frac{\pi^{\frac{1}{2}p(p+n)}}{\Gamma_p(\frac{1}{2}p)\,\Gamma_p(\frac{1}{2}n)}\left[\prod_{i=1}^{p} \lambda_i^{\frac{1}{2}(n-p-1)}\right]\prod_{i<j}(\lambda_i - \lambda_j)\, f(\operatorname{diag}(\lambda_1, \ldots, \lambda_p)).$$

Hence, we find the following theorem.

Theorem 3.5.7. *Assume* $\boldsymbol{X} \sim SS_{n\times p}(f), n \geqslant p$. *Then the density of the eigenvalues* $\lambda_1 > \lambda_2 \cdots > \lambda_p > 0$ *of* $\boldsymbol{W} = \boldsymbol{X}'\boldsymbol{X}$ *is given by* (3.5.11). *Similarly, we can obtain the density of the eigenvalues* $\lambda_1 > \lambda_2 > \cdots > \lambda_p > 0$ *of* $\boldsymbol{B} \sim B_p(\frac{1}{2}n_1, \frac{1}{2}n_2)$ *as follows*

$$\text{(3.5.11)} \qquad \frac{\pi^{\frac{1}{2}p^2}}{\Gamma_p(\frac{1}{2}p)}\frac{\Gamma_p(\frac{1}{2}(n_1+n_2))}{\Gamma_p(\frac{1}{2}n_1)\,\Gamma_p(\frac{1}{2}n_2)}\prod_{i=1}^{p}\left[\lambda_i^{\frac{1}{2}(n_1-p-1)}(1-\lambda_i)^{\frac{1}{2}(n_2-p-1)}\right]\prod_{i<j}(\lambda_i - \lambda_j),$$
$$(1 > \lambda_1 > \lambda_2 \cdots > \lambda_p > 0).$$

3.6. The Generalized Bartlett Decomposition and the Spectral Decomposition of Spherical Matrix Distributions

In this section, we continue to study the relationships among the classes of spherical matrix distributions in the following aspects: coordinate transformations, the generalized Barltett decomposition (cf. Definition 3.6.1), the spectral decompositions, etc. Also, some new subclasses of left–spherical distribution are obtained and their properties are roughly discussed. Some interesting examples are investigated.

3.6.1. Coordinate Transformations

When $\boldsymbol{X} \in \mathfrak{F}_i$, $i = 1, 2, 3, s$, $\boldsymbol{X}$ has stochastic representations (3.1.3), (3.1.4), (3.1.5) and (3.1.7), respectively. Hence $\boldsymbol{U}$, $\boldsymbol{U}_2$, $\boldsymbol{U}_3$ and $(\boldsymbol{U}, \boldsymbol{V})$ play the role of coordinate systems (cf. Section 3.2). As $\mathfrak{F}_3 \subset \mathfrak{F}_2 \subset \mathfrak{F}_1$ and $\mathfrak{F}_3 \subset \mathfrak{F}_s \subset \mathrm{F}_1$, a natural question is how to get stochastic representation for smaller classes by means of the coordinate system of a larger class.

Theorem 3.6.1. *Assume* $\boldsymbol{X} \overset{d}{=} RU_3 \sim VS_{n\times p}(\phi)$. *Then* $\boldsymbol{X} \in \mathfrak{F}_2$ *and*

$$\text{(3.6.1)} \qquad \boldsymbol{X} \overset{d}{=} U_2\boldsymbol{R}^*, \ \boldsymbol{R}^* = \operatorname{diag}(R_1^*, \ldots, R_p^*),$$

where $\boldsymbol{R}^*$ *and* $\boldsymbol{U}_2$ *are independent,* $(R_1^{*2}, \ldots, R_p^{*2}) \sim G_p(\frac{1}{2}n, \ldots, \frac{1}{2}n, \phi)$ *(cf. Section 2.8.1).*

Proof. Let $\boldsymbol{Y} = (\boldsymbol{y}_1, \ldots, \boldsymbol{y}_p) \sim N_{n\times p}(\boldsymbol{O}, \boldsymbol{I}_n \otimes \boldsymbol{I}_p)$. By Lemma 3.2.6 we have

$$\boldsymbol{X} \stackrel{d}{=} RU_3 \stackrel{d}{=} R\boldsymbol{Y}/(tr(\boldsymbol{Y}'\boldsymbol{Y}))^{\frac{1}{2}} = R\left(\frac{\boldsymbol{y}_1}{\|\boldsymbol{y}_1\|}\frac{\|\boldsymbol{y}_1\|}{\|\boldsymbol{Y}\|}, \cdots, \frac{\boldsymbol{y}_p}{\|\boldsymbol{y}_p\|}\frac{\|\boldsymbol{y}_p\|}{\|\boldsymbol{Y}\|}\right),$$

where $\|\boldsymbol{Y}\| = [\mathrm{tr}(\boldsymbol{Y}'\boldsymbol{Y})]^{\frac{1}{2}} = \left(\sum_1^p \|\boldsymbol{y}_j\|^2\right)^{\frac{1}{2}}$. By using Lemma 3.2.6 again, $U_2 = (\boldsymbol{y}_1/\|\boldsymbol{y}_1\|, \ldots, \boldsymbol{y}_p/\|\boldsymbol{y}_p\|)$ is independent of $\boldsymbol{R}$ and $(\|\boldsymbol{y}_1\|^2/\|\boldsymbol{Y}\|^2, \ldots, \|\boldsymbol{y}_p\|^2/\|\boldsymbol{Y}\|^2)$ $\sim D_p(\frac{1}{2}n, \ldots, \frac{1}{2}n)$ which completes the proof. □

Theorem 3.6.2. *Assume* $\boldsymbol{X} \stackrel{d}{=} \boldsymbol{U}_2\boldsymbol{R} \sim MS_{n\times p}(\phi)$. *Then* $\boldsymbol{X} \in F_1$ *and*

$$\boldsymbol{X} \stackrel{d}{=} \boldsymbol{UAR}, \tag{3.6.2}$$

where $\boldsymbol{U}$, $\boldsymbol{A}$, *and* $\boldsymbol{R}$ *are independent,* $\boldsymbol{A} \in UT(p)$ *and*

$$\boldsymbol{A} \stackrel{d}{=} \boldsymbol{T}\mathrm{diag}(\|\boldsymbol{t}_1\|, \ldots, \|\boldsymbol{t}_p\|)^{-1} \tag{3.6.3}$$

with $\boldsymbol{T} = (\boldsymbol{t}_1, \ldots, \boldsymbol{t}_p)$ *and* $\boldsymbol{T}'\boldsymbol{T}$ *being the Bartlett decomposition of* $\boldsymbol{W} \sim W_p(n, \boldsymbol{I}_p)$.

Proof. Let $\boldsymbol{Y} = (\boldsymbol{y}_1, \ldots, \boldsymbol{y}_p) \sim N_{n\times p}(\boldsymbol{O}, \boldsymbol{I}_n \otimes \boldsymbol{I}_p)$. As

$$(\boldsymbol{y}_1/\|\boldsymbol{y}_1\|, \ldots, \boldsymbol{y}_p/\|\boldsymbol{y}_p\|) \stackrel{d}{=} \boldsymbol{U}_2 \stackrel{d}{=} \boldsymbol{U}_1\boldsymbol{A} \text{ and } \boldsymbol{T}'\boldsymbol{T} = \boldsymbol{Y}'\boldsymbol{Y},$$

we have $\quad \boldsymbol{y}_i'\boldsymbol{y}_i = \boldsymbol{t}_i'\boldsymbol{t}_i = \|\boldsymbol{t}_i\|^2, \qquad i = 1, \ldots, p,$

and

$$\boldsymbol{U}_1\boldsymbol{A} \stackrel{d}{=} \boldsymbol{Y}(\boldsymbol{R}^*)^{-1} = (\boldsymbol{y}_1, \ldots, \boldsymbol{y}_p)(\boldsymbol{R}^*)^{-1} = \boldsymbol{QT}(\boldsymbol{R}^*)^{-1},$$

where $\boldsymbol{R}^* = \mathrm{diag}(\|\boldsymbol{t}_1\|, \ldots, \|\boldsymbol{t}_p\|)$ and $\boldsymbol{QT}$ is the orthogonal–triangular decomposition of $\boldsymbol{Y}$ (cf. Section 1.2), i.e., $\boldsymbol{Q}$: $n\times p$, $\boldsymbol{Q}'\boldsymbol{Q} = \boldsymbol{I}_p$, and $\boldsymbol{T} \in UT(p)$. By Lemma 3.1.4, we have $\boldsymbol{T}(\boldsymbol{R}^*)^{-1} \stackrel{d}{=} \boldsymbol{A}$ which completes the proof. □

Remark. Let $\boldsymbol{A} = (a_{ij})$, $\boldsymbol{a}_i = (a_{1i}, \ldots, a_{ii})'$, $\boldsymbol{a}_i^* = (a_{1i}, \ldots, a_{i-1,i})'$ and $\boldsymbol{a}_i^{(2)} = (a_{1i}^2, \ldots, a_{ii}^2)$, $i = 1, \ldots, p$, in (3.6.3). Then by Theorem 3.6.2, we obtain the following facts:

(1) $\boldsymbol{a}_1, \ldots, \boldsymbol{a}_p$ are independent;

(2) $\boldsymbol{a}_k^* \stackrel{d}{=} \boldsymbol{u}_k$, where $\boldsymbol{u}_k$ is the first $(k-1)$-component subvector of $\boldsymbol{u}^{(n)}$, $k = 2, \ldots, p$;

and

(3) $\boldsymbol{a}_k^{(2)} \sim D_k(\frac{1}{2}, \cdots \frac{1}{2}, \frac{1}{2}(n-k+1))$, $k = 2, \ldots, p$.

Theorem 3.6.3. *Assume* $\boldsymbol{X} \stackrel{d}{=} R\boldsymbol{U}_3 \sim VS_{n\times p}(\phi)$. *Then* $\boldsymbol{X} \in \mathfrak{F}_1$ *and*

$$\boldsymbol{X} \stackrel{d}{=} R\boldsymbol{UB}, \tag{3.6.4}$$

where R, $\boldsymbol{U}$,*and* $\boldsymbol{B}$ *are independent,* $\boldsymbol{B} \stackrel{d}{=} \boldsymbol{T}/(\mathrm{tr}(\boldsymbol{T}'\boldsymbol{T}))^{\frac{1}{2}}$ *and* $\boldsymbol{T}$ *is given by Theorem* 3.6.2.

Proof. By Lemma 3.2.6, $U_3 \stackrel{d}{=} Y/(\mathrm{tr}(Y'Y))^{\frac{1}{2}}$ and $T'T/(\mathrm{tr}(T'T)) = Y'Y/(\mathrm{tr}(Y'Y)) \stackrel{d}{=} B'B$. As $B \in UT(p)$ with positive diagonal elements, $B \stackrel{d}{=} T/(\mathrm{tr}(T'T))^{\frac{1}{2}}$.

□

If $X \in \mathcal{F}_s$, then $X = U\Lambda V$((3.1.4)), where U, Λ, V are independent, $V \sim \mathcal{U}_{p,p}$, and $\Lambda = \lambda((X'X)^{\frac{1}{2}})$. Here (U, Λ) plays the role of the coordinate system of the class $\mathcal{F}_s$ and Λ is the "coordinate" of element in $\mathcal{F}_s$. As $\mathcal{F}_3 \subset \mathcal{F}_s$, if $X \in \mathcal{F}_3$ what is the coordinate of element in $\mathcal{F}_s$?

Theorem 3.6.4. *Assume* $X \stackrel{d}{=} RU_3 \sim VS_{n\times p}(\phi)$. *Then* $X \in \mathcal{F}_s$ *and*

$$X \stackrel{d}{=} RU\Lambda V, \tag{3.6.5}$$

where

(1) R, U, Λ, *and* V *are independent, and* R, U, *and* V *have the above meaning;*
(2) $\Lambda^2 = \mathrm{diag}(\lambda_1, \ldots, \lambda_p)$, $\lambda_1 > \lambda_2 > \cdots > \lambda_p > 0$,

$$(\lambda_1, \cdots, \lambda_p) \stackrel{d}{=} (w_1, \ldots, w_p)/(w_1 + \cdots + w_p), \tag{3.6.6}$$

where $w_1, \ldots, w_p$ *are p eigenvalues of* $W \sim W_p(n, I_p)$ *and* (3) $(\lambda_1, \ldots, \lambda_{p-1})$ *has a joint density*

$$\frac{\pi^{\frac{1}{2}p}\Gamma(\frac{1}{2}np)}{\prod_{j=1}^{p}\Gamma(\frac{1}{2}(p-j+1))\,\Gamma(\frac{1}{2}(n-j+1))}\left(\prod_{1}^{p}\lambda_j^{\frac{1}{2}(n-p-1)}\right)\left(\prod_{1\leqslant i<j\leqslant p}(\lambda_i-\lambda_j)\right) \tag{3.6.7}$$

$$\lambda_1 > \cdots \lambda_{p-1} > 0,\ \lambda_p = 1 - \lambda_1 - \cdots - \lambda_{p-1},$$

and $(\lambda_1, \ldots, \lambda_{p-1})$ *is independent of* $w = w_1 + \cdots + w_p$.

Proof. Let $Y \sim N_{n\times p}(O, I_n \otimes I_p)$. Then $Y/(\mathrm{tr}(Y'Y))^{\frac{1}{2}} \stackrel{d}{=} U_3 = U\Lambda V$ and $\lambda(Y'Y/(\mathrm{tr}(Y'Y))) \stackrel{d}{=} \mathrm{diag}(\lambda_1, \ldots, \lambda_p)$. Noting that $\lambda(Y'Y/(\mathrm{tr}(Y'Y))) = \lambda(Y'Y)/\mathrm{tr}(Y'Y)$ and $Y'Y \equiv W \sim W_p(n, I_p)$, $\mathrm{tr}(Y'Y) = w_1 + \cdots + w_p$ and $\lambda(W) = \mathrm{diag}(w_1, \ldots, w_p)$, the first part of the theorem follows. To calculate (3.6.7) we use (3.5.11) with a definite f, $f(\mathrm{diag}(u_1, \ldots, u_p)) = (2\pi)^{-\frac{1}{2}np}\exp(-\frac{1}{2}\Sigma u_j^2)$; then the density of $(w_1, \ldots, w_p)$ is

$$\frac{\pi^{\frac{1}{2}p}}{2^{\frac{1}{2}np}\prod_{j=1}^{p}\Gamma(\frac{1}{2}(p-j+1))\,\Gamma(\frac{1}{2}(n-j+1))}\left(\prod_{i=1}^{p}w_i^{\frac{1}{2}(n-p-1)}\right) \cdot \prod_{i<j}(w_i - w_j)\,e^{-\frac{1}{2}(w_1+\ldots+w_p)} \qquad w_1 > \cdots > w_p > 0. \tag{3.6.8}$$

Take the transformation

$$\begin{cases} \lambda_i = w_i/(w_1 + \cdots + w_p), & i = 1, \ldots, p-1, \\ w = w_1 + \cdots + w_p\ , & \end{cases}$$

whose Jacobian is w^{p-1}, and let $\lambda_p = 1 - \lambda_1 - \cdots - \lambda_{p-1}$. Now (3.6.8) becomes

$$\frac{\pi^{\frac{1}{2}p}\Gamma(\frac{1}{2}np)}{\prod_{j=1}^{p}\Gamma(\frac{1}{2}(p-j+1))\,\Gamma(\frac{1}{2}(n-j+1))}\prod_{i=1}^{p}\lambda_i^{\frac{1}{2}(n-p-1)}\prod_{1\leqslant i<j\leqslant p}(\lambda_i-\lambda_j)$$

$$\cdot\frac{1}{2^{\frac{1}{2}np}\Gamma(\frac{1}{2}np)}w^{\frac{1}{2}np-1}e^{-w/2}\ .$$

The second factor in the above function is the density of the chi-squared distribution with np degrees of freedom. Hence the proof is completed. □

3.6.2. The Generalized Bartlett Decomposition

As we have seen, the Bartlett decomposition plays an important role in many situations (cf. Theorem 3.4.1 and its corollaries, Theorem 3.5.2, Theorem 3.6.2 and Theorem 3.6.3, etc.). So it is quite natural to generalize this concept.

Definition 3.6.1. Let $\boldsymbol{X}$ be an $n\times p$ random matrix and let $\boldsymbol{X}'\boldsymbol{X} = \boldsymbol{T}'\boldsymbol{T}$, where $\boldsymbol{T} = (t_{ij}) \in \boldsymbol{UT}(p)$ with positive diagonal elements. If $\{t_{ij},\ i \leqslant j\}$ are independent, we say that $\boldsymbol{X}'\boldsymbol{X}$ has a *generalized Bartlett decomposition.*

The property of the generalized Bartlett decomposition can be used for characterization of normality. The following results are due to Fang and Wu (1984).

Theorem 3.6.5. *Assume that* $\boldsymbol{X} = (\boldsymbol{x}_1, ..., \boldsymbol{x}_p) \sim MS_{n\times p}(f), n \geqslant p, f(u_1, ..., u_p)$ *is continuous over* R^p, *and* $\boldsymbol{X}'\boldsymbol{X}$ *has a generalized Bartlett decomposition. Then*

(i) $\boldsymbol{x}_1, \boldsymbol{x}_2, ..., \boldsymbol{x}_p$ *are independent;*
(ii) $(\boldsymbol{x}_2, ..., \boldsymbol{x}_p)$ *is normal;*
(iii) *it is not necessary that* $\boldsymbol{x}_1$ *is normal.*

Proof. Let $\boldsymbol{T}'\boldsymbol{T} = \boldsymbol{X}'\boldsymbol{X}$ be the generalized Bartlett decomposition. As

$$\boldsymbol{X}'\boldsymbol{X} = \begin{bmatrix} \boldsymbol{x}_1'\boldsymbol{x}_1 & & & * \\ & \boldsymbol{x}_2'\boldsymbol{x}_2 & & \\ & & \ddots & \\ * & & & \boldsymbol{x}_p'\boldsymbol{x}_p \end{bmatrix} = \boldsymbol{T}'\boldsymbol{T}$$

$$= \begin{bmatrix} t_{11}^2 & & & * \\ & t_{12}^2 + t_{22}^2 & & \\ & & \ddots & \\ * & & & t_{1p}^2 + \cdots + t_{pp}^2 \end{bmatrix}$$

and $\{t_{ij},\ i \leqslant j\}$ are independent, $\{\boldsymbol{x}_i'\boldsymbol{x}_i,\ i=1, ..., p\}$ must be independent. By (3.1.5), it is easily shown that $R_1, \cdots, R_p$ are independent, which then implies that $\boldsymbol{x}_1, ..., \boldsymbol{x}_p$ are independent. From the assumption that $\boldsymbol{X}$ has a density, $\boldsymbol{x}_i$ has a density, too. Its density is $f_i(\boldsymbol{x}_i'\boldsymbol{x}_i)$. Hence the density of $\boldsymbol{X}$ is $\prod_1^p f_i(\boldsymbol{x}_i'\boldsymbol{x}_i)$. It can be shown that $\boldsymbol{T}$ has a density. Denote the density of t_{ij} by $g_{ij}(t_{ij})$, $1 \leqslant i \leqslant j \leqslant p$. By a argument similar to Corollary 1 of Theorem 3.4.1, the density of $\boldsymbol{T}$ has the form (since $\boldsymbol{X} \sim MS(f)$)

$$c\prod_{i=1}^{p} t_{ii}^{n-i} f_1(t_{11}^2)\cdots f_p(t_{1p}^2 + \cdots + t_{pp}^2)$$

for some constant c. Thus

$$\prod_{1 \leqslant i < j \leqslant p} g_{ij}(t_{ij}) = c \prod_{i=1}^{p} t_{ii}^{n-i} f_1(t_{11}^2) \cdots f_p(t_{1p}^2 + \cdots + t_{pp}^2). \tag{3.6.9}$$

Let $f_{ij} = g_{ij}$ for $i \neq j$ and $f_{ii}(t) = g_{ii}(t) t_{ii}^{n-i} c^{1/p}$. By (3.6.9) we have $f_i(y_1^2 + \cdots + y_i^2) = \prod_{j=1}^{i} g_{ji}(y_j)$ which implies $\boldsymbol{x}_i$ is normal by the continuity of f_i and f_{ji}, $i=2, \ldots, p$, i.e. $(\boldsymbol{x}_2, \ldots, \boldsymbol{x}_p)$ is normal.

Now prove (iii). Let $\boldsymbol{Y} \sim N_{n \times p}(\boldsymbol{O}, \boldsymbol{I}_n \otimes \boldsymbol{I}_p)$ and $\boldsymbol{T}^{*\prime}\boldsymbol{T}^*$ be the Bartlett decomposition of $\boldsymbol{Y}'\boldsymbol{Y}$ (Corollary 1 of Theorem 3.5.1). Clearly, $\boldsymbol{Y} = \boldsymbol{U}\boldsymbol{T}^*$, where $\boldsymbol{U} \sim \mathcal{U}_{n,p}$ is independent of $\boldsymbol{T}^*$. Partition $\boldsymbol{T}$, $\boldsymbol{Y}$, and $\boldsymbol{U}$ as

$$\boldsymbol{T} = (\boldsymbol{t}_1 \quad \boldsymbol{T}_2), \qquad \boldsymbol{Y} = (\boldsymbol{y}_1 \quad \boldsymbol{Y}_2), \qquad \text{and} \qquad \boldsymbol{U} = (\boldsymbol{u}_1 \quad \boldsymbol{U}_2),$$

where $\boldsymbol{t}_1$, $\boldsymbol{y}_1$ and $\boldsymbol{u}_1$ are n-dimensional vectors. Noting $\boldsymbol{t}_1 = (t_{11}, 0, \ldots, 0)'$, we have

$$\boldsymbol{y}_1 = t_{11}\boldsymbol{u}_1 \text{ and } \boldsymbol{Y}_2 = \boldsymbol{U}\boldsymbol{T}_2.$$

Take t_{11}^* instead of t_{11} such that t_{11}^* is independent of $\{\boldsymbol{T}_2, \boldsymbol{U}\}$ and $t_{11}^* \sim B(\frac{1}{2}n, \frac{1}{2}m)$ (not χ_n). Let $\boldsymbol{x}_1 = t_{11}^*\boldsymbol{u}_1$, $\boldsymbol{X}_2 = \boldsymbol{Y}_2$ and $\boldsymbol{X} = (\boldsymbol{x}_1 \quad \boldsymbol{X}_2)$. It is easily shown that $\boldsymbol{X}'\boldsymbol{X}$ satisfies the assumption of the theorem, but $\boldsymbol{x}_1 = t_{11}^*\boldsymbol{u}_1$ is not normal. □

Theorem 3.6.6. *Assume that* $\boldsymbol{X} \sim SS_{n \times p}(\phi)$, $\boldsymbol{X}'\boldsymbol{X}$ *has a generalized Bartlett decomposition, and there is an element* x_{ij} *of* $\boldsymbol{X}$ *whose distribution is normal. Then* $\boldsymbol{X}$ *must be normal.*

Proof. By the proof of Theorem 3.6.5, Theorem 2.7.1, and the assumption that x_{ij} is normal, $\boldsymbol{x}_1'\boldsymbol{x}_1, \ldots, \boldsymbol{x}_p'\boldsymbol{x}_p$ are i.i.d and $\boldsymbol{x}_j$ is normal, $j=1, \ldots, p$.

Now we prove that the distribution of $\boldsymbol{X} \in \mathcal{F}_s$ is fully determined by that of $(\boldsymbol{x}_1'\boldsymbol{x}_1, \ldots, \boldsymbol{x}_p'\boldsymbol{x}_p)$. By Theorem 3.1.1, the distribution of $\boldsymbol{X}$ is fully determined by that of $\boldsymbol{X}'\boldsymbol{X}$ because $\boldsymbol{X} \in \mathcal{F}_s \subset \mathcal{F}_1$. The c.f. of $\boldsymbol{X}'\boldsymbol{X}$ is $\phi(\boldsymbol{T}) = E(\exp(i\boldsymbol{T}\boldsymbol{X}'\boldsymbol{X}))$ where $\boldsymbol{T}' = \boldsymbol{T}$. Now write $\boldsymbol{T} = \boldsymbol{P}\Lambda\boldsymbol{P}'$, with $\boldsymbol{P}$ orthogonal and $\Lambda = \operatorname{diag}(\lambda_1, \ldots, \lambda_p)$. Then

$$\operatorname{tr}(\boldsymbol{X}'\boldsymbol{X}\boldsymbol{T}) = \operatorname{tr}(\boldsymbol{X}'\boldsymbol{X}\boldsymbol{P}\Lambda\boldsymbol{P}') = \operatorname{tr}(\boldsymbol{P}'\boldsymbol{X}'\boldsymbol{X}\boldsymbol{P}\Lambda) \stackrel{d}{=} \operatorname{tr}(\boldsymbol{X}'\boldsymbol{X}\Lambda) = \sum_{1}^{p} \lambda_k \boldsymbol{x}_k'\boldsymbol{x}_k$$

because $\boldsymbol{X}$ is spherical. So $\phi(\boldsymbol{T}) = \psi(\Lambda)$, where ψ is the c.f. of $(\boldsymbol{x}_1'\boldsymbol{x}_1, \ldots, \boldsymbol{x}_p'\boldsymbol{x}_p)$ and Λ contains the eigenvalues of $\boldsymbol{T}$. The result follows.

Let $\boldsymbol{y}_j$, $j=1, \ldots, p$ be i.i.d. and $\boldsymbol{y}_1 \stackrel{d}{=} \boldsymbol{x}_1$. Noting that $\boldsymbol{y}_1$ is normal, we have $\boldsymbol{Y} = (\boldsymbol{y}_1, \ldots, \boldsymbol{y}_p) \in \mathcal{F}_s$. Hence $\boldsymbol{X} \stackrel{d}{=} \boldsymbol{Y}$, because $(\boldsymbol{y}_1'\boldsymbol{y}_1, \ldots, \boldsymbol{y}_p'\boldsymbol{y}_p) \stackrel{d}{=} (\boldsymbol{x}_1'\boldsymbol{x}_1, \ldots, \boldsymbol{x}_p'\boldsymbol{x}_p)$. Hence the theorem follows. □

Corollary 1. *Assume that* $\boldsymbol{X} \sim VS_{n \times p}(\phi)$, $p > 1$, $P(\boldsymbol{X} = \boldsymbol{O}) = 0$, *and* $\boldsymbol{X}'\boldsymbol{X}$ *has a generalized Bartlett decomposition. Then* $\boldsymbol{X}$ *must be normal.*

Proof. We can prove that $\boldsymbol{x}_1'\boldsymbol{x}_1, \ldots, \boldsymbol{x}_p'\boldsymbol{x}_p$ are independent, As $\operatorname{vec}(\boldsymbol{X}) = (\boldsymbol{x}_1', \ldots, \boldsymbol{x}_p')' \sim S_{np}(\phi)$ and $P(\operatorname{vec}(\boldsymbol{X}) = \boldsymbol{0}) = P(\boldsymbol{X} = \boldsymbol{O}) = 0$, the corollary follows by Theorem 2.8.1. □

3.6.3. The Spectral Decomposition

For each $p \times p$ symmetric matrix $\boldsymbol{A}$, we have

(3.6.9$_1$) $$A = V\Lambda V' = \sum_{i=1}^{p} \lambda_i v_i v_i',$$

where $V = (v_1, \ldots, v_p) \in O(n)$ is the matrix of the characteristic vector and $\Lambda = \operatorname{diag}(\lambda_1, \ldots, \lambda_p)$, $\lambda_1 \geqslant \lambda_2 \geqslant \cdots \geqslant \lambda_p$ are the characteristic roots of A. The formula (3.6.9$_1$) is called *the spectral decomposition of* A. From the viewpoint of spectral decomposition we can study spherical matrix distributions.

For every $X \in \mathcal{F}_1$, we have $X \overset{d}{=} UA$, where $U \sim \mathcal{U}_{n,p}$ is independent of A, $A \geqslant 0$ and A is left-spherical (Theorem 3.1.2). Let (3.6.9$_1$) be the measurable spectral decomposition of A. Then

(3.6.10) $$X \overset{d}{=} UV\Lambda V' = \sum_{i=1}^{p} \lambda_i (Uv_i v_i') \overset{\Delta}{=} \lambda_1 F_1 + \cdots + \lambda_p F_p,$$

where $F_i = Uv_i v_i'$, $i = 1, \ldots, p$, or

(3.6.11) $$X \overset{d}{=} U\Lambda V'.$$

Here in both (3.6.10) and (3.6.11) U is independent of $\{\Lambda, V\}$ because $X'X \overset{d}{=} V\Lambda^2 V$ and $X \in \mathcal{F}_1$. We call equation (3.6.10) or (3.6.11) the *spectral decomposition of* X.

We should point out that in general it is not necessary that U, Λ, and V are independent and $V \sim \mathcal{U}_{p,p}$. Given a specific subclass of Λ and V, we can obtain a subclass of $\mathcal{F}_1$. In this way, we can construct many new **subclasses of** $\mathcal{F}_1$. First of all, let us have a look at $\mathcal{F}_1$, $\mathcal{F}_2$, $\mathcal{F}_3$ and $\mathcal{F}_s$.

(1) $X \in \mathcal{F}_1$ if and only if X has the decomposition (3.6.11), where U is independent of $\{\Lambda, V\}$;

(2) $X \in \mathcal{F}_s$ if and only if X has the decomposition (3.6.11), where U, Λ, and V are independent and $V \sim \mathcal{U}_{p,p}$;

(3) $X \in \mathcal{F}_3$ if and only if X has the decomposition (3.6.11), where U, Λ and V are independent, $V \sim \mathcal{U}_{p,p}$, $\Lambda \overset{d}{=} \Lambda^* R$ and Λ^{*2} has the distribution (3.6.6);

(4) When $X \in \mathcal{F}_2$, X has the decomposition (3.6.11), but no more information about Λ and V has been obtained. This is an open problem.

Now we consider some subclasses of $\mathcal{F}_1$ through (3.6.10). Note that $\lambda_i F_i$'s are the "cells" of $\mathcal{F}_1$. Then this fact induces us to investigate the following class:

(3.6.12) $$\mathcal{F}_4 = \{\mathcal{L}(X) \mid X \sim LS_{n\times p}(\phi) \quad \text{and } P(\operatorname{rk}(X) = 1) = 1\}.$$

Theorem 3.6.7. $\mathcal{F}_4 = \mathcal{F}_4^{(1)} = \mathcal{F}_4^{(2)} = \mathcal{F}_4^{(3)} = \mathcal{F}_4^{(4)}$,

where

$$\mathcal{F}_4^{(1)} = \{\mathcal{L}(X) \mid X \overset{d}{=} Uyz', \textit{ where } U \textit{ is independent of } \{y, z\} \text{ and } y\text{: } p\times 1, 2\times p\times 1\},$$

$$\mathcal{F}_4^{(2)} = \{\mathcal{L}(X) \mid X \overset{d}{=} yz', y \sim S_n(\phi) \textit{ is independent of } z\text{: } p\times 1\},$$

$$\mathcal{F}_4^{(3)} = \{\mathcal{L}(X) \mid X \overset{d}{=} u^{(n)}z', u^{(n)} \textit{ is independent of } z\text{: } p\times 1\},$$

$$\mathcal{F}_4^{(4)} = \{\mathcal{L}(X) \mid X \overset{d}{=} U1_p z', z(p\times 1) \textit{ is independent of } U\}.$$

Proof. $\mathcal{F}_4 \subset \mathcal{F}_4^{(1)}$ is obvious by noting that $X \in \mathcal{F}_4$ implies $X \stackrel{d}{=} UA$, $P(\mathrm{rk}(A) = 1) = 1$, $A \geqslant 0$ and thus $X \stackrel{d}{=} Uyy'$, where $y(p \times 1)$ is independent of U. If $X \in \mathcal{F}_4^{(1)}$, then $X \stackrel{d}{=} Uyz' \stackrel{d}{=} Ru^{(n)}z' = u^{(n)}(Rz')$, where R, $u^{(n)}$ and z are independent. Thus $X \in \mathcal{F}_4^{(3)}$, i.e., $\mathcal{F}_4^{(2)} \subset \mathcal{F}_4^{(3)}$. As $U1_p \stackrel{d}{=} \sqrt{p}u^{(n)}$, $\mathcal{F}_4^{(3)} \subset \mathcal{F}_4^{(4)}$. Finally, $\mathcal{F}_4^{(4)} \subset \mathcal{F}_4$ by the definitions. □

Lemma 3.6.1. *When* $X \in \mathcal{F}_4$, *if the distribution of* Xa, *where* $a \in R^p$ *is a fixed constant, depends on* a *only through* $a'a$, *then* $X \in \mathcal{F}_s$.

Proof. As $\mathrm{rk}(X) = 1$, a.e., we can write $X'X \stackrel{d}{=} yy'$, $y(p \times 1)$. Now, for every $a \in R^p$, $\Gamma \in O(p)$, we have $Xa \stackrel{d}{=} X\Gamma a$ as $a'a = (\Gamma a)'(\Gamma a)$. Thus $a'X'Xa \stackrel{d}{=} a'yy'a \stackrel{d}{=} a'\Gamma yy'\Gamma a$. Take $\delta \sim \mathcal{U}_{1,1}$ (i.e., $p(\delta = 1) = p(\delta = -1) = 1/2$) independently of y. As $\delta^2 = \delta\delta = 1$, $a'y\delta \cdot \delta y'a \stackrel{d}{=} a'\Gamma' y\delta \cdot \delta y'\Gamma a$. Noting that δ $a'y$ and $\delta a'\Gamma' y$ are all left-spherical (symmetric random variables), by Theorem 3.1.1, we get $\delta \cdot a'y \stackrel{d}{=} \delta a'\Gamma' y$ for each $a \in R^p$ and each $\Gamma \in O(p)$. Then $z = \delta y \sim S_p(\phi)$ by noting that $E(e^{ia'z}) = E(e^{i(\Gamma' a)'z})$ for each $a \in R^p$ and $\Gamma \in O(p)$. Therefore, by $X'X \stackrel{d}{=} yy' = zz'$, $X'X$ is spherical because $z \sim S_p(\phi)$, that is, X is right-spherical and X is spherical. □

By Lemma 3.6.1 we immediately obtain the following theorem.

Theorem 3.6.8. $X \in \mathcal{F}_s \cap \mathcal{F}_4$ *if and only if* $X \in \mathcal{F}_4$ *and the distribution of* Xa *with* $a \in R^p$ *depends on* a *only through* $a'a$. *And we have*

$$\text{(3.6.13)} \qquad \mathcal{F}_4 \cap \mathcal{F}_s = \{\mathcal{L}(X)\colon \ X \stackrel{d}{=} u^{(n)}z',\ u^{(n)} \textit{ is independent of } z \sim S_p(\phi)\}.$$

Let $V \equiv I_p$ in (3.6.11). We get the following subset of $\mathcal{F}_1$.

$$\text{(3.6.14)} \qquad \mathcal{F}_5 = \{X\colon X \stackrel{d}{=} U\Lambda,\ U \text{ is independent of } \Lambda = \mathrm{diag}(\lambda_1, \ldots, \lambda_p)\}.$$

As $X'X \stackrel{d}{=} \Lambda^2$, the normalized characteristic vector matrix of $X'X$ is I_p and its distribution is degenerate. Some interesting examples are listed in Exercises 3.11 and 3.12. In particular, if $\Lambda = RI_p$, i.e., the eigenvalues of $X'X$ are all equal, we get

$$\mathcal{F}_6 = \{\mathcal{L}(X)\colon \ X \stackrel{d}{=} RU,\ U \text{ is independent of } R \geqslant 0\}.$$

Clearly $\mathcal{F}_6 \subset \mathcal{F}_5 \cap \mathcal{F}_s$. In fact, we can prove $\mathcal{F}_6 = \mathcal{F}_5 \cap \mathcal{F}_s$ the proof of which is left to the reader. By means of $\mathcal{F}_6$ we find the following interesting example in which rows and columns of the matrix have normal distribution, but the matrix itself is not normal.

Example 3.6.1. Let $X = (x_1, \ldots, x_p) = (x_{(1)}, \ldots, x_{(n)})' \stackrel{d}{=} RU$ be an $n \times p$ random matrix, where $R \sim \chi_n$ is independent of $\boldsymbol{U}$. Then we have the following facts:

(i) $\boldsymbol{x}_j \sim N_n(\mathbf{0}, \mathbf{I}_n)$, $j=1, \ldots, p$, because $\boldsymbol{x}_j \overset{d}{=} R\boldsymbol{u}_j$, where $R \sim \chi_n$ is independent of $\boldsymbol{u}_j \overset{d}{=} \boldsymbol{u}^{(n)}$. More generally, $\boldsymbol{Xa}$ is normal for every $\mathbf{0} \neq \boldsymbol{a} \in R^p$.

(ii) $\boldsymbol{x}_{(k)} \sim N_p(\mathbf{0}, \sigma^2 \boldsymbol{I}_p)$, $k=1, \ldots, n$ for some $\sigma^2 > 0$. As $\boldsymbol{x}_{(k)} \overset{d}{=} R\boldsymbol{u}_{(k)}$ where $R \sim \chi_n$ is independent of $\boldsymbol{u}_{(k)}$ and $\boldsymbol{u}_{(k)} \overset{d}{=} b\boldsymbol{u}^{(p)}$ with $b \geqslant 0$ being independent of $\boldsymbol{u}^{(p)}$, $b^2 \sim B(\frac{1}{2}p, \frac{1}{2}(n-p))$, we write $\boldsymbol{x}_{(k)} \overset{d}{=} (R_1 b)(R_2 \boldsymbol{u}^{(p)})$, where $R = R_1 R_2$, $R_2 \sim \chi_p$ is independent of $R_1 \sim \chi_{(n-p)}$. As $R_2 \boldsymbol{u}^{(p)}$ is normal, so does $\boldsymbol{x}_{(k)}$. More generally, $\boldsymbol{b}'\boldsymbol{X}$ is normal for every $\boldsymbol{b} \in R^n$.

(iii) $\boldsymbol{X}$ is not normal. Suppose $\boldsymbol{X}$ is normal. Combining (i) and (ii), $\boldsymbol{x}_1, \ldots, \boldsymbol{x}_p$ are i.i.d., i.e., $\boldsymbol{X} \sim N_{n \times p}(\boldsymbol{O}, \boldsymbol{I}_n \otimes \boldsymbol{I}_p)$, and $\boldsymbol{X}$ has a density, but this is impossible when $p > 1$ because $\boldsymbol{U}'\boldsymbol{U} = \boldsymbol{I}_p$.

Perhaps, this example is not convincing because $\boldsymbol{X}$ has no density. Further, we want to find $\boldsymbol{X} \in \mathcal{F}_s$ which has a density with all rows and columns being normal, but $\boldsymbol{X}$ is not normal.

Let $n > 2p$, $\boldsymbol{X} = \begin{pmatrix} \boldsymbol{X}_1 \\ \boldsymbol{X}_2 \end{pmatrix}$ and $\boldsymbol{U} = \begin{pmatrix} \boldsymbol{U}_1 \\ \boldsymbol{U}_2 \end{pmatrix}$ with $\boldsymbol{U}_1$: $(n-p) \times p$ and $\boldsymbol{X}_1$: $(n-p) \times p$. Then $\boldsymbol{X}_1 \overset{d}{=} R\boldsymbol{U}_1$. We can prove that (i) $\boldsymbol{X}_1 \in \mathcal{F}_s$; (ii) both rows and column's marginal distributions are normal; (iii) $\boldsymbol{X}_1$ has a density; (iv) $\boldsymbol{X}_1$ is not normal.

Let $\boldsymbol{X} = (\boldsymbol{x}_{(1)}, \ldots, \boldsymbol{x}_{(n)})' \in \mathcal{F}_1$. In general, it is not necessary that $\boldsymbol{x}_{(1)}$ is spherical. If $\boldsymbol{X} \in \mathcal{F}_s$, we have $\boldsymbol{x}_{(1)}$ to be spherical. One may ask what is the necessary and sufficient condition for $\boldsymbol{x}_{(1)}$ to be spherical? Denote

$$\mathcal{F}_7 = \{\mathcal{L}(\boldsymbol{X}) | \boldsymbol{X} \sim LS_{n \times p}(\phi),\ \boldsymbol{x}_{(1)} \text{ is spherical}\}. \tag{3.6.15}$$

Theorem 3.6.9. *$\boldsymbol{X} \in \mathcal{F}_7$ if and only if $\boldsymbol{X} \sim LS_{n \times p}(\phi)$ and the distribution of $\boldsymbol{Xa}$ depends on $\boldsymbol{a} \in R^p$ only through $\boldsymbol{a}'\boldsymbol{a}$.*

Proof. Suppose that $\boldsymbol{X}_{(1)} \sim S_p(\phi)$ and $\boldsymbol{X} \sim LS_{n \times p}(\psi)$. Then $\psi(bb') = \phi(b'b)$ for each $\boldsymbol{b} \in R^p$. Therefore, for each $\boldsymbol{a} \in R^p$ and $\boldsymbol{t} \in \boldsymbol{R}^n$ we have

$$E(\exp(i\boldsymbol{t}'\boldsymbol{Xa})) = E(\exp(i\boldsymbol{at}'\boldsymbol{X})) = \psi(\boldsymbol{t}'\boldsymbol{taa}') = \phi(\boldsymbol{t}'\boldsymbol{ta}'\boldsymbol{a}),$$

and this shows that the distribution of $\boldsymbol{Xa}$ depends upon $\boldsymbol{a}$ only through $\boldsymbol{a}'\boldsymbol{a}$.

Conversely, suppose $\boldsymbol{X} \sim LS_{n \times p}(\phi)$ and the distribution of $\boldsymbol{Xa}$ depends on $\boldsymbol{a} \in R^p$ only through $\boldsymbol{a}'\boldsymbol{a}$. Then $\boldsymbol{Xa} \overset{d}{=} \boldsymbol{X}\Gamma\boldsymbol{a}$, for each $\Gamma \in O(p)$ and each $\boldsymbol{a} \in R^p$. Hence $\boldsymbol{a}'\Gamma'\boldsymbol{x}_{(1)} \overset{d}{=} \boldsymbol{a}'\boldsymbol{x}_{(1)}$. That means $E(\exp(i\boldsymbol{a}'\boldsymbol{x}_{(1)})) = E(\exp(i(\Gamma\boldsymbol{a})'\boldsymbol{x}_{(1)}))$ for each $\Gamma \in O(p)$ and each $\boldsymbol{a} \in R^p$. Let the c.f. of $\boldsymbol{x}_{(1)}$ be $\phi(\boldsymbol{t})$. Then $\phi(\boldsymbol{t}) = \phi(\Gamma\boldsymbol{t})$ for each $\Gamma \in O(p)$, i.e., $\boldsymbol{x}_{(1)} \sim S_p(\phi)$. □

Clearly, $\mathcal{F}_s \subset \mathcal{F}_7$. There are many properties of $\mathcal{F}_s$ that can be extended to $\mathcal{F}_7$. The following are some of them.

(i) $\mathcal{F}_3 = \mathcal{F}_7 \cap \mathcal{F}_2$ (cf. Corollary of Theorem 3.2.2).

We only need to prove $\mathcal{F}_7 \bigcap \mathcal{F}_2 \subset \mathcal{F}_3$. If $X \in \mathcal{F}_7 \bigcap \mathcal{F}_2$, then $\boldsymbol{x}_{(1)} \sim S_p(\phi)$ by Theorem 3.6.9, and $X \in \mathcal{F}_3$ because $X \in \mathcal{F}_2$ and $\boldsymbol{x}_{(1)} \sim S_p(\phi)$ (cf. Theorem 3.2.2).

(ii) Suppose that $X = (x_{ij}) = (\boldsymbol{x}_1, \ldots, \boldsymbol{x}_p) = (\boldsymbol{x}_{(1)}, \ldots, \boldsymbol{x}_{(n)})' \in \mathcal{F}_7$ and has a finite second moment. Then

(a) $\mathrm{Cov}(\boldsymbol{x}_k, \boldsymbol{x}_l) = \delta_{kl}\sigma^2 \boldsymbol{I}_n$, $k, l = 1, \ldots, p$;

(b) $\mathrm{Cov}(\boldsymbol{x}_{(i)}, \boldsymbol{x}_{(j)}) = \delta_{ij}\sigma^2 \boldsymbol{I}_p$, $i, j = 1, \ldots, n$,

where $\delta_{ij} = 0$ when $i \neq j$ and $\delta_{ii} = 1$ (cf. Lemma 3.2.5).

Proof. We only prove (ii) because the proof of (i) is similar to that of (ii). Noting $\boldsymbol{x}_i \sim S_n(\phi)$, $i = 1, \ldots, p$, since $\boldsymbol{X} \in \mathcal{F}_7$, we have $E(x_{i1}x_{j1}) = 0$, $i \neq j$, and $\boldsymbol{Xa} \overset{d}{=} \boldsymbol{x}_1$, for $\boldsymbol{a} \in R^p$ with $\boldsymbol{a}'\boldsymbol{a} = 1$. Thus $E(\boldsymbol{a}'\boldsymbol{x}_{(i)}\boldsymbol{x}_{(j)}\boldsymbol{a}) = E(x_{i1}x_{j1}) = 0$ and $\boldsymbol{a}'\mathrm{Cov}(\boldsymbol{x}_{(i)}, \boldsymbol{x}_{(j)})\boldsymbol{a} = 0$ for each $\boldsymbol{a} \in R^p$ and $i \neq j$. If we can show that $\mathrm{Cov}(\boldsymbol{x}_{(i)}, \boldsymbol{x}_{(j)})$ is a symmetric matrix, it must be the zero matrix $\boldsymbol{O}$ for $i \neq j$. In fact, $\boldsymbol{X} \overset{d}{=} \boldsymbol{UA}$ and $\mathcal{E}(\boldsymbol{u}_{(i)}\, \boldsymbol{u}'_{(j)}) = \mathcal{E}(\boldsymbol{u}_{(j)}, \boldsymbol{x}'_{(u)})$ for $i \neq j$, where $\boldsymbol{U} = (\boldsymbol{u}_{(1)}, \ldots, \boldsymbol{u}_{(n)})'$. Given $\boldsymbol{A}$, we have $\mathcal{E}(\boldsymbol{A}'\boldsymbol{u}_{(i)}\boldsymbol{u}'_{(j)}\boldsymbol{A}) = \mathcal{E}(\boldsymbol{A}'\boldsymbol{u}_{(j)}\boldsymbol{u}'_{(i)}\boldsymbol{A})$, which implies $\mathcal{E}(\boldsymbol{x}_{(i)}\, \boldsymbol{x}'_{(j)}) = \mathcal{E}(\boldsymbol{x}_{(j)}\, \boldsymbol{x}_{(i)})$ because $\boldsymbol{A}$ is independent of $\boldsymbol{U}$, i.e., $\mathrm{Cov}(\boldsymbol{x}_{(i)}\, \boldsymbol{x}_{(j)})$ is symmetric. □

The material in this subsection is selected from Fang and Chen (1986).

References

Anderson (1984), Anderson and Fang (1982b), (1984), Cambanis, Keener and Simons (1983), Chmielewski (1980), Dawid (1977), (1978), Fang (1986), Fang and Chen (1984), (1986), Fang and Wu (1984), Graham (1981), Jenson and Good (1981), Johnson and Kotz (1972), Kariya (1981), Khatri (1970), Kshirsagar (1961), (1972), Murihead (1982), Xu (1984), Zhang and Fang (1982).

Exercises 3

3.1. If $\boldsymbol{U} \sim \mathcal{U}_{n,p}$ and $\boldsymbol{V}$ are independent and $p(\boldsymbol{V} \in O(n)) = 1$, show that $\boldsymbol{VU} \sim \mathcal{U}_{n,p}$.

3.2. If $\boldsymbol{X} \sim SS_{p \times p}(\phi)$, show that $\boldsymbol{X}' \overset{d}{=} \boldsymbol{X}$.

3.3. If $\boldsymbol{X} \sim SS_{p \times p}(\phi)$, and $\boldsymbol{A}$ is left-spherical such that $\boldsymbol{A}'\boldsymbol{A} \overset{d}{=} \boldsymbol{Y}'\boldsymbol{Y}$, show that $\boldsymbol{A}' \overset{d}{=} \boldsymbol{A}$.

3.4. Let $\boldsymbol{X}$ be an $n \times p$ random matrix with a density

$$f(\boldsymbol{X}) = c|\boldsymbol{I}_p + \boldsymbol{X}'\boldsymbol{X}|^{-\frac{1}{2}(n+p)},$$

where c is a constant. Prove the following assertions:

(i) $\boldsymbol{X} \in \mathcal{F}_s$;

(ii) $(\|\boldsymbol{x}_1\|, \ldots, \|\boldsymbol{x}_p\|)$ and $(\boldsymbol{x}_1 / \|\boldsymbol{x}_p\|, \ldots, \boldsymbol{x}_1 / \|\boldsymbol{x}_p\|)$ are not independent.

(iii) $\boldsymbol{X} \notin \mathcal{F}_2$.

3.5. If $\boldsymbol{X} \in \mathcal{F}_2$, $\boldsymbol{Y} \in \mathcal{F}_2$ and $\boldsymbol{x}_{(1)} \overset{d}{=} \boldsymbol{y}_{(1)}$, prove $\boldsymbol{X} \overset{d}{=} \boldsymbol{Y}$ and show that there is no such property for $\mathcal{F}_1$.

3.6. Show that the set $\mathcal{F}_3$ is a proper subset of $\mathcal{F}_s$ and the latter is a proper subset of $\mathcal{F}_1$.

3.7. (i) Suppose $\boldsymbol{X} \sim N_{n \times p}(\boldsymbol{O}, \boldsymbol{I}_n \otimes \boldsymbol{I}_p)$. Prove $P(\boldsymbol{X}'\boldsymbol{X} > 0) = 1$ if $n > p$.

(ii) Suppose $\boldsymbol{X} \sim LS_{n \times p}(f)$. Prove $P(\boldsymbol{X}'\boldsymbol{X} > 0) = 1$ if $n > p$.

3.8. Let $\boldsymbol{Y} = \boldsymbol{M} + \boldsymbol{XB}$ where $\boldsymbol{X} \sim LS_{n \times q}(\phi)$, $\boldsymbol{B}$: $p \times q$, $\boldsymbol{M}$: $n \times q$ and denote $\boldsymbol{Y} \sim LS_{n \times q}(\boldsymbol{M}, \boldsymbol{B}, \phi)$ or $\boldsymbol{Y} \sim LS_{n \times p}(\boldsymbol{M}, \boldsymbol{B}, f)$ if $\boldsymbol{X} \sim LS_{n \times p}(f)$. Similar notations for MS, VS and SS. Show that

(i) The c.f. of $Y \sim LS_{n\times q}(M, B, \phi)$ is $\exp(iT'M)\phi(BT'TB')$; and the c.f. of $Y \sim SS_{n\times q}(M, B, \phi)$ is $\exp(iT'M)\phi(\lambda(T\Sigma T'))$, where $\Sigma = B'B$. Hence the distribution of Y depends on B only through Σ and we write $SS_{n\times q}(M, \Sigma, \phi)$ instead of $SS_{n\times q}(M, B, \phi)$ and $VS_{n\times q}(M, \Sigma, \phi)$ instead of $VS_{n\times q}(M, \Sigma, \phi)$.

(ii) The density of $Y \sim LS_{n\times q}(M, B, f)$ with $P(|B| = O) = 0$ is $|B|^{-n}f(B'^{-1}(Y - M)'(Y - M)B^{-1})$, and the density of $Y \sim SS_{n\times q}(M, \Sigma, f)$ with $\Sigma > 0$ is $|\Sigma|^{-\frac{1}{2}n}f(\lambda((Y - M)\Sigma^{-1}(Y - M)')) = |\Sigma|^{-\frac{1}{2}n}f(\lambda(\Sigma^{-1}G))$, where $G = (Y - M)'(Y - M)$.

(iii) If $Y \sim SS_{n\times q}(0, \Sigma, f)$ with $\Sigma > 0$, then the density of $Y'Y$ is

$$c_{n;p}|W|^{\frac{1}{2}(n-p-1)}|\Sigma|^{-\frac{1}{2}n}f(\lambda(\Sigma^{-1}W)),$$

where $c_{n,p}$ is defined by (3.4.9). Partition $Y = (Y_1', \ldots, Y_m')'$ where $Y_j: n_j \times q$, $n_j \geqslant m$, $j = 1, \ldots, m$. Find the joint density of $Y_1'Y_1, \ldots, Y_m'Y_m$.

(iv) Under the above assumption, the density of $V = (Y'Y)^{-1}$ is

$$c_{n,p}|\Sigma|^{-\frac{1}{2}n}|V|^{-\frac{1}{2}(n+p+1)}f(\lambda(\Sigma^{-1}V^{-1})).$$

(v) Under the above assumption extend Cochran's Theorem to case (iii).

(vi) If $Y = (Y_1 \quad Y_2) \sim SS_{n\times q}(0, M, f)$ where Y: $n \times r$, partition Σ into $\begin{bmatrix}\Sigma_{11} & \Sigma_{12}\\ \Sigma_{21} & \Sigma_{22}\end{bmatrix}$ where Σ_{11}: $r \times r$. Prove the distribution of $(Y_2 - Y_1\Sigma_{11}^{-1}\Sigma_{12})\Sigma_{22.1}^{-\frac{1}{2}}|Y_1$ is spherical, where $\Sigma_{22.1} = \Sigma_{22} - \Sigma_{21}\Sigma_{11}^{-1}\Sigma_{12}$.

3.9. Let $F \sim MF(p, n_1, n_2)$. Find the density of F^{-1}.

3.10. Find the density of the eigenvalues $\lambda_1 > \lambda_2 > \cdots > \lambda_p$ of $F \sim MF(p, n_1, n_2)$.

3.11. **Put $\Lambda = \mathrm{diag}(R_1, \ldots, R_p)$ into (3.6.5) where $R_i \geqslant 0$, $i = 1, \ldots, p$ and $(R_1^2, \ldots, R_p^2) \sim$ the *generalized Rayleigh distribution*, i.e., $(R_1^2, \ldots, R_p^2) \overset{d}{=} (w_{11}, \ldots, w_{pp})$, which is the diagonal elements of $W \sim W_p(n, \Sigma)$. Then $X \in \mathfrak{F}_5$. Prove that**

(1) the marginal distributions of the columns of X are all normal;

(2) X is not normal.

3.12. Put $\Lambda = \mathrm{diag}(1/d_1, \ldots, 1/d_p)$ into (3.6.14), where $d_i > 0$, $i = 1, \ldots, p$, and $(d_1^2, \ldots, d_p^2) \sim D_p\left(\frac{1}{2}, \ldots, \frac{1}{2}\right)$. Prove that

(i) The c.f.'s of rows of X have the form $\phi(|t_1| + \cdots + |t_p|)$ (cf. Cambanis, Keener and Simons (1983)) and

(ii) The c.f.'s of columns of X have the form $\phi^*(t_1^2 + \cdots + t_n^2)$.

3.13. Suppose $X \in \mathfrak{F}_7$ and $y(p \times 1)$ is independent of X with $P(y'y = 0) = 0$. Prove that the distribution of $y'X'Xy/y'y$ is independent of y.

3.14. Prove that the following statements are equivalent:

(i) $X = (x_1, \ldots, x_p) \in \mathfrak{F}_7$ and $x_1 \sim N(0, I_n)$.

(ii) $X \in \mathfrak{F}_1$ and $a'X'Xa \sim \chi_n^2$ for each $a \in R^p$ with $a'a = 1$.

3.15. If $X \sim N_{n\times p}(M, C \otimes D)$, then the c.f. of X is

$$\exp(iT'M - \frac{1}{2}T'CTD).$$

3.16. If $X \sim N_{n\times p}(M, C \otimes D)$, then

(1) $x_{(i)} \sim N_p(\mu_{(i)}, c_{ii}D)$, $i = 1, \ldots, n$;

(2) $x_j \sim N_n(\mu_j, d_{jj}C)$, $j = 1, \ldots, p$;

(3) $\mathrm{Cov}(x_{(i)}, x_{(j)}) = c_{ij}D$, $i, j = 1, \ldots, n$;

(4) $\mathrm{Cov}(\boldsymbol{x}_i, \boldsymbol{x}_j) = d_{ij}\boldsymbol{C},\ i, j = 1, \ldots, p,$

where $\mu_{(i)}$'s and μ_j's are rows and columns of $\boldsymbol{M}$, respectively.

3.17. Define

$$\Gamma_p(a) = \int_{A>0} \exp(-\boldsymbol{A})|\boldsymbol{A}|^{a-\frac{1}{2}(p+1)} d\boldsymbol{A}, \qquad \mathrm{Re}(a) > \frac{1}{2}(p-1).$$

Show that $\Gamma_p(a) = \pi^{\frac{1}{4}p(p-1)} \prod_{i=1}^{p} \Gamma(a - \frac{1}{2}(i-1))$. $\Gamma_p(a)$ is called the multivariate Gamma function.

3.18. Let $\boldsymbol{X}$ be an $n \times p$ random matrix and have moments of order 2. Prove that

(1) $\mathcal{E}(\boldsymbol{X}) = 0$ and $\mathcal{D}(\mathrm{vec}\boldsymbol{X}') = \boldsymbol{I}_n \otimes \boldsymbol{V}$ where $\boldsymbol{V} = \mathcal{E}(\boldsymbol{x}_{(1)}\boldsymbol{x}'_{(1)})$ if $\boldsymbol{X} \in \mathcal{F}_1$;

(2) $\mathcal{D}(\mathrm{vec}\boldsymbol{X}') = \sigma^2 \boldsymbol{I}_n \otimes \boldsymbol{I}_p = \sigma^2 \boldsymbol{I}_{np}$ where $\sigma^2 = Ex_{11}^2$ if $\boldsymbol{X} \in \mathcal{F}_s$.

3.19. Let $\boldsymbol{X} = (x_{ij}) \in \mathcal{F}_s$ be an $n \times p$ matrix. Then $\boldsymbol{X} \in \mathcal{F}_3$ iff $(x_{11}, \ldots, x_{pp})'$ is a spherical.

3.20. Prove that the c.f. of $\boldsymbol{U}_1$ is

$$E[\exp(\mathrm{i}\boldsymbol{T}'\boldsymbol{U}_1)] = {}_0F_1\Big(\frac{1}{2}n;\ -\frac{1}{4}\boldsymbol{T}'\boldsymbol{T}\Big),$$

where

$${}_0F_1(a; \boldsymbol{A}) = \sum_{k=0}^{\infty} \frac{1}{[a]_k k!} \boldsymbol{A}^k$$

and

$$[a]_k = a(a+1) \cdots (a+k-1).$$

CHAPTER IV
ESTIMATION OF PARAMETERS

In the present chapter we are concerned with the estimators of the mean vector μ and the covariance matrix Σ in spherical matrix distributions. Hence in this chapter we are restricted to point estimate. We shall study the maximum likelihood estimates (MLE's) of μ and Σ, and their functions, such as correlation coefficients and multivariate correlation. In Section 4.3 we discuss many properties of MLE's. In the last section, we deal with the minimax and admissible estimates as well as Bayes estimates. For convenience we appoint that $LS_{n\times p}(\mu, \Sigma, f)$ replaces the $LS_{n\times p}(\mathbf{1}\mu', \Sigma, f)$ that is defined before. Analogously, $SS_{n\times p}(\mu, \Sigma, f)$ replaces the $SS_{n\times p}(\mathbf{1}\mu', \Sigma, f)$ and $VS_{n\times p}(\mu, \Sigma, f)$ replaces the $VS_{n\times p}(\mathbf{1}\mu', \Sigma, f)$. Besides, we assume $\Sigma > 0$ throughout the chapter.

4.1. MLE's of Mean Vector and Covariance Matrix

4.1.1. MLE's of μ and Σ in $VS_{n\times p}(\mu, \Sigma, f)$

Suppose $\mathbf{X} \sim VS_{n\times p}(\mu, \Sigma, f)$ with $\Sigma > 0$, that is, $\mathbf{X}$ has a density (cf. Section 3.2.4)

$$|\Sigma|^{-n/2} f[\operatorname{tr}(\mathbf{X} - \mathbf{1}\mu')\Sigma^{-1}(\mathbf{X} - \mathbf{1}\mu')']. \tag{4.1.1}$$

Let

$$\boldsymbol{G} \hat{=} (\mathbf{X} - \mathbf{1}\mu')'(\mathbf{X} - \mathbf{1}\mu'). \tag{4.1.2}$$

Then (4.1.1) can be rewritten as follows:

$$|\Sigma|^{-n/2} f(\operatorname{tr}\Sigma^{-1}\boldsymbol{G}). \tag{4.1.3}$$

Put $\boldsymbol{S} \hat{=} \sum_{i=1}^{n} (\boldsymbol{x}_{(i)} - \bar{\boldsymbol{x}})(\boldsymbol{x}_{(i)} - \bar{\boldsymbol{x}})'$, or equivalently

$$\boldsymbol{S} = (\mathbf{X} - \mathbf{1}\bar{\boldsymbol{x}}')'(\mathbf{X} - \mathbf{1}\bar{\boldsymbol{x}}') = \mathbf{X}'(\boldsymbol{I}_n - \frac{1}{n}\mathbf{1}\mathbf{1}')\mathbf{X}, \tag{4.1.4}$$

where $\boldsymbol{x}_{(1)}, \ldots, \boldsymbol{x}_{(n)}$ are the $\boldsymbol{n}$ row vectors of $\mathbf{X}$, and $\bar{\boldsymbol{x}} = \frac{1}{n}\sum_{i=1}^{n} \boldsymbol{x}_{(i)}$. We often call $\frac{1}{n}\boldsymbol{S}$ and $\bar{\boldsymbol{x}}$ the *sample covariance matrix* and the *sample mean* vector respectively. It is well–known that in univariate statistics there is a useful relationship between $\sum_{i=1}^{n}(X_i - \theta)^2$ and $\sum_{i=1}^{n}(X - \bar{x})^2$ where $\bar{x} = \frac{1}{n}\sum_{i=1}^{n} X_i$, which is $\sum_{i=1}^{n}(X_i - \theta)^2 = \sum_{i=1}^{n}(X_i - \bar{x})^2 + n(\bar{x} - \theta)^2$. Similarly, we have

$$\begin{aligned} \boldsymbol{G} &= (\mathbf{X} - \mathbf{1}\mu')'(\mathbf{X} - \mathbf{1}\mu') = (\mathbf{X} - \mathbf{1}\bar{\boldsymbol{x}}' + \mathbf{1}\bar{\boldsymbol{x}}' - \mathbf{1}\mu')'(\mathbf{X} - \mathbf{1}\bar{\boldsymbol{x}}' + \mathbf{1}\bar{\boldsymbol{x}}' - \mathbf{1}\mu') \\ &= (\mathbf{X} - \mathbf{1}\bar{\boldsymbol{x}}')'(\mathbf{X} - \mathbf{1}\bar{\boldsymbol{x}}) + n(\bar{\boldsymbol{x}} - \mu)(\bar{\boldsymbol{x}} - \mu)' \\ &= \boldsymbol{S} + n(\bar{\boldsymbol{x}} - \mu)(\bar{\boldsymbol{x}} - \mu)', \end{aligned} \tag{4.1.5}$$

because

$$\boldsymbol{O} = (\mathbf{X} - \mathbf{1}\bar{\boldsymbol{x}}')'(\mathbf{1}\bar{\boldsymbol{x}}' - \mathbf{1}\mu') \tag{4.1.6}$$

and

(4.1.7) $$(\mathbf{1}\bar{\boldsymbol{x}}' - \mathbf{1}\mu')'(\mathbf{1}\bar{\boldsymbol{x}}' - \mathbf{1}\mu') = n(\bar{\boldsymbol{x}} - \mu)(\bar{\boldsymbol{x}} - \mu)'.$$

Puting (4.1.5) into (4.1.3), we can rewrite (4.1.1) again as follows:

(4.1.8) $$|\Sigma|^{-n/2} f(\mathrm{tr}\Sigma^{-1}\boldsymbol{S} + n(\bar{\boldsymbol{x}} - \mu)'\Sigma^{-1}(\bar{\boldsymbol{x}} - \mu)).$$

Now we are interested in the MLE (maximum likelihood estimate) of the parameter (μ, Σ) for a given f(·). Imposing some simple conditions on f(·) we shall find out the LME of (μ, Σ). Let us start with several lemmata.

Lemma 4.1.1. *$\boldsymbol{G}$ and $\boldsymbol{S}$ are positive definite with probability one iff $n>p$, where $\boldsymbol{G}$ and $\boldsymbol{S}$ are defined by* (4.1.2) *and* (4.1.4) *respectively.*

Proof. By (4.1.5) it is sufficient to prove that $\boldsymbol{S}$ is positive definite with probability one iff $n>p$. If $n>p$, first, we want to show that $P(\boldsymbol{S}>0)=1$ for the case of $\boldsymbol{X} \sim N_{n\times p}(\mathbf{1}\mu', I_n\otimes\Sigma)$. Consider the transformation $\boldsymbol{Z} = (\boldsymbol{z}_{(1)}, ..., \boldsymbol{z}_{(n)})' = \Gamma\boldsymbol{X}$, where $\Gamma\in O(n)$ with the n-th row vector being $(1/\sqrt{n}, ..., 1/\sqrt{n})'$. As $\Gamma(\mathbf{1}\mu') = (\mathbf{0}', n^{1/2}\mu)'$, we have

(4.1.9) $$\boldsymbol{Z} \sim N_{n\times p}((\mathbf{0}', n^{1/2}\mu)', \boldsymbol{I}_n\otimes\Sigma).$$

So $\boldsymbol{z}_{(1)}, ..., \boldsymbol{z}_{(n)}$ are independent. Now

(4.1.10) $$\begin{aligned}\boldsymbol{S} &= (\boldsymbol{X} - \mathbf{1}\bar{\boldsymbol{x}}')'(\boldsymbol{X} - \mathbf{1}\bar{\boldsymbol{x}}') = \boldsymbol{X}'(\boldsymbol{I}_n - \frac{1}{n}\mathbf{1}\mathbf{1}')\boldsymbol{X} = \boldsymbol{Z}'\Gamma(\boldsymbol{I}_n - \frac{1}{n}\mathbf{1}\mathbf{1}')\Gamma'\boldsymbol{Z}\\ &= \boldsymbol{Z}'\boldsymbol{Z} - \boldsymbol{z}_{(n)}\boldsymbol{z}'_{(n)} = \sum_{i=1}^{n-1}\boldsymbol{z}_{(i)}\boldsymbol{z}'_{(i)} \,\hat{=}\, \boldsymbol{B}'\boldsymbol{B},\end{aligned}$$

where $\boldsymbol{B} = (\boldsymbol{z}_{(1)}, ..., \boldsymbol{z}_{(n-1)})'$. To show that $P(\boldsymbol{S}>0)=1$ when $n>p$, it is sufficient to show that $\boldsymbol{B}$ has rank p with probability one. Obviously, by adding more columns to $\boldsymbol{B}$ we cannot diminish its rank. Thus we need only to prove that $\boldsymbol{B}$ has rank p with probability one when $n-1=p$. For any set of $p-1$ p-vectors $(\boldsymbol{a}_1, ..., \boldsymbol{a}_{p-1})$ in R^p let $L(\boldsymbol{a}_1, ..., \boldsymbol{a}_{p-1})$ be the linear subspace spanned by $\boldsymbol{a}_1, ..., \boldsymbol{a}_{p-1}$. Noting $\Sigma>0$, for any given $\boldsymbol{a}_1, ..., \boldsymbol{a}_{p-1}$, we have

(4.1.11) $$P(\boldsymbol{z}_{(1)}\in L(a_1, ..., a_{p-1})) = 0.$$

Now as $\boldsymbol{z}_{(1)}, ..., \boldsymbol{z}_{(p)}$ are i.i.d., we have, by equation (4.1.11),

$$\begin{aligned}&P(\boldsymbol{z}_{(1)}, ..., \boldsymbol{z}_{(p)} \text{ are linearly dependent})\\ &\leqslant \sum_{i=1}^{p} P(\boldsymbol{z}_{(i)}\in L(\boldsymbol{z}_{(1)}, ..., \boldsymbol{z}_{(i-1)}, \boldsymbol{z}_{(i+1)}, ..., \boldsymbol{z}_{(p)}))\\ &= p.P(\boldsymbol{z}_{(1)}\in L(\boldsymbol{z}_{(2)}, ..., \boldsymbol{z}_{(p)}))\\ &= p.E[P(\boldsymbol{z}_{(1)}\in L(\boldsymbol{z}_{(2)}, ..., \boldsymbol{z}_{(p)}))\,|\,\boldsymbol{z}_{(2)}, ..., \boldsymbol{z}_{(p)}]\\ &= p.E(0) = 0,\end{aligned}$$

which shows that $P(\boldsymbol{S}>0)=1$ when $n>p$ in the case of $\boldsymbol{X} \sim N_{n\times p}(\mathbf{1}\mu', \boldsymbol{I}_n\otimes\Sigma)$. (The above proof is due to Dykstra (1970).) Now we assume $\boldsymbol{X} \sim VS_{n\times p}(\mu, \Sigma, f)$ and try to prove $P(\boldsymbol{S}>0)=1$ when $n>p$. Since

(4.1.12) $$\boldsymbol{X} \overset{d}{=} \mathbf{1}\mu' + RU\Sigma^{1/2},$$

where $\boldsymbol{U}$: $n\times p$ and vec $\boldsymbol{U} \overset{d}{=} \boldsymbol{u}^{(np)}$, hence $\boldsymbol{S} = \boldsymbol{X}'(\boldsymbol{I}_n - \boldsymbol{P}_n)\boldsymbol{X} \overset{d}{=} (\mathbf{1}\mu' + RU\Sigma^{1/2})'(\boldsymbol{I}_n - \boldsymbol{P}_n)(\mathbf{1}\mu' + RU\Sigma^{1/2}) = R^2\Sigma^{1/2}\boldsymbol{U}'(\boldsymbol{I}_n - \boldsymbol{P}_n)\boldsymbol{U}\Sigma^{1/2} \overset{d}{=} R^2\Sigma^{1/2}\boldsymbol{Y}'(\boldsymbol{I}_n - \boldsymbol{P}_n)\boldsymbol{Y}\Sigma^{1/2}/\mathrm{tr}\boldsymbol{Y}'\boldsymbol{Y}$, where $\boldsymbol{P}_n = \frac{1}{n}\mathbf{1}\mathbf{1}'$ and $\boldsymbol{Y} \sim N_{n\times p}(\boldsymbol{O}, \boldsymbol{I}_n\otimes\boldsymbol{I}_p)$. Since $P(\boldsymbol{Y}'(\boldsymbol{I}_n - \boldsymbol{P}_n)\boldsymbol{Y}>0)=1$

and $P(R^2>0)=1$, we immediately get $P(\boldsymbol{S}>0)=1$, as desired.

To show the "only if" part, from (4.1.9) it is obvious. In fact, if $n\leqslant p$ (equivalently, $n-1<p$) $\boldsymbol{S}$ has rank $\mathrm{rk}(\boldsymbol{S})<p$, *for each* $\boldsymbol{Z}$, equivalently for each $\boldsymbol{X}$. □

Lemma 4.1.2. *Suppose that $f(\cdot)$ is a nonincreasing and continuous function such that $cf(\boldsymbol{x}'\boldsymbol{x})$, $\boldsymbol{x}\in R^m$, is a density in R^m for some constant c. Then the function*

$$h(x) = x^{m/2} f(x), \qquad x\geqslant 0,$$

has a maximum at some finite $x_0>0$. Moreover if f is differentiable x_0 is essentially a solution of

$$f'(x) + (m/2x)f(x) = 0.$$

Proof. Since $cf(\boldsymbol{x}'\boldsymbol{x})$ is a density, we have

$$\int_{R^m} f(\boldsymbol{x}'\boldsymbol{x})d\boldsymbol{x} = [\pi^{m/2}/\Gamma(m/2)]\int_0^{\infty} r^{m/2-1} f(r)dr, \tag{4.1.13}$$

by formula (2.4.13). From (4.1.13) and monotonicity of $f(\cdot)$,

$$2^{-m/2}(2x)^{m/2} f(2x) = x^{m/2} f(2x) \leqslant x^{m/2-1}\int_x^{2x} f(r)dr \leqslant \int_x^{2x} f(r)r^{m/2-1}\,dr \to 0$$

as $x\to\infty$, i.e., $h(x) = x^{m/2} f(x)\to 0$ as $x\to\infty$. By virtue of this fact, $h(0) = 0$ and $h(x) \geqslant 0$, for each $x\geqslant 0$. The first assertion of the lemma follows. Now assume that f is differentiable. Then, necessarily,

$$0 = h'(x_0) = x_0^{m/2}\,[(m/2x_0)f(x_0) + f'(x_0)],$$

which completes the proof. □

From Lemma 4.1.2, when $m=np$, the function

$$f^*(\lambda) = \lambda^{-np/2} f(p/\lambda)$$

arrives at its maximum at some finite $\lambda_0>0$. In the sequel we shall denote the λ_0 by $\lambda_{\max}(f)$.

Theorem 4.1.1. *Suppose that $\boldsymbol{X}\sim VS_{n\times p}(\mu, \Sigma, f)$ with $n>p$ and $f(\cdot)$ being nonincreasing and continuous. Then the maximum likelihood estimate of (μ, Σ) is*

$$(\hat{\mu}, \hat{\Sigma}) = (\bar{\boldsymbol{x}}, \lambda_{\max}(f)\boldsymbol{S}).$$

Proof. From the supposition the likelihood function is

$$L(\mu, \Sigma) = |\Sigma|^{-n/2} f(\mathrm{tr}\Sigma^{-1}\boldsymbol{S} + n(\bar{\boldsymbol{x}}-\mu)'\Sigma^{-1}(\bar{\boldsymbol{x}}-\mu))$$

by (4.1.8). By monotonicity of $f(\cdot)$ for any given $\Sigma>0$, $L(\mu, \Sigma)$ as a function of $\mu\in R^p$ arrives at its maximum at $\mu = \bar{\boldsymbol{x}}$ and the concentrated likelihood is

$$L(\bar{\boldsymbol{x}}, \Sigma) = |\Sigma|^{-n/2} f(\mathrm{tr}\Sigma^{-1}\boldsymbol{S}). \tag{4.1.14}$$

By Lemma 4.1.1, $\boldsymbol{S}>0$ with probability one. Put $\tilde{\Sigma} = \boldsymbol{S}^{-1/2}\Sigma\boldsymbol{S}^{-1/2}$ (cf. Section 1.2 for the definition of $\boldsymbol{S}^{-1/2}$). Then (4.1.14) becomes

$$L(\bar{\boldsymbol{x}}, \Sigma) = |\boldsymbol{S}^{\frac{1}{2}}\tilde{\Sigma}\boldsymbol{S}^{\frac{1}{2}}|^{-n/2} f(\mathrm{tr}\tilde{\Sigma}^{-1}) = |\boldsymbol{S}|^{-n/2}|\tilde{\Sigma}|^{-n/2} f(\mathrm{tr}\tilde{\Sigma}^{-1}).$$

Let $\lambda_1 \geqslant \cdots \geqslant \lambda_p>0$ be the eigenvalues of $\tilde{\Sigma}^{-1}$ and $\bar{\lambda}^{-1} = (\lambda_1 + \cdots + \lambda_p)/p$. Then

$$L(\bar{\boldsymbol{x}}, \Sigma) = |\boldsymbol{S}|^{-n/2}\prod_1^p \lambda_i^{n/2} f(\sum_1^p \lambda_i)$$

$$= |\boldsymbol{S}|^{-n/2} \prod_{1}^{p} (\lambda_i^{1/p})^{np/2} f(p\bar{\lambda}^{-1})$$

$$\leqslant |\boldsymbol{S}|^{-n/2} \bar{\lambda}^{-np/2} f(p/\bar{\lambda}) \text{ (by the arithmetic mean-geometric mean inequality)}$$

$$\leqslant |\boldsymbol{S}|^{-n/2} \lambda_{max}^{-np/2}(f) f(p/\lambda_{\max}(f)).$$

The above inequality holds for all $\lambda_1 \geqslant \cdots \geqslant \lambda_p > 0$ with equality when $\lambda_1 = \cdots = \lambda_p = (1/p)\lambda_{\max}^{-1}(f)$. The theorem follows from Lemma 4.1.2. □

Theorem 4.1.1 and Lemma 4.1.2 are due to Anderson & Fang (1982c).

There is a fact that the MLE of (μ, Σ) is possibly not unique though $f(\cdot)$ is strictly monotone (cf. Exercise 4.11).

4.1.2 Examples

Example 4.1.1. (Normal) Suppose $\boldsymbol{X} \sim N_{n\times p}(\mathbf{1}\mu', \boldsymbol{I}_n \otimes \Sigma)$, i.e., $\boldsymbol{X}$'s rows, $\boldsymbol{x}_{(1)}, \ldots, \boldsymbol{x}_{(n)}$, are i.i.d and $\boldsymbol{x}_{(1)} \sim N_p(\mu, \Sigma)$. Thus $\boldsymbol{X}$ has likelihood

$$L(\mu, \Sigma) = |\Sigma|^{-n/2} (2\pi)^{-np/2} \exp\{(-1/2)\Sigma^{-1}\boldsymbol{G}\},$$

where $\boldsymbol{G}$ is defined by (4.1.2). According to the notations in Section 4.1.1, we have $f(x) = (2\pi)^{-np/2} e^{-x/2}$, $x \geqslant 0$. It is easy to see that $\lambda_{\max}(f) = 1/\text{n}$. Hence by Theorem 4.1.1 we obtain the MLE's, $\hat{\mu}$ and $\hat{\Sigma}$, of μ and Σ. They are

$$\hat{\mu} = \bar{\boldsymbol{x}} = \frac{1}{n}\sum_{i=1}^{n} \boldsymbol{x}_{(i)} \tag{4.1.15}$$

and

$$\hat{\Sigma} = \frac{1}{n}\boldsymbol{S} = \frac{1}{n}\sum_{i=1}^{n} (\boldsymbol{x}_{(i)} - \bar{\boldsymbol{x}})(\boldsymbol{x}_{(i)} - \bar{\boldsymbol{x}})', \tag{4.1.16}$$

respectively. The maximum likelihood value is

$$L(\hat{\mu}, \Sigma) = (2\pi)^{-np/2} |\boldsymbol{S}|^{-n/2} n^{np/2} e^{-np/2}. \tag{4.1.17}$$

Example 4.1.2. Let $\boldsymbol{X} \sim VS_{n\times p}(\mu, \Sigma, f)$ where $f(\cdot) = c(1+\cdot)^{-(np+1)/2}$. Now f satisfies the conditions of Theorem 4.1.1 and $\lambda_{\max}(f) = 1/np$. Hence the MLE of (μ, Σ) is $(\bar{\boldsymbol{x}}, \frac{1}{np}\boldsymbol{S})$ which is different from that in the case of normal.

Example 4.1.3. (Compound normal) Let $\boldsymbol{X} \sim VS_{n\times p}(\mu, \Sigma, f)$ and f be given by

$$f(u) = \int_0^\infty e^{-ru} dG(r), \ u \geqslant 0,$$

where G(·) is a distribution function so that G([0, ∞)) = 1. Clearly f(u) satisfies the conditions of Theorem 4.1.1. Then consider the equation $f(x) + (np/2x) f(x) = 0$, i.e.,

$$x\int_0^\infty re^{-rx} dG(r) = (np/2)\int_0^\infty e^{-rx} dG(r),$$

equivalently, $\int_0^\infty (xr - np/2) e^{-rx} dG(r) = 0$. In general it is difficult to get an analytic solution of the equation.

4.1.3. MLE's of μ and Σ in $LS_{n\times p}(\mu, \Sigma, f)$ and $SS_{n\times p}(\mu, \Sigma, \mathbf{f})$

Suppose $\boldsymbol{X} \sim LS_{n\times p}(\mu, \Sigma, f)$, i.e., $\boldsymbol{X}$ has a density (cf. Section 3.2.4)

$$|\Sigma|^{-n/2} f(\Sigma^{-1/2}(\boldsymbol{X} - \mathbf{1}\mu')'(\boldsymbol{X} - \mathbf{1}\mu')\Sigma^{-1/2}). \tag{4.1.18}$$

Similarly to (4.1.8), we can rewrite (4.1.18) as follows:

$$|\Sigma|^{-n/2} f(\Sigma^{-1/2}[\boldsymbol{S} + n(\bar{\boldsymbol{x}} - \mu)(\bar{\boldsymbol{x}} - \mu)']\Sigma^{-1/2}). \tag{4.1.19}$$

In this paragraph suppose that f is:

(a) nonincreasing, i.e., $\Sigma_1 \geqslant \Sigma_2 \geqslant 0$ implies $f(\Sigma_1) \leqslant f(\Sigma_2)$;

(b) continuous, i.e., $f(\Sigma) \to f(\Sigma_0)$ as $\Sigma \to \Sigma_0$ along nonnegative definite matrices, for each $\Sigma_0 \geqslant 0$.

Lemma 4.1.3. $\boldsymbol{S} > 0$ *with probability one iff* $n > p$.

Lemma 4.1.4. $|\Sigma|^{n/2} f(\Sigma)$ *arrives at its maximum at some positive definite matrix,* Σ^* *say.*

The proofs of Lemma 4.1.3 and Lemma 4.1.4 are similar to those of Lemma 4.1.1 and Lemma 4.1.2 and are left to the reader. (See Zhang, Fang and Chen, 1985.)

Theorem 4.1.2. μ *and* Σ *have the maximum likelihood estimators* $\hat{\mu} = \bar{\boldsymbol{x}}$ *and* $\hat{\Sigma} = \boldsymbol{S}^{1/2}\boldsymbol{G}^*\boldsymbol{S}\boldsymbol{G}^*\boldsymbol{S}^{-1/2}$ *where* $\boldsymbol{G}^* = (\boldsymbol{S}^{1/2}\Sigma^*\boldsymbol{S}^{1/2})^{-1/2}$ *with* Σ^* *being defined by Lemma 4.1.4.*

Proof. Since f is nonincreasing in the sense of (a), for a sample $\boldsymbol{X}$, the likelihood function

$$L(\mu, \Sigma) = f(\Sigma^{-1/2}[\boldsymbol{S} + n(\bar{\boldsymbol{x}} - \mu)(\bar{\boldsymbol{x}} - \mu)']\Sigma^{-1/2})|\Sigma|^{-n/2}$$

as a function of μ, i.e., for any given $\Sigma > 0$, arrives at its maximum at $\mu = \bar{\boldsymbol{x}}$ and the concentrated likelihood is

$$L(\bar{\mathbf{x}}, \Sigma) = |\Sigma|^{-n/2} f(\Sigma^{-1/2}\boldsymbol{S}\Sigma^{-1/2}).$$

Let $\boldsymbol{W} = \Sigma^{-1/2}\boldsymbol{S}\Sigma^{-1/2}$. Then $\boldsymbol{w} > 0$ with probability one by Lemma 4.1.3. Thus we have

$$L(\bar{\boldsymbol{x}}, \Sigma) = |\boldsymbol{W}|^{n/2}|\boldsymbol{S}|^{-n/2} f(\boldsymbol{W}). \tag{4.1.20}$$

Using Lemma 4.1.4, we show that the right-hand side of equation (4.1.20) as a function of $\boldsymbol{W}$ arrives at its maximum at $\Sigma^* > 0$. Now $\Sigma^* = \hat{\Sigma}^{-1/2}\boldsymbol{S}\hat{\Sigma}^{-1/2}$. Put $\boldsymbol{G}^* \hat{=} (\boldsymbol{S}^{-1/2}\hat{\Sigma}^{1/2}\boldsymbol{S}^{-1/2})^{-1/2}$, that is, $\hat{\Sigma} = \boldsymbol{S}^{1/2}\boldsymbol{G}^*\boldsymbol{S}\boldsymbol{G}^*\boldsymbol{S}^{1/2}$, and the proof is complete.□

Let $\boldsymbol{X} \sim SS_{n\times p}(\mu, \Sigma, f)$, that is, $\boldsymbol{X}$ has a density

$$\begin{aligned} &|\Sigma|^{-n/2} f(\lambda[(\boldsymbol{X} - \mathbf{1}\mu')\Sigma^{-1}(\boldsymbol{X} - \mathbf{1}\mu')']) \\ &= |\Sigma|^{-n/2} f(\lambda[\Sigma^{-1}\boldsymbol{S} + n\Sigma^{-1}(\bar{\boldsymbol{x}} - \mu)(\bar{\boldsymbol{x}} - \mu)']), \end{aligned} \tag{4.1.21}$$

where $\lambda(\boldsymbol{A}) = \mathrm{diag}(\lambda_1, \ldots, \lambda_p)$, and $\lambda_1 \geqslant \cdots \geqslant \lambda_p$ are p eigenvalues of $\boldsymbol{A}$. Assume that the function f satisfies conditions (a) and (b). It may be shown that when $f(\lambda(\boldsymbol{A}))$ is a symmetric function of $\lambda_1, \ldots, \lambda_p$, where $\lambda(\mathrm{A}) = \mathrm{diag}(\lambda_1, \ldots, \lambda_p)$, $\Sigma^* = \alpha_0\boldsymbol{I}_p$ and is given by Lemma 4.1.4. Via Theorem 4.1.2, the MLE's of μ and Σ are $\hat{\mu} = \bar{\boldsymbol{x}}$ and $\hat{\Sigma} = (1/\alpha_0)\boldsymbol{S}$ because $\boldsymbol{G}^* = (\boldsymbol{S}^{1/2}\Sigma^*\boldsymbol{S}^{1/2})^{-1/2} = \alpha_0^{1/2}\mathrm{S}^{-1/2}$. The maximum value of likelihood is

$$L(\hat{\mu}, \hat{\Sigma}) = \alpha_0^{n/2}|\boldsymbol{S}|^{-n/2} f(\alpha_0\boldsymbol{I}_p). \tag{4.1.22}$$

4.1.4. MLE's of Parametric Functions

Let x be a sample from some population, $\boldsymbol{F}_\theta$ say, $\theta \in \Theta$, where θ is a parameter and Θ a parametric space. Let the $\boldsymbol{F}_\theta$ have a density $f(x|\theta)$. Thus the likelihood function is $L(\theta) = f(x|\theta)$. We are required to estimate some parametric function, $g(\theta)$ say, through the sample x. Let

$$\Theta^* \hat{=} \{\theta^* | g(\theta) = \theta^*, \theta \in \Theta\} \text{ and } C(\theta^*) \hat{=} \{\theta \in \Theta | g(\theta) = \theta^*\}.$$

Put $M(\theta^*) = \sup_{\theta\in C(\theta^*)} L(\theta)$, $\theta^*\in\Theta^*$. Maybe it is reasonable to call $M(\theta^*)$ the induced likelihood from $L(\theta)$.

Definition 4.1.1. $\hat{\theta}^*$ is called a *maximum induced likelihood estimator* of $g(\theta)$ if $\hat{\theta}^*\in\Theta^*$ such that $M(\tilde{\theta}^*) \geqslant M(\theta^*)$ for each $\theta^*\in\Theta^*$. Simply, we call $\hat{\theta}^*$ a maximum likelihood estimator of $g(\theta)$.

The following lemma gives an important property, usually called the *invariance property* of the method of maximum likelihood in statistical estimation. The word "invariance" refers to such a problem: If $g(\theta)$ is a one-to-one function from Θ onto Θ^*, then we may get two maximum likelihood estimators $\hat{\theta}$ and $\hat{\theta}^*$ of θ and $g(\theta)$, respectively. It may be natural to demand that $\hat{\theta}^* = g(\hat{\theta})$ if both $\hat{\theta}^*$ and $\hat{\theta}$ exist. The following lemma shows that this is true for an arbitrary mapping $g(\cdot)$, i.e., $g(\cdot)$ may not be one-to-one.

Lemma 4.1.5. *Let $\theta\in\Theta$ (often being an interval in R^k) and let $g(\cdot)$ be an arbitrary transformation mapping Θ to Θ^* (being an interval in R^r). Then $g(\hat{\theta})$ is a maximum induced likelihood estimator of $g(\theta)$, provided that the maximum likelihood estimator $\hat{\theta}$ of θ exists.*

Proof. Since $\hat{\theta}\in\Theta$, $g(\hat{\theta})\in\Theta^*$. Clearly, $\{C(\theta^*)\,|\,\theta^*\in\Theta^*\}$ is a partition of Θ and $\hat{\theta}$ belongs to one and only one cell of this partition, $C(\hat{\theta}^*)$ say. So we have

$$L(\hat{\theta}) = \sup_{\theta\in C(\hat{\theta}^*)} L(\theta) = M(\hat{\theta}^*) \leqslant \sup_{\theta^*\in\Theta^*} M(\theta^*) = \sup_{\theta^*\in\Theta^*}\ \sup_{\theta\in C(\theta^*)} L(\theta)$$
$$= \sup_{\theta\in\Theta} L(\theta) = L(\hat{\theta}),$$

that is, $M(\hat{\theta}^*) = \sup_{\theta^*\in\Theta^*} M(\theta^*)$. As $\hat{\theta}\in C(\hat{\theta}^*)$ we get $\hat{\theta}^* = g(\hat{\theta})$ by $C(\hat{\theta}^*)$'s definition. □

Let $\boldsymbol{X} = (X_{ij}) \sim VS_{n\times p}(\mu, \Sigma, \phi)$, and be nondegenerate. When $\boldsymbol{X} = (\boldsymbol{x}_{(1)}, \ldots, \boldsymbol{x}_{(n)})'$ has the second order moment, $\mathscr{D}(\boldsymbol{x}_{(k)}) = -2\phi'(0)\Sigma \mathrel{\hat{=}} \sigma\Sigma = (\sigma\sigma_{ij})$, $k=1, \ldots, n$ (cf. Theorem 2.6.5). Hence the correlation coefficient between X_{ki} and X_{kj} is

$$\text{(4.1.23)}\qquad \rho_{ij} = \mathrm{Corr}(X_{ki}, X_{kj}) = \frac{\mathrm{Cov}(X_{ki}, X_{kj})}{[\mathrm{Var}(X_{ki})\mathrm{Var}(X_{kj})]^{1/2}} = \frac{\sigma_{ij}}{(\sigma_{ii}\sigma_{jj})^{1/2}},$$

$i,j=1, \ldots, p$; $k=1, \ldots, n$ (See Definition 2.2.2). From Corollary 3 of Lemma 2.6.1, partitioning $\boldsymbol{x}_{(k)}$ as $\boldsymbol{x}_{(k)} = (\boldsymbol{x}'_{(k1)}, \boldsymbol{x}'_{(k2)})'$, $k=1, \ldots, n$, *where* $\boldsymbol{x}_{(k1)}: m\times 1$, we have

$$\varepsilon(\boldsymbol{x}_{(k1)}\,|\,\boldsymbol{x}_{(k2)}) = \mu^{(1)} + \Sigma_{12}\Sigma_{22}{}^{-1}(\boldsymbol{x}_{(k2)} - \mu^{(2)})$$

and

$$\mathscr{D}(\boldsymbol{x}_{(k1)}\,|\,\boldsymbol{x}_{(k2)}) = \sigma(\boldsymbol{x}_{(k2)})\Sigma_{11.2},$$

where $\mu = \begin{pmatrix}\mu(1)\\ \mu(2)\end{pmatrix}$, $\Sigma = \begin{pmatrix}\Sigma_{11} & \Sigma_{12}\\ \Sigma_{21} & \Sigma_{22}\end{pmatrix}$, $\Sigma_{11.2} = \Sigma_{11} - \Sigma_{12}\Sigma_{22}^{-1}\Sigma_{21}$ and $\sigma(\boldsymbol{x}_{(k2)})>0$. $\beta_{1.2} \mathrel{\hat{=}} \Sigma_{12}\Sigma_{22}^{-1}$ is called the regression coefficients of $\boldsymbol{x}_{(k1)}$ on $\boldsymbol{x}_{(k2)}$. When $\boldsymbol{X} \sim N_{n\times p}(\boldsymbol{1}\mu', \boldsymbol{I}_n\otimes\Sigma)$ and $\sigma(\boldsymbol{x}_{(k2)}) = 1$, $\Sigma_{11.2}$ is called the *conditional covariance matrix.*

Theorem 4.1.3 *Let $\boldsymbol{X} \sim VS_{n\times p}(\mu, \Sigma, f)$ with $n>p$ and f being nonincreasing and continuous. Then the maximum likelihood estimators of ρ_{ij}, $\beta_{1.2}$, $\Sigma_{11.2}$ are respectively*

$$\hat{\rho}_{ij} = \frac{S_{ij}}{(S_{ii}S_{jj})^{1/2}},\quad i, j=1, \ldots, p,$$

$$\hat{\beta}_{1.2} = S_{21}S_{11}^{-1},$$

$$\hat{\Sigma}_{11.2} = \lambda_{\max}(f)\,(S_{11} - S_{12}S_{22}^{-1}S_{21}),$$

where $S = (S_{ij}) = \begin{pmatrix} S_{11} & S_{12} \\ S_{21} & S_{22} \end{pmatrix}$ *is defined by* (4.1.4).

Proof. This follows from Theorem 4.1.1 and Lemma 4.1.5. □

Let $X = (X_1, ..., X_n)'$ be a random vector with covariance matrix $\Sigma > 0$. Partition x and Σ as

$$x = \begin{pmatrix} X_1 \\ x^{(2)} \end{pmatrix}, \ \Sigma = \begin{pmatrix} \sigma_{11} & \sigma_{12} \\ \sigma_{21} & \Sigma_{22} \end{pmatrix},$$

where $x^{(2)}$:$(n-1) \times 1$ and Σ_{22}:$(n-1) \times (n-1)$.

Definition 4.1.2. The *multiple correlation coefficient between* X_1 and $X_2, ..., X_n$ is the maximum correlation between X_1 and any linear function $a'x^{(2)}$ of $X_2, ..., X_n$.

Lemma 4.1.6. *With the above notations, of all linear combinations* $a'x^{(2)}$, *the combination that minimizes the variance of* $X_1 - a'x^{(2)}$ *and maximizes the correlation between* X_1 *and* $a'x^{(2)}$, *is the linear combination* $\beta' x^{(2)}$, *where* $\beta' = \sigma_{12}\Sigma_{22}^{-1}$, *if* $\mathrm{Var}(X_1) > 0$.

Proof. Without loss of generality assume $\mu = 0$. (Otherwise, $x - \mu$ is replaced by x.) Since $\beta' = \sigma_{12}\Sigma_{22}^{-1}$, $\mathcal{E}(X_1 x^{(2)\prime}) = \sigma_{12}$ and $\mathcal{E}(x^{(2)}x^{(2)\prime}) = \Sigma_{22}$, we have, for each a,

$$\begin{aligned} &\mathcal{E}[(X_1 - \beta' x^{(2)})(\beta - a)' x^{(2)}] \\ &= \mathcal{E}[(X_1 - \sigma_{12}\Sigma_{22}^{-1}x^{(2)})\,x^{(2)\prime}(\beta - a)] \\ &= (\sigma_{12} - \sigma_{12}\Sigma_{22}^{-1}\Sigma_{22})(\beta - a) = 0. \end{aligned} \tag{4.1.24}$$

Via the above fact we have, for each a,

$$\begin{aligned} &E(X_1 - a'x^{(2)})^2 \\ &= E(X_1 - \beta' x^{(2)})^2 + E(\beta' x^{(2)} - a' x^{(2)})^2 \\ &= E(X_1 - \beta' x^{(2)})^2 + (\beta - a)'\Sigma_{22}(\beta - a) \\ &\geqslant E(X_1 - \beta' x^{(2)})^2. \end{aligned} \tag{4.1.25}$$

Moreover, in (4.1.25) the equality holds iff $a = \beta$, and the first assertion of the lemma follows.

Now we show that the maximum correlation between X_1 and $a'x^{(2)}$ is given by $a = \beta$. By the Schwarz inequality,

$$\begin{aligned} &\frac{E(X_1 a' x^{(2)})}{[\sigma_{11}E(a'x^{(2)})^2]^{1/2}} \\ &= \frac{\sigma_{12}a}{(\sigma_{11}a'\Sigma_{22}a)^{1/2}} = \frac{\beta'\Sigma_{22}a}{(\sigma_{11}a'\Sigma_{22}a)^{1/2}} \\ &\leqslant \left(\beta'\Sigma_{22}^{1/2}\Sigma_{22}^{1/2}\beta \cdot \frac{a'\Sigma_{22}^{1/2}\Sigma_{22}^{1/2}a}{\sigma_{11}a'\Sigma_{22}a}\right)^{1/2} \end{aligned} \tag{4.1.26}$$

$$= (\sigma_{11}^{-1}\beta'\Sigma_{22}\beta)^{1/2}, \text{ for each } \boldsymbol{a}.$$

The equality holds in (4.1.26) iff $\Sigma_{22}^{1/2}\beta = \Sigma_{22}^{1/2}\boldsymbol{a}$, i.e., $\beta = \boldsymbol{a}$, which proves the lemma. □

Corollary. *Let $\boldsymbol{x} \sim EC_p(\mu, \Sigma, \phi)$ be nondegenerate with the 2nd moment and $\Sigma > 0$. Then the multiple correlation coefficient between X_1 and $\boldsymbol{x}^{(2)}$ is*

$$R = \frac{(\sigma_{12}\Sigma_{22}^{-1}\sigma_{21})^{1/2}}{(\sigma_{11})^{1/2}}.$$

Proof. By Theorem 2.6.5, we obtain that the covariance matrix of $\boldsymbol{x}$ is $-2\phi'(0)\Sigma \mathrel{\hat{=}} \sigma\Sigma$ (say). The corollary follows from Lemma 4.1.6. □

Theorem 4.1.4. *Let $\mathbf{X} \sim VS_{n\times p}(\mu, \Sigma, f)$ with $n>p$ and f being nonincreasing and continuous and let $\mathbf{X} = (\boldsymbol{x}_{(1)}, \dots, \boldsymbol{x}_{(n)})'$, $\boldsymbol{x}_{(k)} = \begin{pmatrix} \boldsymbol{x}_{(k1)} \\ \boldsymbol{x}_{(k2)} \end{pmatrix}$, where $\boldsymbol{x}_{(k1)}$:$q\times 1$, and X_{kj} be a component of $\boldsymbol{x}_{(k1)}$, $k=1, \dots, n$. Write*

$$\Sigma = \begin{pmatrix} \Sigma_{11} & \Sigma_{12} \\ \Sigma_{21} & \Sigma_{22} \end{pmatrix}, \quad \boldsymbol{S} = \sum_{i=1}^{n} (\boldsymbol{x}_{(i)} - \bar{\boldsymbol{x}})(\boldsymbol{x}_{(i)} - \bar{\boldsymbol{x}})' = \begin{pmatrix} \boldsymbol{S}_{11} & \boldsymbol{S}_{12} \\ \boldsymbol{S}_{21} & \boldsymbol{S}_{22} \end{pmatrix} = (\mathrm{S}_{ij}),$$

where Σ_{11} and $\boldsymbol{S}_{11}$ are $q\times q$ matrices. Then

(i) *the multiple correlation coefficient between X_{kj} and $\boldsymbol{x}_{(k2)}$, (given k, considering the vector $\boldsymbol{x}_{(k)}$), is*

$$R = \frac{(\boldsymbol{b}'\Sigma_{12}\Sigma_{22}^{-1}\Sigma_{21}\boldsymbol{b})^{1/2}}{(\sigma_{jj})^{1/2}}, \qquad \text{for all } 1 \leqslant k \leqslant n, \tag{4.1.27}$$

where $\boldsymbol{b}' = \boldsymbol{e}_j(q)$. In particular, when $q=1$,

$$R_0 = \frac{(\Sigma_{12}\Sigma_{22}^{-1}\Sigma_{21})^{1/2}}{(\sigma_{11})^{1/2}}. \tag{4.1.28}$$

(ii) *MLE's of R and R_0 (as given above) are, respectively,*

$$\hat{R} = \frac{(\boldsymbol{b}'\boldsymbol{S}_{12}\boldsymbol{S}_{22}^{-1}\boldsymbol{S}_{21}\boldsymbol{b})^{1/2}}{(\mathrm{S}_{ij})^{1/2}} \tag{4.1.29}$$

and

$$\hat{R}_0 = \frac{(\boldsymbol{S}_{12}\boldsymbol{S}_{22}^{-1}\boldsymbol{S}_{21})^{1/2}}{(\boldsymbol{S}_{11})^{1/2}}. \tag{4.1.30}$$

Proof. Since $\boldsymbol{x}_{(1)} \overset{d}{=} \cdots \overset{d}{=} \boldsymbol{x}_{(n)}$, the theorem follows from corollary of Lemma 4.1.6 and Theorem 4.1.1. □

It is worth noting that R defined by (4.1.27) is independent of the density $f(\cdot)$.

4.2. The Distributions of Some Estimators

4.2.1. Joint Density

Let $\mathbf{X} \sim LS_{n\times p}(\mu, \Sigma, f)$ with $n>p$, i.e., $\mathbf{X}$ have a density

$$\begin{aligned} &|\Sigma|^{-n/2} f(\Sigma^{-1/2}(\mathbf{X} - \mathbf{1}\mu')'(\mathbf{X} - \mathbf{1}\mu')\Sigma^{-1/2}) \\ &= |\Sigma|^{-n/2} f(\Sigma^{-1/2}[\boldsymbol{S} + n(\bar{\boldsymbol{x}} - \mu)(\bar{\boldsymbol{x}} - \mu)']\Sigma^{-1/2}), \end{aligned} \tag{4.2.1}$$

where $S = \sum_{i=1}^{n} (x_{(i)} - \bar{x})(x_{(i)} - \bar{x})' = X'(I_n - P_n)X$ and $P_n = \frac{1}{n}\mathbf{1}\mathbf{1}'$. By virtue of (4.1.10), $S = \sum_{i=1}^{n-1} z_{(i)} z_{(i)}' = B'B$ where $Z = (z_{(1)}, \ldots, z_{(n)})' = \Gamma X$, with $\Gamma \in O(n)$ and its nth row being $(n^{-1/2}, \ldots, n^{-1/2})$. It is easy to see that Z has a density

$$|\Sigma|^{-n/2} f(\Sigma^{-1/2}[B'B + n(n^{1/2}z_{(n)} - \mu)(n^{1/2}z_{(n)} - \mu)']\Sigma^{-1/2}). \tag{4.2.2}$$

Now we can get the joint density of $(\bar{x}, S)$.

Theorem 4.2.1. *Let* $X \sim LS_{n\times p}(\mu, \Sigma, f)$ *with* $n>p$, *i.e.,* X *has a density* (4.2.1). *Then the joint density of* $(\bar{x}, S)$ *is*

$$\left[n^{p/2}\pi^{(n-1)p/2} / \Gamma_p\left(\frac{n-1}{2}\right)\right] |S|^{(n-p)/2-1} |\Sigma|^{-n/2} f(\Sigma^{-1/2}[S + n(\bar{x} - \mu)(\bar{x} - \mu)']\Sigma^{-1/2}),\quad S>0 \tag{4.2.3}$$

where $\Gamma_p(n/2) = \pi^{p(p-1)/4} \prod_{j=1}^{p} \Gamma((n-j+1)/2)$ *is the multivariate gamma function (c.f. Exercise* 3.17).

Proof. By (3.4.2) and (4.2.2), for any nonnegative Borel function g, we have

$$E[g(\bar{x}, S)] = E[g(n^{1/2}z_{(n)}, B'B)]$$

$$= \int g(n^{1/2}z_{(n)}, B'B)|\Sigma|^{-1/2} f(\Sigma^{-1/2}[B'B + n(n^{1/2}z_{(n)} - \mu)(n^{1/2}z_{(n)} - \mu)']\Sigma^{-1/2})dZ$$

$$= \left[\pi^{(n-1)p/2} / \Gamma_p\left(\frac{n-1}{2}\right)\right] \int_{S>0} g(n^{1/2}z_{(n)}, S)|S|^{(n-p)/2-1}|\Sigma|^{-n/2} f(\Sigma^{-1/2}[S + n(n^{1/2}z_{(n)} - \mu)(n^{1/2}z_{(n)} - \mu)']\Sigma^{-1/2})dSdz_{(n)}. \tag{4.2.4}$$

Again let $\bar{x} = n^{1/2}z_{(n)}$. And from (4.2.4), we have

$$E[g(\bar{x}, S)] = \left[n^{p/2}\pi^{(n-1)p/2} / \Gamma_p\left(\frac{n-1}{2}\right)\right] \int_{S>0} g(\bar{x}, S)|S|^{(n-p)/2-1} |\Sigma|^{-n/2} f(\Sigma^{-1/2}[S + n(\bar{x} - \mu)(\bar{x} - \mu)']\Sigma^{-1/2})dSd\bar{x}, \tag{4.2.5}$$

which shows that $(\bar{x}, S)$ has a joint density (4.2.3). □

Corollary 1. *Let* $X \sim VS_{n\times p}(\mu, \Sigma, f)$ *with* $n>p$. *Then* $(\bar{x}, S)$ *has a joint density*

$$\left[n^{p/2}\pi^{(n-1)p/2} / \Gamma_p\left(\frac{n-1}{2}\right)\right] |S|^{(n-p)/2-1} |\Sigma|^{-n/2} f(\mathrm{tr}\Sigma^{-1}S + n(\bar{x} - \mu)'\Sigma^{-1}(\bar{x} - \mu)),\ S > 0,\ \bar{x} \in R^p. \tag{4.2.6}$$

Corollary 2. *Let* $X \sim N_{n\times p}(\mathbf{1}\mu', I_n \otimes \Sigma)$ *with* $n>p$ *and* $\Sigma>0$. *Then*

(i) $\bar{x}$ *and* S *are independent, and*

(ii) $\bar{x} \sim N(\mu, \frac{1}{n}\Sigma)$ *and* $S \sim W_p(n-1, \Sigma)$.

Proof. Using Corollary 1 for $f(\mathrm{tr}(\cdot)) = (2\pi)^{-np/2}\exp\{-\frac{\cdot}{2}\}$, we directly show (i) and (ii). □

4.2.2. Marginal Density

In the preceding section, Corollary 2 of Theorem 4.2.1 gives the marginal densities of $\bar{\boldsymbol{x}}$ and $\boldsymbol{S}$ in the normal case. Now we would like to derive them from (4.2.6) under the assumption of $\boldsymbol{X} \sim VS_{n\times p}(\mu, \Sigma, f)$.

Theorem 4.2.2. *Let* $\boldsymbol{X} \sim VS_{n\times p}(\mu, \Sigma, f)$ *with* $n>p$. *Then*

(i) $\bar{\boldsymbol{x}} \overset{d}{=} \mu + n^{-1/2}R^*\Sigma^{1/2}u^{(p)}$ *where* R^* *and* $u^{(p)}$ *are independent,* $R^* \overset{d}{=} Rb_{p/2,(n-1)p/2}$, *with* $b_{p/2,(n-1)p/2}$ *being independent of* $R>0$ *and* R *having a density*

$$[2\pi^{np/2}/\Gamma(np/2)]r^{np-1}f(r),\ r>0. \tag{4.2.7}$$

(ii) $\boldsymbol{S}$ *has a density*

$$\left[2\pi^{p/2}/\Gamma(p/2)\Gamma_p\left(\frac{n-1}{2}\right)\right]|\Sigma|^{-(n-1)/2}|\boldsymbol{S}|^{(n-p)/2-1}\int_0^\infty r^{p-1} \times f(r^2+\mathrm{tr}\Sigma^{-1}\boldsymbol{S})dr. \tag{4.2.8}$$

Proof. To show (i), first, let $\boldsymbol{X} \sim VS_{n\times p}(\boldsymbol{O}, \boldsymbol{I}_p, \phi)$. Then $\mathrm{vec}\boldsymbol{X} \overset{d}{=} \mathrm{vec}\boldsymbol{X}' \overset{d}{=} R\,\boldsymbol{u}^{(np)}$ where $R \leftrightarrow \phi \in \Phi_{np}$ (cf. Section 2.5) is independent of $\boldsymbol{u}^{(np)}$. Hence $\bar{\boldsymbol{x}} = \frac{1}{n}\boldsymbol{X}'\boldsymbol{1} = \frac{1}{n}(\boldsymbol{1}'\otimes\boldsymbol{I}_p)\mathrm{vec}\boldsymbol{X}' \overset{d}{=} \frac{1}{n}(\boldsymbol{1}'\otimes\boldsymbol{I}_p)(R\boldsymbol{u}^{(np)})$. By Corollary 1 of Theorem 2.6.1

$$\bar{\boldsymbol{x}} \sim \mathrm{EC}_p\left(0, \frac{1}{n}\boldsymbol{I}_p, \phi\right), \text{ with } \phi\in\Phi_{np},$$

because

$$\left(\frac{1}{n}\boldsymbol{1}'\otimes\boldsymbol{I}_p\right)\left(\frac{1}{n}\boldsymbol{1}'\otimes\boldsymbol{I}_p\right)' = \frac{1}{n}(1\otimes\boldsymbol{I}_p) = \frac{1}{n}\boldsymbol{I}_p.$$

This means that

$$\bar{\boldsymbol{x}} \overset{d}{=} n^{-1/2}R^*\boldsymbol{u}^{(p)},$$

where $R^* \overset{d}{=} Rb_{p/2,(n-1)p/2}$ as desired (cf. Section 2.5.1). Now we turn back to the case of $\boldsymbol{X} \sim VS_{n\times p}(\mu, \Sigma, f)$. By considering the transformation $\Sigma^{-1/2}(\bar{\boldsymbol{x}}-\mu)$ the first assertion follows from the above result.

The second assertion can be found by integrating (4.2.6) with respect to $\bar{\boldsymbol{x}}$ and using Formula (2.4.15). □

4.2.3. Independence of $\bar{\boldsymbol{x}}$ and $\boldsymbol{S}$

Corollary 2 of Theorem 4.2.1 shows that $\bar{\boldsymbol{x}}$ and $\boldsymbol{S}$ are independent when $\boldsymbol{X} \sim N_{n\times p}(\boldsymbol{1}\mu', \boldsymbol{I}_n\otimes\Sigma)$. We would like to state that it arises only when $\boldsymbol{X}$ are normal. Hence the independency of $\bar{\boldsymbol{x}}$ and $\boldsymbol{S}$ is a characterization of normality.

Theorem 4.2.3. *Suppose* $\boldsymbol{X} \sim VS_{n\times p}(\mu, \Sigma, f)$ *with* $n>p$. *Then* $\bar{\boldsymbol{x}}$ *and* $\boldsymbol{S}$ *are independent iff* $\boldsymbol{X}$ *is normal.*

Proof. We need only to prove the necessity. Noting that $\bar{\boldsymbol{x}}$ and $\boldsymbol{S}$ are independent iff $\Sigma^{-1/2}(\bar{\boldsymbol{x}}-\mu)$ and $\Sigma^{-1/2}\boldsymbol{S}\Sigma^{-1/2}$ are independent, we may assume that $\mu=\boldsymbol{0}$ and $\Sigma=\boldsymbol{I}_p$. It is well known that there exists a $\Gamma\in O(n)$ such that

$$\boldsymbol{S}=\sum_{i=1}^{n-1}\boldsymbol{z}_{(i)}\boldsymbol{z}_{(i)}' \text{ and } \bar{\boldsymbol{x}}=n^{1/2}\boldsymbol{z}_{(n)},$$

where $\boldsymbol{Z}=(\boldsymbol{z}_{(1)},\ldots,\boldsymbol{z}_{(n)})'=\Gamma\boldsymbol{X}$. Now that $\bar{\boldsymbol{x}}$ and $\boldsymbol{S}$ are independent implies that tr $\boldsymbol{S}$ $=\boldsymbol{z}_{(1)}'\boldsymbol{z}_{(1)}+\cdots+\boldsymbol{z}_{(n-1)}'\boldsymbol{z}_{(n-1)}$ and $\bar{\boldsymbol{x}}'\bar{\boldsymbol{x}}=n\boldsymbol{z}_{(n)}'\boldsymbol{z}_{(n)}$ are independent. By Theorem 2.8.1, $\boldsymbol{Z}$ is normal, equivalently, $\boldsymbol{X}$ is normal, which proves the theorem. □

Remark. For the "only if" part of Theorem 4.2.3, it is not necessary to suppose $\boldsymbol{X}$ has a density, and we need only to put the condition $P(\boldsymbol{X}'\boldsymbol{X}>0)=1$ on $\boldsymbol{X}$ (See Exercise 4.5).

4.2.4. Distribution of Sample Correlation Coefficients

Let $\boldsymbol{X}\sim VS_{n\times p}(\mu,\Sigma,f)$, $\Sigma=(\sigma_{ij})$. Then the correlation coefficient matrix is defined as

$$\boldsymbol{R}=(\rho_{ij}), \tag{4.2.9}$$

where $\rho_{ij}=\sigma_{ij}/(\sigma_{ii}\sigma_{jj})^{1/2}$, $i,j=1,\ldots,p$. (See (4.1.23)). From Theorem 4.1.3, we know that when f is nonincreasing and continuous and $n>p$, the MLE of $\boldsymbol{R}$ is

$$\hat{\boldsymbol{R}}=(\hat{\rho}_{ij}), \tag{4.2.10}$$

where $\hat{\rho}_{ij}=S_{ij}/(S_{ii}S_{jj})^{1/2}$ and $\boldsymbol{S}=(S_{ij})$ defined by (4.1.4). We may rewrite (4.2.10) as follows :

$$\hat{\boldsymbol{R}}=\operatorname{diag}(S_{11}^{-1/2},\ldots,S_{pp}^{-1/2})\,\boldsymbol{S}\,\operatorname{diag}(S_{11}^{-1/2},\ldots,S_{pp}^{-1/2}). \tag{4.2.11}$$

Theorem 4.2.4. *Let $\boldsymbol{X}\sim VS_{n\times p}(\mu,\boldsymbol{I}_n,f)$ with $n>p$. Then $\hat{\boldsymbol{R}}$ has a density*

$$\left[\pi^{np/2}[\Gamma((n-1)/2)]^p\Big/\Gamma_p[(n-1)/2]\Gamma\left(\frac{np}{2}\right)\right]$$

$$\times|\hat{\boldsymbol{R}}|^{(n-p-2)/2}\int_0^\infty r^{\frac{p}{2}-1}f(r)dr.$$

Proof. Consider the transformation

$$\boldsymbol{S}=\operatorname{diag}(S_{11}^{1/2},\ldots,S_{pp}^{1/2})\,\hat{\boldsymbol{R}}\,\operatorname{diag}(S_{11}^{1/2},\ldots,S_{pp}^{1/2}).$$

Then the Jacobian of the transformation is

$$J(\boldsymbol{S}\to\hat{\boldsymbol{R}},S_{11},\ldots,S_{pp})=\prod_{i<j}(S_{ii}S_{jj})^{1/2}=\prod_{i=1}^{p}S_{ii}^{(p-1)/2}.$$

By (4.2.8) we get the joint density of $\hat{\boldsymbol{R}}$, $S_{11},\ldots,S_{pp}$ as follows:

$$c\left[\prod_{i=1}^{p}S_{ii}^{(n-3)/2}\right]|\hat{\boldsymbol{R}}|^{(n-p-2)/2}\int_0^\infty r^{p-1}f\left(r^2+\sum_{i=1}^{p}S_{ii}\right)dr, \tag{4.2.12}$$

where $c=2\pi^{np/2}/\Gamma(p/2)\Gamma_p((n-1)/2)$. Integrating (4.2.12) with respect to $(S_{11},\ldots,S_{pp})$ and using Lemma 2.4.4, we complete the proof. □

Corollary. *If $\boldsymbol{X}\sim N_{n\times p}(\boldsymbol{1}\mu',\boldsymbol{I}_n\otimes\boldsymbol{I}_p)$, $\hat{\boldsymbol{R}}$ and $(S_{11},\ldots,S_{pp})$ are independent.*

Proof. Here $f(t) = (2\pi)^{-np/2}e^{-t/2}$. Thus, for some constant c,

$$f\left(r^2 + \sum_{i=1}^{p} S_{ii}\right) = c f(r^2) f\left(\sum_{i=1}^{p} S_{ii}\right).$$

Then the corollary follows from (4.2.12). □

Note. We speak of $\hat{\boldsymbol{R}}$'s density (or $\boldsymbol{S}$'s density) which means the density of independent matrix elements of $\hat{\boldsymbol{R}}$ (or $\boldsymbol{S}$). For example, $\hat{\boldsymbol{R}}$'s density means $\{\hat{\rho}_{12}, \ldots, \hat{\rho}_{1p}; \ldots, \hat{\rho}_{p-1,p-1}\}$'s joint density.

4.3. Properties of $\hat{\mu}$ and $\hat{\Sigma}$

In section 4.1, we have investigated the maximum likelihood estimates of mean vector and covariance matrix in spherical matrix distributions. In the present section, we try to discuss some of their simple properties, for example, unbiasedness, sufficiency, consistency, and so on. Further discussion of their properties will be found in Section 4.4.

4.3.1. Unbiasedness

Let $\boldsymbol{x}$ be a sample from a population P_θ, $\theta \in \Theta$ and let statistic $\boldsymbol{t}(\boldsymbol{x})$ be used to estimate $g(\theta)$ which is a function of parameter θ. We call $\boldsymbol{t}(\boldsymbol{x})$ an *unbiased estimate* of $g(\theta)$ if $\boldsymbol{E}_\theta(t(x)) = g(\theta)$ for each $\theta \in \Theta$, where $\boldsymbol{E}_\theta$ denotes the expectation under parametric value θ, but the subscript θ is often omitted if this does not cause any confusion.

Lemma 4.3.1 *Let* $\mathbf{X} = (\mathbf{x}_{(1)}, \ldots, \boldsymbol{x}_{(n)})' \sim LS_{n\times p}(\mu, \Sigma, \phi)$ *with the finite 2nd moment. Then*

(i) $\mathcal{E}(\bar{\boldsymbol{x}}) = \mu$ *and*
(ii) $\mathcal{E}(\boldsymbol{S}) = (n-1)(-2\phi'(0)\Sigma)$.

Proof. As $\boldsymbol{x}_{(i)} \sim EC_p(\mu, \Sigma, \phi)$, $i = 1, \ldots, n$, we have

$$\mathcal{E}(\bar{\boldsymbol{x}}) = \mathcal{E}\left(\frac{1}{n}\sum_{i=1}^{n} \boldsymbol{x}_{(i)}\right) = \frac{1}{n}\sum_{i=1}^{n} \mathcal{E}(\boldsymbol{x}_{(i)}) = \mu.$$

By (4.1.10), we have

$$\boldsymbol{S} = \sum_{i=1}^{n-1} \boldsymbol{z}_{(i)}\boldsymbol{z}_{(i)}',$$

where $\boldsymbol{z}_{(i)} \sim EC_p(\mathbf{0}, \Sigma, \phi)$, $i = 1, \ldots, n-1$. Hence

$$\mathcal{E}(\boldsymbol{S}) = \sum_{i=1}^{n-1} \mathcal{E}(\boldsymbol{z}_{(i)}\boldsymbol{z}_{(i)}') = (n-1)(-2\phi'(0)\Sigma),$$

by Theorem 2.6.5. The proof is completed. □

Corollary 1. *Let* $\boldsymbol{X} \sim VS_{n\times p}(\mu, \Sigma, f)$ *with the finite 2nd moment and let f be nonincreasing and continuous. Then the* $\hat{\mu} = \bar{\boldsymbol{x}}$ *and* $[2(1-n)\lambda_{\max}^{-1}(f)\phi'(0)]^{-1}\hat{\Sigma} = [2(1-n)\phi'(0)]^{-1}\boldsymbol{S}$ *are unbiased estimators of* μ *and* Σ *respectively.*

Proof. It follows from Lemma 4.3.1. □

Corollary 2. *Let* $\boldsymbol{X} \sim N_{n\times p}(\mathbf{1}\mu', \boldsymbol{I}_n \otimes \Sigma)$. *Then* $\left(\hat{\mu}, \frac{n}{n-1}\hat{\Sigma}\right) = \left(\bar{\boldsymbol{x}}, \frac{1}{n-1}\boldsymbol{S}\right)$ *is an unbiased estimator of* (μ, Σ).

Proof. By Corollary 1, we need only to verify $-\phi'(0)=\frac{1}{2}$ which is obvious from $\phi(t)=e^{t/2}$. □

4.3.2. Sufficiency

Sufficiency is a very important concept in statistical inference. If a statistic $T(\boldsymbol{X})$ induces a likelihood function equivalent to that of $\boldsymbol{X}$ it will give all the necessary statistical information and is called a sufficient statistic. Here our statement is only descriptive. And the rigorous définition can be found in many statistical textbook. The most useful theorem used to verify a statistic to be sufficient or not is the famous Fisher-Neyman Factorization Theorem: Let $\boldsymbol{X}$ have a density $f(\boldsymbol{X}|\theta)$, $\theta\in\Theta$. Then a statistic $T(\boldsymbol{X})$ is sufficient iff these can be found two nonnegative functions $g(t|\theta)$ (not necessarily a density) and $h(\boldsymbol{X})$ such that

$$f(\boldsymbol{X}|\theta) = g(T(\boldsymbol{X})|\theta)\, h(\boldsymbol{X}), \tag{4.3.1}$$

where $g(t|\theta)$ depends on $\boldsymbol{X}$ only through $T(\boldsymbol{X})$ and h is independent of θ. The reader may look up Halmos and Savage (1949) for a general proof involving some deeper theorems of measure theory.

Let $\boldsymbol{X} \sim LS_{n\times p}(\mu, \Sigma, f)$. Then from (4.1.19) we immediately know that $(\bar{\boldsymbol{x}}, \boldsymbol{S})$ is a sufficient statistic of (μ, Σ) by the Fisher-Neyman Factorization Theorem, i.e., $\boldsymbol{X}$'s density has the form of (4.3.1), here $h(\boldsymbol{X})=1$, $\boldsymbol{T}(\boldsymbol{X})=(\bar{\boldsymbol{x}}, \boldsymbol{S})$. If μ is known, $\sum_{i=1}^{n}(\boldsymbol{x}_{(i)}-\mu)(\boldsymbol{x}_{(i)}-\mu)'$ is sufficient for Σ and but $\boldsymbol{S}$ is not. If Σ is known generally $\bar{\boldsymbol{x}}$ is not sufficient for μ, but it is sufficient when $\boldsymbol{X} \sim N_{n\times p}(\boldsymbol{1}\mu', \boldsymbol{I}_n\otimes\Sigma)$.

4.3.3. Completeness

Let $T(\boldsymbol{X})$ be a statistic with an induced density from $\boldsymbol{X}$, $f_T(t|\theta)$, $\theta\in\Theta$. $T(\boldsymbol{X})$ is called a *complete statistic* if for each real Borel function $g(\boldsymbol{T})$, $E_\theta[g(T)]=0$, for every $\theta\in\Theta$, implies $P_\theta(g(T)=0)=1$, for every $\theta\in\Theta$. If $T(\boldsymbol{X})$ is also sufficient, we call it a *complete sufficient statistic.*

Lemma 4.3.2 *Suppose* $\boldsymbol{X} \sim N_{n\times p}(\boldsymbol{1}\mu', \boldsymbol{I}_n\otimes\Sigma)$ *with* $n>p$ *and* $\Sigma>0$. *Then* $(\bar{\boldsymbol{x}}, \boldsymbol{S})$ *is a complete sufficient statistic for* (μ, Σ).

Proof. That $(\bar{\boldsymbol{x}}, \boldsymbol{S})$ is sufficient has been seen in Section 4.3.2. We want to show that $(\bar{\boldsymbol{x}}, \boldsymbol{S})$ is complete. Given any real Borel function $g(\bar{\boldsymbol{x}}, \boldsymbol{S})$, by (4.2.6), we have

$$E[g(\bar{\boldsymbol{x}}, \boldsymbol{S})] = c\int_{S>0} g(\bar{\boldsymbol{x}}, \boldsymbol{S})\,|\Sigma|^{-n/2}\,|\boldsymbol{S}|^{(n-p-2)/2}\exp\left\{-\frac{1}{2}(\Sigma^{-1}\boldsymbol{S})\right\} \exp\left\{-\frac{1}{2}n(\bar{\boldsymbol{x}}-\mu)'\Sigma^{-1}(\bar{\boldsymbol{x}}-\mu)\right\}d\bar{\boldsymbol{x}}\,d\boldsymbol{S}. \tag{4.3.2}$$

Write $\Sigma^{-1}=\boldsymbol{I}_p-2\boldsymbol{Q}$ where $\boldsymbol{Q}'=\boldsymbol{Q}$ such that $\boldsymbol{I}_p-2\boldsymbol{Q}>0$. Let $\mu=(\boldsymbol{I}-2\boldsymbol{Q})^{-1}\boldsymbol{a}$. If $E[g(\bar{\boldsymbol{x}}, \boldsymbol{S})]=0$, for each (μ, Σ), then from (4.3.2) we get

$$0 = \int_{S>0} g(\bar{\boldsymbol{x}}, \boldsymbol{S})\,|\boldsymbol{I}_p-2\boldsymbol{Q}|^{n/2}\,|\boldsymbol{S}|^{(n-p-2)/2}\exp\left\{-\frac{1}{2}(\boldsymbol{I}_p-2\boldsymbol{Q})(\boldsymbol{S}+n\bar{\boldsymbol{x}}\bar{\boldsymbol{x}}')-2n\boldsymbol{a}\bar{\boldsymbol{x}}'\right.$$

$$+ n(\boldsymbol{I}_p - 2\boldsymbol{Q})^{-1}\boldsymbol{a}\boldsymbol{a}' \Big\} d\bar{x}\, dS,$$

or

$$(4.3.3) \qquad \int_{S>0} g(\bar{x}, S + n\bar{x}\bar{x}' - n\bar{x}\bar{x}')|S|^{(n-p-2)/2}$$
$$\cdot \exp\left\{-\frac{1}{2}(S + n\bar{x}\bar{x}') + Q(S + n\bar{x}\bar{x}') + na\bar{x}'\right\} d\bar{x}\, dS = 0$$

which is a function in $\boldsymbol{Q}$ and $\boldsymbol{a}$. We now identify (4.3.3) as the Laplace transform of

$$g(\bar{x}, S + n\bar{x}\bar{x}' - \bar{x}\bar{x}'n)|S|^{(n-p-2)/2} \exp\left\{-\frac{1}{2}(S + n\bar{x}\bar{x}')\right\}$$

with respect to variables $n\bar{x}$, $S + n\bar{x}\bar{x}'$. Since this is identically equal to zero for each pair (a, Q), where Q satisfies $I_p - 2Q > 0$, we conclude that $g(\bar{x}, S) = 0$ except possibly for a set of values of $(\bar{x}, S)$ with probability zero. Hence the theorem is proved. □

Let

$$\mathfrak{F} = \{|\Sigma|^{-n/2} f((X - \mathbf{1}\mu')\Sigma^{-1}(X - \mathbf{1}\mu')') \mid f(X'X) \text{ is a density and } \mu \in R^p, \Sigma > 0\}$$

be a class of densities.

Theorem 4.3.1. *The statistic* $(\bar{x}, S)$ *is complete and sufficient for the family* $\mathfrak{F}$.

Proof. Let $g(\bar{x}, S)$ be a statistic such that

$$\int g(\bar{x}, S)|\Sigma|^{-n/2} f((X - \mathbf{1}\mu')\Sigma^{-1}(X - \mathbf{1}\mu')')dX = 0$$

for each $|\Sigma|^{-n/2} f((X - \mathbf{1}\mu')\Sigma^{-1}(X - \mathbf{1}\mu')') \in \mathfrak{F}$. Since the normal density is in $\mathfrak{F}$, by virtue of Lemma 4.3.2 we have

$$\int_{g(\bar{x}, S)=0} \exp\left\{-\frac{1}{2}(X - \mathbf{1}\mu')\Sigma^{-1}(X - \mathbf{1}\mu')'\right\} dX = 1,$$

for all $\mu \in R^p$, $\Sigma > 0$. Now, $|\Sigma|^{-n/2} f((X - \mathbf{1}\mu')\Sigma^{-1}(X - \mathbf{1}\mu')')$ is absolutely continuous with respect to the normal density if it belongs to $\mathfrak{F}$. Hence

$$\int_{g(\bar{x}, S)=0} f((X - \mathbf{1}\mu')\Sigma^{-1}(X - \mathbf{1}\mu')')\, dX = 1,$$

which shows that $(\bar{x}, S)$ is complete. The sufficiency of $(\bar{x}, S)$ is obvious. So the theorem follows. □

4.3.4. Consistency

Let X_n be a sample of size n from a population, $F_{n,\theta}$, say, $n \geq 1$, with a parameter θ to be estimated, $\theta \in \Theta$, and let $g(\theta)$ be a function of θ and $T_n(X_n)$ be a statistic used to estimate $g(\theta)$. $\{T_n\}$ is called *weakly consistent* for $g(\theta)$, if, for any given $\varepsilon > 0$,

$$\lim_{n\to\infty} P(|T_n - g(\theta)| < \varepsilon) = 1, \qquad \text{for all } \theta \in \Theta,$$

and $\{T_n\}$ called *strongly consistent* if

$$P(\lim_{n\to\infty} T_n = g(\theta)) = 1, \qquad \text{for all } \theta\in\Theta .$$

When $\boldsymbol{T}_n$ and $\boldsymbol{g}(\theta)$ are vectors (or matrices), $|\boldsymbol{T}_n - \boldsymbol{g}(\theta)|$ is regarded as an Euclidean norm.

First, we should introduce the following concepts in spherical matrix distributions. $\boldsymbol{X}$ is called row-presentable if $\boldsymbol{V}>0$ has a distribution, π (say), and, given $\boldsymbol{V}$, $\boldsymbol{X} \sim N_{n\times p}(\boldsymbol{O}, \boldsymbol{I}_n \otimes \boldsymbol{V})$. (See Dawid, 1978). It is easy to see that $\boldsymbol{X}$ is row-presentable iff $\boldsymbol{X} \stackrel{d}{=} \boldsymbol{Y}\boldsymbol{A}$, where $\boldsymbol{Y} \sim N_{n\times p}(\boldsymbol{O}, \boldsymbol{I}_n \otimes \boldsymbol{I}_p)$, and $\boldsymbol{Y}$ is independent of $\boldsymbol{A}$: $p\times p$. When $p=1$, by Corollary 1 of Theorem 2.5.7, $\boldsymbol{x}$ is row-presentable iff $\boldsymbol{x} \sim S_n(\phi)$ with $\phi\in\Phi_\infty$.

Now let $\boldsymbol{X}_n \stackrel{d}{=} \boldsymbol{1}\mu' + \boldsymbol{Y}_n\boldsymbol{A}$, $n=p+1, p+2, \ldots$, where $\boldsymbol{Y}_n \sim N_{n\times p}(\boldsymbol{O}, \boldsymbol{I}_n \otimes \boldsymbol{I}_p)$, $\boldsymbol{A}$ is independent of $\boldsymbol{Y}_n = (\boldsymbol{y}_{(1)}, \ldots, \boldsymbol{y}_{(n)})'$ and let μ and Σ be constant. Let $\boldsymbol{X}_n = (\boldsymbol{x}_{(1)}, \ldots, \boldsymbol{x}_{(n)})'$. Then we have

$$\bar{\boldsymbol{x}}_n \hat{=} \frac{1}{n}\sum_{i=1}^{n} \boldsymbol{x}_{(i)} \stackrel{d}{=} \frac{1}{n}\left[\sum_{i=1}^{n} (\boldsymbol{A}'\boldsymbol{y}_{(i)} + \mu)\right] = \boldsymbol{A}'\left(\frac{1}{n}\sum_{i=1}^{n} \boldsymbol{y}_{(i)}\right) + \mu \to \mu \text{ as } \quad n\to\infty$$

with probability one because $\frac{1}{n}\sum_{i=1}^{n} \boldsymbol{y}_{(i)} \to \boldsymbol{0}$, a.e., as $n\to\infty$ by virtue of Kolmogorov's strong law of large numbers. Hence we get $P(\lim_{n\to\infty} \bar{\boldsymbol{x}}_n = \mu) = 1$, i.e., $\bar{\boldsymbol{x}}_n$ is strongly consistent for μ. To study the consistency of $\boldsymbol{S}/n$ for Σ, we may, without loss of generality assume that $\mu = \boldsymbol{0}$. Similarly to (4.1.10) we have

$$\frac{1}{n}\boldsymbol{S}_n \stackrel{d}{=} \frac{1}{n}\boldsymbol{A}'\boldsymbol{Y}'_n(\boldsymbol{I}_n - \boldsymbol{P}_n)\boldsymbol{Y}_n\boldsymbol{A} \stackrel{d}{=} \boldsymbol{A}'\left(\frac{1}{n}\sum_{i=1}^{n-1} \boldsymbol{y}_{(i)}\boldsymbol{y}'_{(i)}\right)\boldsymbol{A}.$$

Thus $\boldsymbol{S}_n/n \to \boldsymbol{A}'\boldsymbol{A}$ as $n\to\infty$ with probability one because $\frac{1}{n}\sum_{i=1}^{n-1} \boldsymbol{y}_{(i)}\boldsymbol{y}'_{(i)} \to \boldsymbol{I}_p$ as $n \to\infty$. We sum up the above results as follows.

Theorem 4.3.2. *Under the above assumption, we have*

(i) $P(\lim_{n\to\infty} \bar{\boldsymbol{x}}_n = \mu) = 1$, *and*

(ii) $P\left(\lim_{n\to\infty} \frac{1}{n}\boldsymbol{S} = \boldsymbol{A}'\boldsymbol{A}\right) = 1.$

4.4. Minimax and Admissible Characters of $\hat{\mu}$ and $\hat{\Sigma}$

Now we would like to investigate the parametric estimation from a decision theoretic approach which is connected with the conceptions of Bayesian method, minimax and admissibility. We assume that the reader has been familiar with it in univariate statistics. For convenience we give their definitions and state some usual results without rigorous proofs. Let $(\mathcal{X}, \mathcal{B}, P_\theta)$ be a probability space with $\theta\in\Theta$, where Θ is called the *parametric space* (an interval in R^m) and $\mathcal{X}$ is called the *sample space*. Let $\mathcal{D}$ be the set of all possible estimators of θ. A nonnegative function $L(\theta, \mathrm{d})$, $\theta\in\Theta$, $\mathrm{d}\in\mathcal{D}$, defined on $\Theta \times \mathcal{D}$, represents the loss of erroneously estimating θ by d. Let

$$(4.4.1) \qquad R(\theta, d) \hat{=} E_\theta[L(\theta, d)] = \int L(\theta, d(\boldsymbol{x}))f(\boldsymbol{x}\,|\,\theta)d\boldsymbol{x},$$

where $f(\boldsymbol{x}|\theta)$ denotes the density of random vector $\boldsymbol{x}$ taking values in $\mathcal{X}$, corresponding to P_θ with respect to the Lebesgue measure $d\boldsymbol{x}$. $R(\theta,d)$ is called the *risk function* of the estimator $d(\boldsymbol{x})$ for θ. Let $h(\theta)$, $\theta\in\Theta$, denote the prior density on Θ, which is an experimenter's belief in parameter θ before experimenting. The *posterior density* of θ, given $\boldsymbol{x}$, (i.e., the experimenter has obtained data $\boldsymbol{x}$) is

$$h(\theta \mid \boldsymbol{x}) = \frac{f(x|\theta)\, h(\theta)}{\int_\Theta f(x|\theta)\, h(\theta)\, d\theta}.$$

The prior risk (Bayes risk of d with respect to the prior $h(\theta)$) is given by

$$R_h(d) = \int_\Theta R(\theta,\, d)\, h(\theta)\, d\theta. \tag{4.4.2}$$

We can interchange the order of integration in $R_h(d)$, provided that $R_h(d) < \infty$, and obtain

$$R_h(d) = \int \{ \int L(\theta,\, d(\boldsymbol{x}))\, f(\boldsymbol{x}|\theta) d\boldsymbol{x} \}\, h(\theta)\, d\theta$$

$$= \int \tilde{f}(\boldsymbol{x}) \{ \int L(\theta,\, d(\boldsymbol{x}))\, d\theta \} d\boldsymbol{x} \tag{4.4.3}$$

where $\tilde{f}(\boldsymbol{x}) \,\hat{=}\, \int f(\boldsymbol{x}|\theta)\, h(\theta)\, d\theta$. The quantity (*posterior risk*)

$$\int L(\theta,\, d(\boldsymbol{x}))\, h(\,\theta\,|\boldsymbol{x}) d\theta \tag{4.4.4}$$

is called the expected loss of erroneously estimating by $d(\boldsymbol{x})$ when $\boldsymbol{x}$ has been observed.

Definition 4.4.1. A statistic $d_0(\boldsymbol{x})$ used to estimate θ is called a *Bayes estimate* of θ associated with the prior density $h(\theta)$ and the loss function $L(\theta,\, d)$, if $d_0(\boldsymbol{x})$ minimizes the posterior risk (4.4.4), that is, $d_0(\boldsymbol{x})$ is defined as

$$\int L(\theta,\, d_0(x)) h\,(\theta|x)\, d\theta = \inf_{d\in\mathfrak{D}} \int L(\theta,\, d)\, h(\theta|x)\, d\theta.$$

Note. (i) Under some simple regular conditions, there exists a Bayes estimate, but it is not necessarily unique. However, if $L(\theta,d)$ is strictly convex in d for given θ, then the Bayes estimate is essentially unique if it exists.

(ii) It is easy to see that a Bayes estimate also minimizes the prior risk (4.4.2).

In practice we usually take $L(\theta,\, \boldsymbol{d})$ to be the square loss function

$$L(\theta,\, \boldsymbol{d}) = (\theta - \boldsymbol{d})'\,(\theta - \boldsymbol{d}) \tag{4.4.5}$$

or

$$L(\theta,\, \boldsymbol{d}) = (\theta - \boldsymbol{d})'\, \boldsymbol{D}^{-1}\,(\theta - \boldsymbol{d}), \tag{4.4.6}$$

where $\boldsymbol{D}$ is a positiue definite matrix.

Lemma 4.4.1. *Let* $L(\theta,\, \boldsymbol{d})$ *be taken as* (4.4.6). *Then the unique Bayes estimate of* θ *is* $\mathcal{E}(\theta|\boldsymbol{x})$, *where*

$$\mathcal{E}(\theta|\boldsymbol{x}) \,\hat{=}\, \int \theta h(\theta|\boldsymbol{x}) d\theta$$

is the posterior expectation of θ.

This follows from the following:

$$\int (\theta - \mathcal{E}(\theta|\boldsymbol{x}))' \boldsymbol{D}^{-1}(\mathcal{E}(\theta|\boldsymbol{x}) - \boldsymbol{d}) h(\theta|\boldsymbol{x}) d\theta = 0.$$

Definition 4.4.2. An estimate $d^*(\boldsymbol{x})$ of θ is called *minimax* if

$$\sup_{\theta \in \Theta} R(\theta, \boldsymbol{d}^*) = \inf_{\boldsymbol{d} \in \mathcal{D}} \sup_{\theta \in \Theta} R(\theta, \boldsymbol{d}).$$

In other words, the minimax estimate protects against the largest possible risk when θ varies over Θ. There is a useful connection between the Bayes and minimax estimates.

Lemma 4.4.2. *Suppose that $d_0(\boldsymbol{x})$ is a Bayes estimate associated with some prior density, $h(\theta)$ say, and that the risk function $R(\theta, \boldsymbol{d}_0)$ does not depend on θ. Then $d_0(\boldsymbol{x})$ is a minimax estimate. Further, let $\{h_k(\theta), k \geqslant 1\}$ be a sequence of prior densities on Θ and let $\{\hat{\theta}_k, k \geqslant 1\}$ be the corresponding sequence of Bayes estimates with prior risks $\{R(h_k, \hat{\theta}_k), k \geqslant 1\}$. If there exists an estimate $\hat{\theta}^*$ for which*

$$\sup_{\theta \in \Theta} R(\theta, \hat{\theta}^*) \leqslant \lim_{k \to \infty} \sup R(h_k, \hat{\theta}_k),$$

then $\hat{\theta}^*$ is a minimax estimate.

Now we come to introduce the admissibility. An estimate $d_1(\boldsymbol{x})$ is said to be at least as good as another estimate $d_2(\boldsymbol{x})$ if

$$R(\theta, d_1) \leqslant R(\theta, d_2), \qquad \text{for each } \theta \in \Theta, \tag{4.4.7}$$

and $d_1(\boldsymbol{x})$ said to be better than or strictly dominates $d_2(\boldsymbol{x})$ if (4.4.7) holds with strict inequality for at least one $\theta \in \Theta$.

Definition 4.4.3. An estimate $d^*(\boldsymbol{x})$ is said to be *admissible* if, for any estimate $d(\boldsymbol{x})$, that $R(\theta, d^*) \leqslant R(\theta, d)$, for each $\theta \in \Theta$, implies that $R(\theta, d^*) = R(\theta, d)$ on Θ. An estimate is said to be *inadmissible* if it is not admissible.

Lemma 4.4.3. (Blyth, 1951) *If the risk function $R(\theta, d)$ is continuous in θ for each $d \in \mathcal{D}$, and if the $\hat{\theta}_h(\boldsymbol{x})$ is a Bayes estimate corresponding to a prior density $h(\theta)$ which is positive at all $\theta \in \Theta$, then the Bayes estimate $\hat{\theta}_h(x)$ is admissible.*

Proof. By negation, if $\hat{\theta}_h(x)$ is inadmissible then there exists an estimate $\hat{\theta}^*(x)$ such that (i) $R(\theta, \hat{\theta}^*) \leqslant R(\theta, \hat{\theta}_h)$, for each $\theta \in \Theta$, and (ii) $R(\theta_0, \hat{\theta}^*) < R(\theta_0, \theta_h)$, for some $\theta_0 \in \Theta$. Since $R(\theta, d)$ is continuous in θ for each d, there exists a neighborhood, $N(\theta_0)$ say, around θ_0 over which the inequality (ii) holds for all $\theta \in N(\theta_0)$. Noting $h(\theta) > 0$ for all $\theta \in \Theta$, we have

$$R(h, \hat{\theta}^*) = \int_\Theta R(\theta, \hat{\theta}^*)\, h(\theta)\, d\theta = \left(\int_{N(\theta_0)} + \int_{\Theta - N(\theta_0)} \right) R(\theta, \hat{\theta}^*) h(\theta)\, d\theta$$

$$< \int_{N(\theta_0)} R(\theta, \hat{\theta}_h)\, h(\theta)\, d\theta + \int_{\Theta - N(\theta_0)} R(\theta, \hat{\theta}_h{}^*)\, h(\theta)\, d\theta$$

$$= \int_\Theta R(\theta, \hat{\theta}^*)\, h(\theta)\, d\theta = R(h, \hat{\theta}_h),$$

contradicting the assumption that $\hat{\theta}_h$ is a Bayes estimate corresponding to h. □

4.4.1. Inadmissibility of $\bar{\boldsymbol{x}}$ as an Estimate of μ

Suppose that $\boldsymbol{x}_1, \ldots, \boldsymbol{x}_n$ are n independent observations from the population $N_p(\mu, \boldsymbol{I}_p)$. Stein (1956) and James and Stein (1961) pointed out that $\bar{\boldsymbol{x}}$ is an inadmissible estimate of μ when $p>2$, which opened up a whole area of research and led to the development of a new type of estimate. They showed that under the quadratic loss (4.4.5), the estimate

$$\hat{\mu} = \left(1 - \frac{p-2}{n\bar{\boldsymbol{x}}'\bar{\boldsymbol{x}}}\right)\bar{\boldsymbol{x}} \tag{4.4.8}$$

is a better estimate than $\bar{\boldsymbol{x}}$ when $p>2$, and (4.4.8) is called the *James-Stein estimate.* Now we try to establish the inadmissibility of $\bar{\boldsymbol{x}}$ in spherical matrix distributions. It was studied by many authors, such as Strawderman (1974), Brandwein and Strawderman (1978), Brandwein (1979), and Fan and Fang (1986). They considered the more general form as follows

$$\delta_a(\boldsymbol{x}) = (1 - a/\boldsymbol{x}'\boldsymbol{x})\,\boldsymbol{x} \tag{4.4.9}$$

and showed that $\delta_a(\boldsymbol{x})$ is a better estimate than $\boldsymbol{x}$ for certain a's. First we need some lemmas. The following lemma shows that we need only to consider the case of $n=1$.

Lemma 4.4.4. *Let* $\boldsymbol{X} = (\boldsymbol{x}_{(1)}, \ldots, \boldsymbol{x}_{(n)})' \sim SS_{n\times p}(\mu', \boldsymbol{I}_p, \phi)$. *Then*

$$\bar{\boldsymbol{x}} \overset{\mathrm{d}}{=} \frac{1}{n}(\boldsymbol{x}_{(1)} - \mu) + \mu. \qquad (\textit{Note that } \boldsymbol{x}_{(1)} - \mu \sim \boldsymbol{S}_p(\psi).)$$

Proof. Let $\boldsymbol{Y} \sim SS_{n\times p}(\boldsymbol{0}, \boldsymbol{I}_p, \phi)$. Then $\boldsymbol{X} \overset{\mathrm{d}}{=} \boldsymbol{1}\mu' + \boldsymbol{Y}$. Therefore

$$\bar{\boldsymbol{x}} = \frac{1}{n}\boldsymbol{X}'\boldsymbol{1} \overset{\mathrm{d}}{=} \frac{1}{n}(\boldsymbol{1}\mu' + \boldsymbol{Y})'\boldsymbol{1} = \frac{1}{n}\boldsymbol{Y}'\boldsymbol{1} + \mu.$$

Taking $\Gamma \in O(n)$ such that $\Gamma\boldsymbol{1} = (n^{1/2}, 0, \ldots, 0)'$, we have

$$\bar{\boldsymbol{x}} \overset{\mathrm{d}}{=} \frac{1}{n}\boldsymbol{Y}'\Gamma'\Gamma\boldsymbol{1} + \mu \overset{\mathrm{d}}{=} \frac{1}{n}\boldsymbol{Y}'\Gamma\boldsymbol{1} + \mu = n^{-1/2}\boldsymbol{y}_{(1)} + \mu \overset{\mathrm{d}}{=} n^{-1/2}(\boldsymbol{x}_{(1)} - \mu) + \mu,$$

where $\boldsymbol{Y} = (\boldsymbol{y}_{(1)}, \ldots, \boldsymbol{y}_{(n)})'$. The lemma follows. □

Lemma 4.4.5. *Suppose $r(t)$ is a non-negative continuous non-decreasing function. Then*

$$\int_{-1}^{1} \frac{\alpha v r(1 + 2\alpha v + \alpha^2)}{1 + 2\alpha v + \alpha^2}(1 - v^2)^{\frac{p-3}{2}}\, dv > -\frac{2}{p-1}\int_{-1}^{1} \frac{r(1+2\alpha v+\alpha^2)}{1+2\alpha v+\alpha^2}(1-v^2)^{\frac{p-3}{2}}\, dv$$

provided $p \geqslant 4$.

Proof. From the following argument, without loss generality, we can assume that the first-order derivative function of $r(t)$ always exists, as $r(t)$ is a non-decreasing continuous function. By using integration by parts and

$$\alpha^2(1-v^2) < 1 + 2\alpha v + \alpha^2$$

we obtain that

$$\int_{-1}^{1} \frac{\alpha v\, r(1 + 2\alpha v + \alpha^2)}{1 + 2\alpha v + \alpha^2}(1 - v^2)^{\frac{p-3}{2}}\, dv$$

$$= -\int_{-1}^{1}\frac{2}{p-1}\frac{\alpha^2 r(1+2\alpha v+\alpha^2)}{(1+2\alpha v+\alpha^2)^2}(1-v^2)^{\frac{p-1}{2}}dv + \frac{2\alpha^2}{p-1}\int_{-1}^{1}\frac{r'(1+2\alpha v+\alpha^2)}{1+2\alpha v+\alpha^2}$$

$$\times(1-v^2)^{\frac{p-1}{2}}dv > -\frac{2}{p-1}\int_{-1}^{1}\frac{r(1+2\alpha v+\alpha^2)}{1+2\alpha v+\alpha^2}(1-v^2)^{\frac{p-3}{2}}dv.$$

The proof is completed. □

Lemma 4.4.6. *For $p\geqslant 4$, the function defined by*

$$f(\alpha) = \int_{-1}^{1}\frac{1}{1+2\alpha v+\alpha^2}(1-v^2)^{\frac{p-3}{2}}dv$$

is an non-increasing function of α over $[0, \infty)$.

Proof. When $p\geqslant 5$ differentiating $f(\alpha)$ with respect to α and using integration by parts we have

$$f'(\alpha) = -2\int_{-1}^{1}\frac{\alpha}{(1+2\alpha v+\alpha^2)^2}(1-v^2)^{(p-3)/2} - 2\int_{-1}^{1}\frac{v(1-v^2)^{(p-3)/2}}{(1+2\alpha v+\alpha^2)^2}dv$$

$$= -2\alpha\int_{-1}^{1}\frac{\alpha^2+2\alpha v+1-4(1-v^2)/(p-1)}{(1+2\alpha v+\alpha^2)^3}(1-v^2)^{(p-3)/2}dv\leqslant 0.$$

When $p=4$, direct calculation shows that

$$f'(\alpha) = -2\int_{-1}^{1}\frac{\alpha+v}{(1+2\alpha v+\alpha^2)^2}(1-v^2)^{\frac{1}{2}}dv$$

$$= -\frac{1}{\alpha}\int_{-1}^{1}\frac{1-\alpha v-2\alpha v^2}{1+2\alpha v+\alpha^2}(1-v^2)^{-\frac{1}{2}}dv$$

$$= -\frac{1}{\alpha^2}\int_{-1}^{1}\frac{\alpha+v}{1+2\alpha v+\alpha^2}(1-v^2)^{-\frac{1}{2}}dv$$

$$= \begin{cases} -\pi/\alpha^3 & \alpha>1 \\ 0 & \alpha<1 \end{cases}$$

Thus, it is concluded $f'(\alpha)\leqslant 0$, for $\alpha\neq 1$, provided $p\geqslant 4$, and the desired conclusion is obtained due to the continuity of $f(\alpha)$. □

Theorem 4.4.1. *Let $\boldsymbol{x}\sim EC_p(\mu, \mathbf{I}_p, f)$ and $E_0(\|\boldsymbol{x}\|^{-2})$ be finite, where E_0 denotes the expectation when $\mu=\mathbf{0}$. Then the estimator $\delta_a(\boldsymbol{x})$ beats the usual estimator $\boldsymbol{x}$ under quadratic loss* (4.4.5), *provided $p\geqslant 4$ and*

$$0\leqslant a\leqslant 2(p-3)/(p-1)E_0(\|\boldsymbol{x}\|^{-2}).$$

Proof. Let $\boldsymbol{y}=\boldsymbol{x}-\mu$. The difference in risks between $\boldsymbol{x}$ and $\delta_a(\boldsymbol{x})$ is

$$R(\boldsymbol{x}, \mu)-R(\delta_a(\boldsymbol{x}), \mu) = E\|\boldsymbol{x}-\mu\|^2 - E\|\delta_a(\boldsymbol{x})-\mu\|^2$$

$$= a\{2E[(\boldsymbol{y}+\mu)'\boldsymbol{y}/\|\boldsymbol{y}+\mu\|^2] - aE\|\boldsymbol{y}+\mu\|^{-2}\}.$$

Without loss of generality we can assume $\mu=(\|\mu\|, 0, \ldots, 0)'$ because of spherically symmetry of $\boldsymbol{y}$ so that

$$R(\boldsymbol{x}, \mu) - R(\delta_a(\boldsymbol{x}), \mu) = aE\frac{2\|\boldsymbol{y}\|^2 + 2Y_1\|\mu\| - a}{(Y_1 + \|\mu\|)^2 + Y_2^2 + \cdots + Y_p^2}.$$

Denote $\alpha = \|\mu\| / R$. *As the conditional distribution of* $\boldsymbol{y} / \|\boldsymbol{y}\|$ given $\|\boldsymbol{y}\|^2 = R^2$ is the uniform distribution on the unit sphere, by Lemma 4.4.5 and Lemma 4.4.6 we have

$$E\left\{\frac{2\|\boldsymbol{y}\|^2 + 2Y_1\|\mu\| - a}{(Y_1 + \|\mu\|)^2 + Y_2^2 + \cdots + Y_p^2}\,\middle|\, \|\boldsymbol{y}\|^2 = R^2\right\}$$
$$= \frac{1}{B(\frac{1}{2}, \frac{p-1}{2})}\int_{-1}^{1}\frac{2+2\alpha v - aR^{-2}}{1+2\alpha v + \alpha^2}(1-v^2)^{(p-3)/2}\,dv$$
$$> \frac{1}{B(\frac{1}{2}, \frac{p-1}{2})}\int_{-1}^{1}\frac{2-4/(p-1) - aR^{-2}}{1+2\alpha v + \alpha^2}(1-v^2)^{(p-3)/2}\,dv$$
$$= \frac{1}{B(\frac{1}{2}, \frac{p-1}{2})}(2(p-3)/(p-1) - aR^{-2})f(\|\mu\|/R)$$
$$\geqslant \frac{1}{B(\frac{1}{2}), \frac{p-1}{2}}(2(p-3)/(p-1) - aR^{-2})f(\|\mu\|/R_0),$$

where $R_0 = ((p-1)a/2(p-3))^{\frac{1}{2}}$. The above inequality holds as $f(\|\mu\|/R)$ defined in Lemma 4.4.6 is a non-decreasing function of R and $2(p-3)/(p-1) - aR^{-2} \geqslant (<) 0$ whenever $R \geqslant (<) R_0$. Combining the above facts we have

$$R(\boldsymbol{x}, \mu) - R(\delta_a(\boldsymbol{x}), \mu)$$
$$> \frac{1}{B(\frac{1}{2}, \frac{p-1}{2})}f(\|\mu\|/R_0)(2(p-3)/(p-1) - aE_0\|x\|^{-2}) \geqslant 0$$

and the assertion follows. □

Remark. The fact established in this theorem was first proved by Brandwein and Strawderman (1978). But their proof is too long to read. The present proof is due to Fan and Fang (1986). In the normal case the similar result can easily be obtained.

Lemma 4.4.7. *Let* $x \sim N(\theta, 1)$. *If* $h(t)$ *is differentiable and* $E[h'(x)] < \infty$, *then*

(4.4.10) $$E[h(x)(x-\theta)] = E(h'(x)).$$

Proof. Taking the integration by parts, we have

$$E[h(x)(x-\theta)] = (2\pi)^{-1/2}\int_{-\infty}^{\infty} h(x)(x-\theta)e^{-(x-\theta)^2/2}\,dx$$
$$= -(2\pi)^{-1/2}\int_{-\infty}^{\infty} h(x)\,d[e^{-(x-\theta)^2/2}]$$
$$= (2\pi)^{-1/2}\int_{-\infty}^{\infty} h'(x)e^{-(x-\theta)^2/2}\,dx$$
$$= E[h'(x)].$$
□

Theorem 4.4.2. *Suppose that* $\boldsymbol{x} \sim N(\mu, \boldsymbol{I}_p)$ *and* $p \geqslant 3$. *Then* $\delta_a(\boldsymbol{x})$ *is better than* $\boldsymbol{x}$ *under the quadratic loss, provided* $0 < a \leqslant 2(p-2)$.

Proof. The difference of risks between $\boldsymbol{x}$ and $\delta_a(\boldsymbol{x})$ is

$$R(\mu, \boldsymbol{x}) - R(\mu, \delta_a) = p - E\left(\left\|\left(1-\frac{a}{\boldsymbol{x}'\boldsymbol{x}}\right)\boldsymbol{x} - \mu\right\|^2\right)$$

$$= p - E(\|\boldsymbol{x}-\mu\|^2) - 2aE(\boldsymbol{x}'(\boldsymbol{x}-\mu)/\boldsymbol{x}'\boldsymbol{x}) + a^2E(1/\boldsymbol{x}'\boldsymbol{x})$$

$$= 2aE[\boldsymbol{x}'(\boldsymbol{x}-\mu)/\boldsymbol{x}'\boldsymbol{x}] - a^2E(1/\boldsymbol{x}'\boldsymbol{x}).$$

As

$$E[\boldsymbol{x}'(\boldsymbol{x}-\mu)/\boldsymbol{x}'\boldsymbol{x}] = \sum_{i=1}^{p} E[x_i(x_i-\mu_i)/\boldsymbol{x}'\boldsymbol{x}],$$

and for $1 \leqslant i \leqslant p$

$$E\left[\frac{x_i(x_i-\mu_i)}{\boldsymbol{x}'\boldsymbol{x}}\right] = E\left[\frac{\partial}{\partial x_i}\left(\frac{x_i}{\boldsymbol{x}'\boldsymbol{x}}\right)\right] = E\left[\frac{\boldsymbol{x}'\boldsymbol{x} - 2x_i^2}{(\boldsymbol{x}'\boldsymbol{x})^2}\right], \text{ (Lemma 4.4.5)}$$

thus

$$E[\boldsymbol{x}'(\boldsymbol{x}-\mu)/\boldsymbol{x}'\boldsymbol{x}] = \sum_{i=1}^{p} E\left[\frac{\boldsymbol{x}'\boldsymbol{x} - 2x_i^2}{(\boldsymbol{x}'\boldsymbol{x})^2}\right] = E[(p-2)/\boldsymbol{x}'\boldsymbol{x}].$$

Noting $E(1/\boldsymbol{x}'\boldsymbol{x}) = (p-2)^{-1}$, we get

$$R(\mu, \boldsymbol{x}) - R(\mu, \delta_a) = a(2 - a/(p-2)) > 0.$$

provided that $0 < a < 2(p-2)$. □

But Stein (1956) showed that $\boldsymbol{x}$ is an admissible estimate of μ for $\boldsymbol{x} \sim N(\mu, \boldsymbol{I}_p)$, $p = 1,2$.

Though we proved that $(1-a/\boldsymbol{x}'\boldsymbol{x})\boldsymbol{x}$ is better than $\boldsymbol{x}$ so $\boldsymbol{x}$ is inadmissible, $(1-a/\boldsymbol{x}'\boldsymbol{x})\boldsymbol{x}$ itself is not admissible. Actually $(1-a/\boldsymbol{x}'\boldsymbol{x})^+\boldsymbol{x}$ is better than $(1-a/\boldsymbol{x}'\boldsymbol{x})\boldsymbol{x}$ where $t^+ = \max(0, t)$ (See Exercise 4.10). Strawderman (1971) found a class of proper Bayes estimates, all of which are better than $\boldsymbol{x}$ for a normal mean vector. Thus they are all admissible by Lemma 4.4.3. The following theorem of Cohen (1966) characterizes all the admissible linear estimates of the mean vector of multivariate normal distributions.

Theorem 4.4.3. (Cohen, 1966). *Let $\boldsymbol{x} \sim N(\mu, \boldsymbol{I}_p)$. Then $\boldsymbol{Ax}$ is an admissible estimate of μ where $\boldsymbol{A}$ is a known $p \times p$ matrix, iff $\boldsymbol{A}$ is symmetric and its eigenvalues $\lambda_i (i=1, \ldots, p)$ satisfy the inequality*

$$0 \leqslant \lambda_i \leqslant 1, \text{ for all } i = 1, \ldots, p,$$

with equality to 1 for at most two of the eigenvalues.

For a proof of the theorem see Cohen (1966). Note that for $\boldsymbol{x}$, $\boldsymbol{A} = \boldsymbol{I}_p$ and all the eigenvalues are equal to 1. Then $\boldsymbol{x}$ is inadmissible if $p \geqslant 3$. In particular, $\boldsymbol{0}$, a constant vector, is an admissible estimate of μ. However, any statistician would never like to use $\boldsymbol{0}$ as an estimate of μ because estimate $\boldsymbol{0}$ does not give any information about observation $\boldsymbol{x}$. This fact partly shows that "admissibility" only reflects a certain aspect of the properties of an estimate. A statistician would like to get rid of an inadmissible estimate, and but he, maybe, would not use rashly an admissible estimate.

4.4.2. Discussion on Estimation of Σ

In Section 4.1 we have got the maximum likelihood estimates of μ and Σ for spherical matrix distributions. When μ is unknown, the maximum likelihood estimate of Σ is $\hat{\Sigma} = \lambda_{\max}(f)\boldsymbol{S}$, where $\boldsymbol{S} = \boldsymbol{X}'(\boldsymbol{I}_n - \frac{1}{n}\boldsymbol{11}')\boldsymbol{X}$ and $\boldsymbol{X} \sim VS_{n\times p}(\mu, \Sigma. f)$ with $n > p$ and f

satisfying certain condition. From Corollary 1 of Theorem 4.1.1, $\frac{\sigma}{n-1}S$ with $\sigma = (-2\phi'(0))^{-1} > 0$ is an unbiased estimate of Σ. Now we present two kinds of loss functions considered in the literature by James and Stein (1961), Olkin and Selliah (1977), and Haff (1980) as follows.

(4.4.11)
$$L_1(\Sigma, \boldsymbol{D}) = \operatorname{tr}(\Sigma^{-1}\boldsymbol{D}) - \log(|\Sigma^{-1}\boldsymbol{D}|) - p$$

and

(4.4.12)
$$L_2(\Sigma, \boldsymbol{D}) = \operatorname{tr}(\Sigma^{-1}\boldsymbol{D} - \boldsymbol{I}_p)^2.$$

Both loss functions are nonnegative and are zero when $\Sigma = \boldsymbol{D}$. Certainly there are many other possible loss functions with these properties; the two above, however, have the attractive feature that they are relatively easy to work with. We shall study the risks of estimates of Σ under the loss function L_1 and L_2, respectively. Let us start with L_1.

Theorem 4.4.4. *Let* $\boldsymbol{X} \sim LS_{n\times p}(\mu, \Sigma, \phi)$ *with* $n>p$ *and the 2nd finite moment. Then the best estimate (involving smallest risk) of* Σ, *having the form* $\alpha\boldsymbol{S}(\alpha > 0)$ *under the loss function* (4.4.11), is $[1/(n-1)(-2\phi'(0))]\ \boldsymbol{S}$ *which is unbiased.*

Proof. The risk of the estimate $\alpha\boldsymbol{S}$ is

$$R_1(\Sigma, \alpha\boldsymbol{S}) = E[\alpha \operatorname{tr}(\Sigma^{-1}\boldsymbol{S}) - \log|\alpha\Sigma^{-1}\boldsymbol{S}| - p$$

$$= \alpha \operatorname{tr}\Sigma^{-1}\mathcal{E}(\boldsymbol{S}) - p\log\alpha - E\left[\log\frac{|\boldsymbol{S}|}{|\Sigma|}\right] - p$$

(4.4.13)
$$= \alpha p(n-1)(-2\phi'(0)) - p\log\alpha - E\left[\log\frac{|\boldsymbol{S}|}{|\Sigma|}\right] - p,$$

as $\mathcal{E}(\boldsymbol{S}) = (n-1)(-2\phi'(0))\Sigma$. The proof is completed by noting that the value of α which minimizes the right side of (4.4.13) is $\alpha = [(n-1)(-2\phi'(0))]^{-1}$. □

Corollary. *Let* $\boldsymbol{X} \sim N_{n\times p}(\boldsymbol{1}\mu', \boldsymbol{I}_n \otimes, \Sigma)$ *with* $n>p$. *Then the best estimate of* Σ *having the form* $\alpha\boldsymbol{S}$, $\alpha > 0$ *under the loss* $L_1(\Sigma, \boldsymbol{D})$ *is* $\boldsymbol{S}/(n-1)$.

If we look outside the class of estimates of the form $h(\boldsymbol{S}) = \alpha\boldsymbol{S}$, $\alpha > 0$, we can find some better estimates than the sample covariance matrix $(1/n)\boldsymbol{S}$. Then there is another approach of letting $\boldsymbol{S} = \boldsymbol{T}'\boldsymbol{T}$ where $\boldsymbol{T} \in \mathrm{UT}(P)$ with positive diagonal elements and considering

(4.4.14)
$$h(\boldsymbol{S}) = \boldsymbol{T}'\Delta\boldsymbol{T}$$

with $\Delta = \operatorname{diag}(\delta_1, \ldots, \delta_p) > 0$. Noting that $h(\mathrm{S})$ defined by (4.4.14) satisfies

(4.4.15)
$$h(\boldsymbol{L}'\boldsymbol{S}\boldsymbol{L}) = \boldsymbol{L}'h(\boldsymbol{S})\boldsymbol{L}$$

for each $\boldsymbol{L} \in \mathrm{UT}(p)$ with positive diagonal elements, we may consider using $h(\boldsymbol{S})$ to estimate Σ with $h(\boldsymbol{S})$ satisfying (4.4.15). However, the following lemma shows that both of $h(\boldsymbol{S})$'s satisfying (4.4.14) and (4.4.15) are the same.

Lemma 4.4.6. $h(\boldsymbol{S})$ *satisfying* (4.4.15) *has necessarily the form* (4.4.14).

Proof. Taking $\boldsymbol{S} = \boldsymbol{I}_p$ in (4.4.15) gives

(4.4.16)
$$h(\boldsymbol{L}'\boldsymbol{L}) = \boldsymbol{L}'h(\boldsymbol{I}_p)\boldsymbol{L}'.$$

Now letting

$$L = \begin{pmatrix} \pm 1 & & 0 \\ & \ddots & \\ 0 & & +1 \end{pmatrix},$$

then $L'L = I_p$ and (4.4.16) becomes

$$h(I_p) = L'h(I_p)L. \tag{4.4.17}$$

Note that (4.4.17) holds for each such L and implies that $h(I_p)$ is diagonal and $h(I_p) = \operatorname{diag}(\delta_1, \ldots, \delta_p) = \Delta$, say. Now write $S = T'T$, $T \in \mathrm{UT}(p)$ with positive diagonal elements. Then $h(S) = T'h(I_p)T = T'\Delta T$. □

Lemma 4.4.7. *Let $X \sim SS_{n\times p}(\mu, \Sigma, f)$ with $n > p$ and have the 2nd finite moment. Then $R_1(\Sigma, h(S))$ does not depend on Σ where $h(S)$ is given by* (4.4.14).

Proof.

$$\begin{aligned} R_1(\Sigma, h(S)) &= E[\operatorname{tr}(\Sigma^{-1}h(S)) - \log|\Sigma^{-1}h(S)| - p] \\ &= E[\operatorname{tr}(h(T'ST)) - \log|h(T'ST)| - p], \end{aligned}$$

where $\Sigma^{-1} = T'T$, $T \in UT(p)$ with positive diagonal elements. The density of X is

$$|\Sigma|^{-n/2} f[\lambda((X - 1\mu')\Sigma^{-1}(X - 1\mu')')].$$

So, by $S = X'(I_n - P_n)X$, $P_n = (1/n)11'$, we get

$$\begin{aligned} R_1(\Sigma, h(S)) = |\Sigma|^{-n/2} \int &[\operatorname{tr}(h(T'X'(I_n - P_n)X)) - \log|h(T'X'(I_p - P_n)XT)| - p] \\ &\cdot f(\lambda[(X - 1\mu')\Sigma^{-1}(X - 1\mu')'])dX \\ = |\Sigma|^{-n/2} \int &[\operatorname{tr}(h(TX'(I_n - P_n)XT)) - \log|h(T'X'(I_n - P_n)XT)| - p] \\ &\cdot f(\lambda(X'\Sigma^{-1}X))dX. \end{aligned}$$

Putting $Y = TX$, this becomes

$$\begin{aligned} R_1(\Sigma, h(S)) &= \int [\operatorname{tr}(h(Y'(I_n - P_n)y) - \log h(Y'(I_n - P_n)Y) - p] \\ &\quad \cdot f(\lambda(Y'Y))dY \\ &= R_1(I_p, h(S)), \end{aligned}$$

which completes the proof. □

Theorem 4.4.5. *Let X satisfy the conditions of Lemma* 4.4.7. *Then the best (smallest risk) estimate of Σ in the class of estimates satisfying* (4.4.15) *under the loss function $L_1(\Sigma, D)$ is*

$$h^*(S) = T'\Delta^* T,$$

where $T' = (t_{(1)}, \ldots, t_{(p)})$ was defined before and $\Delta^* = \operatorname{diag}(E(\|t_{(1)}\|^2), \ldots, E(\|t_{(p)}\|^2))$.

Proof. By Lemma 4.4.6 and Lemma 4.4.7, we have

$$\begin{aligned} R_1(\Sigma, h(S)) = R_1(I_p, T'\Delta T) &= E[\operatorname{tr}(T'\Delta T) - \log|T'\Delta T| - p] \\ &= E[\operatorname{tr}(T'\Delta T) - \log|\Delta| - \log|S| - p] \end{aligned}$$

$$= E[\mathrm{tr}(\sum_{i=1}^{p} \delta_i \boldsymbol{t}_{(i)}\boldsymbol{t}'_{(i)}) - \sum_{i=1}^{p} \log \delta_i - \log|\boldsymbol{S}| - p]$$
$$= E[\sum_{i=1}^{p} \delta_i \boldsymbol{t}'_{(i)}\boldsymbol{t}'_{(i)}] - \sum_{i=1}^{p} \log \delta_i - E(\log|\boldsymbol{S}|) - p$$
$$= \sum_{i=1}^{p} \delta_i E(\boldsymbol{t}'_{(i)}\boldsymbol{t}_{(i)}) - \sum_{i=1}^{p} \log \delta_i - E(\log|\boldsymbol{S}|) - p.$$

This attains its minimum value when $\delta_i = E(\boldsymbol{t}'_{(i)}\boldsymbol{t}_{(i)})$, $i = 1, \ldots, p$. □

Corollary 1. *In Theorem* 4.4.5., $\Delta^* = \mathrm{diag}(\delta_1^*, \ldots, \delta_p^*)$, *where* $\delta_i^* = 1/(n + p - 2i)$, $i = 1, \ldots, p$, *provided* $\boldsymbol{X} \sim N_{n \times p}(\boldsymbol{1}\mu, \boldsymbol{I}_n \otimes \Sigma)$.

Proof. By using the Bartlett decomposition (cf. Corollary 1 of Theorem 3.4.1) and noting that $\boldsymbol{t}'_{(i)}\boldsymbol{t}_{(i)} \sim \chi^2_{n+p-2i}$, the corollary follows. □

Corollary 2. *Let* $\boldsymbol{X} \sim SS_{n \times p}(\mu, \Sigma, f)$ *with* $n > p$ *and let* $\boldsymbol{X}$ *have the 2nd finite moment. Then under the loss function* $L_1(\Sigma, \boldsymbol{D})$, *any estimate of the form* $\alpha\,\boldsymbol{S}$, $\alpha > 0$, *of* Σ *is inadmissible. In particular,* $(n-1)^{-1}(-2\phi'(0))^{-1}\boldsymbol{S}$ *is inadmissible.*

Proof. By noting that the estimate $\boldsymbol{S}$ satisfies the requirement (4.4.15), the corollary then follows from Theorem 4.4.5. □

Finally, we will discuss the problem of estimating Σ using the loss function $L_2(\Sigma, \boldsymbol{D}) = \mathrm{tr}(\Sigma^{-1}\boldsymbol{D} - \boldsymbol{I}_p)^2$.

Theorem 4.4.6. *Let* $\boldsymbol{X} \sim SS_{n \times p}(\mu, \Sigma, f)$ *with* $n > p$ *and have the 2nd finite moment. Under the loss function* $L_2(\Sigma, \boldsymbol{D})$, *the best (smallest risk) estimate of* Σ *in the class of estimates satisfying* (4.4.15), *is*

$$h_0(\boldsymbol{S}) = \boldsymbol{T}'\Delta_0\boldsymbol{T},$$

where $\boldsymbol{T}$ *was defined before and* $\Delta_0 = \mathrm{diag}(\delta_1, \ldots, \delta_p)$ *and* $\{\delta_i, 1 \leqslant i \leqslant p\}$, *are given by the solution of* (4.4.18).

Proof. Similar to Lemma 4.4.7, we have $R_2(\Sigma, h(\boldsymbol{S})) = R_2(\boldsymbol{I}_p, h(\boldsymbol{S}))$. Hence, by Lemma 4.4.6,

$$R_2(\Sigma, h(\boldsymbol{S})) = R_2(\boldsymbol{I}_p, h(\boldsymbol{S})) = E[L_2(\boldsymbol{I}_p, h(\boldsymbol{S})] = E[\mathrm{tr}(h(\boldsymbol{S}) - \boldsymbol{I}_p)^2]$$
$$= E[\mathrm{tr}(\boldsymbol{T}'\Delta\boldsymbol{T}\boldsymbol{T}'\Delta\boldsymbol{T})] - 2E[\mathrm{tr}(\boldsymbol{T}'\Delta\boldsymbol{T})] + p$$
$$= E[\sum_{i,j} \delta_i\delta_j(\boldsymbol{t}'_{(i)}\boldsymbol{t}_{(j)})^2] - 2E[\sum_{i=1}^{p} \delta_i\boldsymbol{t}'_{(i)}\boldsymbol{t}_{(i)}] + p$$
$$= \sum_{i,j} \delta_i[E(\boldsymbol{t}'_{(i)}\boldsymbol{t}_{(j)})^2]\delta_j - 2\sum_{i=1}^{p} \delta_i E(\boldsymbol{t}'_{(i)}\boldsymbol{t}_{(i)}) + p$$
$$\mathrel{\hat{=}} \delta'\boldsymbol{B}\delta - 2\delta'\boldsymbol{b} + p.$$

We get

$$\frac{\partial R_2(\Sigma, h(\boldsymbol{S}))}{\partial \delta} = 2\boldsymbol{B}\delta - 2\boldsymbol{b} = 0,$$

and

(4.4.18) $$\boldsymbol{B}\delta = \boldsymbol{b},$$

which is desired. □

Corollary. *Under the assumption of Theorem 4.4.6., we have*
(i) *any eatimate of the form* $\alpha\mathbf{S}$, $\alpha > 0$, *of* Σ *is inadmissible, and*
(ii) *if* $\boldsymbol{X} \sim N_{n\times p}(\mathbf{1}\mu', \boldsymbol{I}_p \otimes \Sigma)$, $\boldsymbol{B} = (b_{ij})$, $\boldsymbol{b}' = (b_1, ..., b_p)$, *then* $b_{ii} = (n + p - 2i)(n + p - 2i + 2)$, $b_i = n + p - 2i$, $i = 1, ..., p$, $b_{ij} = n + p - 2j$, $i < j$.

Proof. (i) directly follows from Theorem 4.4.6. and (ii) follows by applying Bartlett decomposition. □

4.4.3. Minimax Estimates of μ

First, we study the minimax estimation about μ, given Σ. So we can assume $\Sigma = \boldsymbol{I}_p$ and as stated before we may work with models $\boldsymbol{X} \sim EC_p(\mu, \boldsymbol{I}_p, f)$.

Lemma 4.4.8. *Let* $\boldsymbol{X} \sim EC_p(\mu, \boldsymbol{I}_p, f)$ *with the 2nd finite moment. Then* $\boldsymbol{x}$ *is a minimax estimate of* μ *under the quadratic loss* (4.4.5).

Proof. This may follow from Kiefer (1957). Alternatively, by virtue of Chen (1964), there is a minimax estimate of μ with form

$$\hat{\mu}_0 = \boldsymbol{x} + h_0(\boldsymbol{x}), \tag{4.4.19}$$

where h_0 satisfies $h_0(\boldsymbol{x} + \boldsymbol{a}) = h_0(\boldsymbol{x})$ for each $\boldsymbol{a} \in R^p$. Hence if we can show that $\boldsymbol{x}$ is minimax in the class of the estimates having transitive invariance, in which each estimate has necessarily the form (4.4.19), then the lemma follows clearly. Now, $h(\boldsymbol{x} + \boldsymbol{c}) = h(\boldsymbol{x})$, for each $\boldsymbol{c} \in R^p$ and $\boldsymbol{x} \in R^p$, iff $h(\boldsymbol{x}) = h(\boldsymbol{0})$ for each $\boldsymbol{x} \in R^p$. Therefore, satisfying (4.4.19) has necessarily the form $\hat{\mu} = \boldsymbol{x} + \boldsymbol{c}$ for some constant $\boldsymbol{c} \in R^p$. However, for any $\boldsymbol{c} \in R^p$, we have

$$\begin{aligned} R(\mu, \boldsymbol{x}) - R(\mu, \boldsymbol{x} + \boldsymbol{c}) &= E[(\boldsymbol{x} - \mu)'(\boldsymbol{x} - \mu)] - E[(\boldsymbol{x} + \boldsymbol{c} - \mu)'(\boldsymbol{x} + \boldsymbol{c} - \mu)] \\ &= \boldsymbol{c}'\boldsymbol{c} \geqslant 0. \end{aligned}$$

This means that $\boldsymbol{x}$ is minimax in the class of the estimates having transitive invariance, which completes the proof. □

Note that if $\hat{\theta}_1$ is better than $\hat{\theta}_2$. Then that $\hat{\theta}_2$ is minimax implies that $\hat{\theta}_1$ is minimax. So $\delta_{\boldsymbol{a}}(\boldsymbol{x})$ is minimax by Theorem 4.4.2.

Finally, we are going the devote a little space to the study of *M*-estimates which is associated with robustness. We know that if $X_1, ..., X_n$ are a sample of size n from a population, $f(\frac{x-\theta}{\sigma})$, say, then the MLE's of θ and σ are the solutions of the equations

$$\sum_{i=1}^{n} \frac{f'\left(\dfrac{X_i - \theta}{\sigma}\right)}{f\left(\dfrac{X_i - \theta}{\sigma}\right)} = 0 \tag{4.4.20}$$

and

$$\sum_{i=1}^{n} \left[\left(\frac{X_i - \theta}{\sigma}\right) \frac{f'\left(\dfrac{X_i - \theta}{\sigma}\right)}{f\left(\dfrac{X_i - \theta}{\sigma}\right)} - 1\right] = 0 \tag{4.4.21}$$

In analogy to the maximum likelihood estimate solution and, in order to avoid strong dependence on a special form of $f(\boldsymbol{x})$, a general class of *M-estimates of* μ and Σ are defined as the solutions of systems of equations of the form

$$(4.4.22)\qquad \begin{cases} \dfrac{1}{n}\sum_{i=1}^{n} u_1[\{(\boldsymbol{X}_i-\mu)'\Sigma^{-1}(\boldsymbol{X}_i-\mu)\}^{1/2}](\boldsymbol{X}_i-\mu)=\boldsymbol{0},\\ \dfrac{1}{n}\sum_{i=1}^{n} u_2[\{(\boldsymbol{X}_i-\mu)'\Sigma^{-1}(\boldsymbol{X}_i-\mu)\}](\boldsymbol{X}_i-\mu)(\boldsymbol{X}_i-\mu)'=\Sigma, \end{cases}$$

where u_1 and u_2 are functions satisfying a set of general assumptions which ensure the existence and uniqueness of solution of equations (4.4.22). Huber (1964) proposed the M- estimates for which

$$u_1(t)=\begin{cases} -k, & t\leqslant -k,\\ t, & -k<t<k,\\ k, & t\geqslant k, \end{cases}$$

and

$$u_2(t)=u_1^2(t)-\frac{1}{(2\pi)^{1/2}}\int_{-\infty}^{\infty} u_1^2(x)e^{-t_x^2/2}dx.$$

Many developments have recently been attained in the M-estimate. If $\boldsymbol{x}=(X_1,\ldots,X_n)'\sim EC_n(\mu,\Sigma,\phi)$, (Note that $X_1,\ldots,X_n$ are not necessarily independent.), we also call the solutions of (4.4.22) the M–estimates. Bariloche (1976) has shown that for $\boldsymbol{X}\sim EC_n(\mu,\Sigma,\phi)$, there is a unique solution of (4.4.22) under some regular conditions.

References

Anderson and Fang (1982c), Bariloche (1976), Blyth (1951), Brandwein (1979), Brandwein, and Strawderman (1978), Chen (1964), Cohen (1966), Dawid (1978), Dykstra (1970), Fan and Fang (1986), Haff (1980), Halmos and Savage (1949), Huber (1964), James and Stein (1961), Kiefer (1957), Olkin and Selliah (1977), Stein (1956), Strawderman (1971, 1972 1974), Zhang, Fang and Chen (1985), Zhang and Bian (1984).

Exercises 4

4.1. Let $\boldsymbol{X}\sim VS_{n\times p}(\mu,\Sigma,f)$ with $n>p$ and let f be nonincreasing and differentiable and let $\boldsymbol{X}$ have the 1st moment. Find the MLE of $\mathcal{E}(\boldsymbol{X}_1|\boldsymbol{X}_2)$, (given $\boldsymbol{X}_2$), where $\boldsymbol{X}=(\boldsymbol{X}_1,\boldsymbol{X}_2)$.

4.2. Let $\boldsymbol{X}$ satisfy the conditions of Exercise 4.1 but $\mu=\boldsymbol{C}\beta$ where $\boldsymbol{C}: p\times r$, a known constant matrix, β is a parametric vector and $\mathbf{rk}(\boldsymbol{C})=r$. Find the maximum likelihood estimates of β and Σ.

4.3. Prove that Theorem 4.1.4. is still true if $\boldsymbol{X}\sim SS_{n\times p}(\mu,\Sigma,f)$, f satisfying the conditions (a) $f(\lambda(\Sigma_1))\geqslant f(\lambda(\Sigma_2))$ if $\Sigma_1\leqslant\Sigma_2$ and (b) $f(\lambda(\Sigma))\to f(\lambda(\Sigma_0))$ as $0\leqslant\Sigma\to\Sigma_0$ for each $\Sigma_0\geqslant 0$.

4.4. Let $\boldsymbol{X}$ satisfy the conditions in Theorem 4.1.1. Then prove that

(i) the maximum likelihood estimate of μ when Σ known is $\hat{\mu}=\bar{\boldsymbol{x}}$.

(ii) the maximum likelihood estimate of Σ when μ is known, $\mu=\mu_0$ say, is

$$\hat{\Sigma}_0=\lambda_{\max}(f)\sum_{i=1}^{n}(\boldsymbol{x}_{(i)}-\mu_0)(\boldsymbol{x}_{(i)}-\mu_0)'.$$

4.5. Let $\boldsymbol{X} \sim VS_{n\times p}(\phi)$ with $n > p$ and let $P(\boldsymbol{X}'\boldsymbol{X} > 0) = 1$. Then $\bar{\boldsymbol{x}}$ and $\boldsymbol{S}$ are independent iff $\boldsymbol{X}$ is normal. (Hint: use Theorem 2.8.1. for $\boldsymbol{Xa}$ where $\boldsymbol{a} \in R^p$ is any given constant vector.) Is it true if $\boldsymbol{X} \sim LS_{n\times p}(f)$?

4.6. Let $\boldsymbol{X} \sim N(\mu, \sigma^2 \boldsymbol{I}_p)$ and $\mu \sim N(\boldsymbol{a}, \tau^2 \boldsymbol{I}_p)$. Calculate the posterior density $p(\mu|\boldsymbol{x})$ and show that $\mathcal{E}(\mu|\boldsymbol{x}) = \left(\frac{1}{\sigma^2}\boldsymbol{x} + \frac{1}{\tau^2}\mu\right) \Big/ \left(\frac{1}{\sigma^2} + \frac{1}{\tau^2}\right)$. What information can you get through comparing $\mathcal{E}(\mu|\boldsymbol{x})$ with μ, the latter being $\boldsymbol{x}$'s expectation given μ?

4.7. Let $\boldsymbol{x} \sim N(\mu, \boldsymbol{I}_p)$ and $p \geqslant 5$. Take the prior density $\mu|\lambda \sim N(\boldsymbol{O}, \frac{1-\lambda}{\lambda}\boldsymbol{I}_p)$, $\lambda \sim (1-a)\lambda^{-a}$, $0 \leqslant \lambda \leqslant 1$, $(0 < a < 1)$, i.e., μ has a prior density

$$\int_0^1 (2\pi\frac{1-\lambda}{\lambda})^{-1/2}(1-a)\lambda^{-a}\exp\{-\frac{1}{2}\left(\frac{1-\lambda}{\lambda}\right)\mu'\mu\}d\lambda, \qquad \mu \in R^p.$$

Verify that the Bayes estimate corresponding to the above prior, $\mathcal{E}(\mu|\boldsymbol{x}) = (1 - E(\lambda|\boldsymbol{x}))\boldsymbol{x}$, satisfies the conditions in Corollary of Theorem 4.4.2 so that $\mathcal{E}(\mu|\boldsymbol{x})$ is an estimate with being both better than $\boldsymbol{x}$ and admissible. (Strawderman, 1972. He also showed that when $p = 3,4$, there is no Bayes estimate for μ.)

4.8. Let $\boldsymbol{X} \sim VS_{n\times p}(\boldsymbol{O}, \Sigma, \phi)$, $n > p$, $\Sigma > 0$ and $P(\boldsymbol{X} = \boldsymbol{O}) = 0$. Show that $\boldsymbol{S}$ has a density and represent it through $\mathrm{tr}\boldsymbol{X}'\boldsymbol{X}$'s distribution function.

4.9. Use Lemma 4.4.2. to show that if $\boldsymbol{x} \sim N(\mu, \boldsymbol{I}_p)$ then $\boldsymbol{x}$ is minimax. (Hint: Consider a prior density $\mu \sim N(\boldsymbol{0}, \frac{1}{k}\boldsymbol{I}_p)$, and then let $k \to \infty$.)

4.10. Let $\boldsymbol{x} \sim N(\mu, \boldsymbol{I}_p)$. Then $(1 - \frac{a}{\boldsymbol{x}'\boldsymbol{x}})^+\boldsymbol{x}$ is better than $(1 - \frac{a}{\boldsymbol{x}'\boldsymbol{x}})\boldsymbol{x}$, where $c^+ = \max(c, \boldsymbol{0})$ and $0 < a \leqslant 2(p-2)$, $p > 2$.

4.11. Set an example to show that there are at least two solutions satisfying the equation

$$f'(x) + (N/2x)f(x) = 0$$

where f satisfies the conditions in Lemma 4.1.2 (See Anderson and Fang, 1982c).

4.12. Suppose that $g(y)$ is continuous $(0 \leqslant y < \infty)$ such that $g(\boldsymbol{x}'\boldsymbol{x})$ is a density for some $\boldsymbol{x} \in R^N$ and that $\mathcal{E}(\boldsymbol{x}'\boldsymbol{x}) < \infty$. Then $h(y) = y^{N/2}g(y)$ has a maximum at some finite $y_g > 0$.

CHAPTER V

TESTING HYPOTHESES

In this chapter, we would like to discuss the statistical tests. In Section 5.1, distribution-free statistics are studied, which forms a preparation for deriving the distributions of the statistics used as tests. Then using the likelihood ratio method, testing means and covariance matrices are investigated. It is found that most of the likelihood ratio tests in spherical matrix distributions are the same as those in normal distributions. In the last section the robustness of tests is treated and goodness of fit tests is studied in which several procedures of testing elliptically contoured distributions are suggested.

5.1. Distribution-Free Statistics

In Chapter 3, we have found that many statistics have the same distribution in the classes of spherical matrix distributions. A natural problem is what kind of statistics has the same distribution in the whole class. Now, we want to discuss this problem. Suppose $\mathfrak{F}$ is a set of some random variables (vectors, or matrices) and $t(\boldsymbol{x})$ a statistic. It is called *distribution-free on* $\mathfrak{F}$ if

$$t(\boldsymbol{x}) \overset{\mathrm{d}}{=} t(\boldsymbol{y}), \quad \text{for each} \quad \boldsymbol{x} \in \mathfrak{F}, \quad \boldsymbol{y} \in \mathfrak{F}.$$

Sometimes a distribution-free statistic on $\mathfrak{F}$ is said to be *invariant on* $\mathfrak{F}$. And the words "on $\mathfrak{F}$" can be omitted if we are not concerned with $\mathfrak{F}$. Let (cf. Section 3.1)

$$\mathfrak{F}_1^+ = \{\boldsymbol{X} \in \mathfrak{F}_1 | P\{|\boldsymbol{X}'\boldsymbol{X}| = 0\} = 0\},$$

$$\mathfrak{F}_2^+ = \{\boldsymbol{X} \in \mathfrak{F}_2 | P\{\boldsymbol{x}_i = \boldsymbol{0}\} = 0,\ i = 1, \dots, p\} \text{ and}$$

$$\mathfrak{F}_3^+ = \{\boldsymbol{X} \in \mathfrak{F}_3 | P\{\boldsymbol{X} = \boldsymbol{O}\} = 0\}.$$

Theorem 5.1.1. *Suppose that $t(\boldsymbol{X})$ is a statistic. Then*

(a) *$t(\boldsymbol{X})$ is distribution-free on $\mathfrak{F}_1^+$ iff*

(5.1.1) $$t(\boldsymbol{X}\boldsymbol{A}) \overset{\mathrm{d}}{=} t(\boldsymbol{X})$$

for each constant matrix $\boldsymbol{A} \in UT(p)$ with positive diagonal elements and each $\boldsymbol{X} \in \mathfrak{F}_1^+$.

(b) *$t(\boldsymbol{X})$ is distribution-free on $\mathfrak{F}_2^+$ iff*

(5.1.2) $$t(\boldsymbol{X}\boldsymbol{R}) \overset{\mathrm{d}}{=} t(\boldsymbol{X})$$

for each constant diagonal matrix $\boldsymbol{R} = \mathrm{diag}(r_1, \dots, r_p) > 0$ and each $\boldsymbol{X} \in \mathfrak{F}_2^+$.

(c) *$t(\boldsymbol{X})$ is distribution–free on $\mathfrak{F}_3^+$ iff*

(5.1.3) $$t(a\boldsymbol{X}) \overset{\mathrm{d}}{=} t(\boldsymbol{X})$$

for each real $a > 0$ *and each* $\boldsymbol{X} \in \mathcal{F}_3^+$.

Proof. We only prove (a); others are similar. The "only if" part is trivial. Now suppose $\boldsymbol{X} \in \mathcal{F}_1^+$ and $\boldsymbol{X} \stackrel{d}{=} \boldsymbol{UA}$ where $\boldsymbol{U}$ and $\boldsymbol{A}$ are independent, $\boldsymbol{U}$ is the random matrix having uniform distribution on the Stiefel manifold, and $\boldsymbol{A} \in UT(p)$. Then for each Borel function $h \geqslant 0$, we have (by using the assumption that $t(\boldsymbol{XB}) \stackrel{d}{=} t(\boldsymbol{X})$ for each constant $\boldsymbol{B} \in UT(p)$ with positive diagonal elements)

$$\begin{aligned} E[h(t(\boldsymbol{X}))] &= E[h(t(\boldsymbol{UA}))] = E_A\{E[h(t(\boldsymbol{UA})|\boldsymbol{A}]\} \\ &= E_A\{E[h(t(\boldsymbol{U}))]\} = E[h(t(\boldsymbol{U}))], \end{aligned}$$

i.e., $t(\boldsymbol{X}) \stackrel{d}{=} t(\boldsymbol{U})$, for each $\boldsymbol{X} \in \mathcal{F}_1^+$. □

Now we consider the following more general classes:

(5.1.4) $$\mathcal{F}_i^+(\boldsymbol{M}, \Sigma) = \{\boldsymbol{Y} | \boldsymbol{Y} \stackrel{d}{=} \boldsymbol{M} + \boldsymbol{X}\Sigma^{1/2}, \ \boldsymbol{X} \in \mathcal{F}_i^+\}, \qquad i = 1, 2, 3,$$

where $\Sigma > 0$ and $\boldsymbol{M}$ are constant matrices. Let ω_m and ω_c be sets of some $n \times p$ matrices $\boldsymbol{M}$ and some positive definite matrices Σ, respectively. It is always assumed that $\boldsymbol{O} \in \omega_m$ and $\boldsymbol{I}_p \in \omega_c$. If a statistic $t(\boldsymbol{X})$ satisfies

(5.1.5) $t(\boldsymbol{X} + \boldsymbol{T}) \stackrel{d}{=} t(\boldsymbol{X})$ for each $\boldsymbol{X} \in \mathcal{F}_i^+(\boldsymbol{M}, \Sigma)$, $\boldsymbol{M} \in \omega_m$, $\boldsymbol{T} \in \omega_m$ and $\Sigma \in \omega_c$,

and

(5.1.6) $t(\boldsymbol{XC}) \stackrel{d}{=} t(\boldsymbol{X})$ for each $\boldsymbol{X} \in \mathcal{F}_i^+(\boldsymbol{M}, \Sigma)$, $\boldsymbol{B} \in \omega_c$, $\boldsymbol{C} \in \omega_c$ and $\Sigma \in \omega_c$,

then the distribution of $t(\boldsymbol{X})$ is invariant on $\mathcal{F}_i^+(\boldsymbol{M}, \Sigma)$ with $\boldsymbol{M} \in \omega_m$ and $\Sigma \in \omega_c$. When ω_c is the whole set of $p \times p$ positive definite matrices δ_p, it can be shown that the condition (5.1.6) implies the conditions (5.1.1), (5.1.2), and (5.1.3), for $i = 1, 2, 3$, seperately. In the theory of testing hypotheses, ω_m and ω_c have different definitions for different cases.

It is an known fact that if $\boldsymbol{X} \in N_{n \times p}(\boldsymbol{M}, \boldsymbol{I}_n \otimes \Sigma)$ then $\boldsymbol{X} \in \mathcal{F}_i^+(\boldsymbol{M}, \Sigma)$ for $i = 1, 2, 3$. In particular, if $\boldsymbol{Z} \sim N_{n \times p}(\boldsymbol{O}, \boldsymbol{I}_n \otimes \boldsymbol{I}_p)$, then $\boldsymbol{Z} \in \mathcal{F}_i^+$, $i = 1, 2. 3$. By using this fact, we obtain immediately the following corollaries:

Corollary 1. *If a statistic* $t(\boldsymbol{X})$ *satisfies* (5.1.5.) *and* (5.1.6) *for* $\boldsymbol{X} \in \mathcal{F}_i^+(\boldsymbol{M}, \Sigma)$, $\boldsymbol{M} \in \omega_m$ *and* $\Sigma \in \omega_c = \delta_p$, *then*

$$t(\boldsymbol{X}) \stackrel{d}{=} t(\boldsymbol{Z}), \textit{ for } \boldsymbol{X} \in \mathcal{F}_i^+(\boldsymbol{M}, \Sigma), \ \boldsymbol{M} \in \omega_m \textit{ and } \Sigma \in \delta_p,$$

where $\boldsymbol{Z} \sim N_{n \times p}(\boldsymbol{O}, \boldsymbol{I}_n \otimes \boldsymbol{I}_p)$ *and* $i = 1, 2\ 3$.

Corollary 2. *If a statistic* $t(\boldsymbol{X})$, $\boldsymbol{X} \in \mathcal{F}_3^+(\boldsymbol{M}, \boldsymbol{I}_p)$, *satisfies* (5.1.3) *and* (5.1.5) *for each* $\boldsymbol{M} \in \omega_m$, *then*

$$t(\boldsymbol{X}) \stackrel{d}{=} t(\boldsymbol{Z}) \textit{ for } \boldsymbol{X} \in \mathcal{F}_3^+(\boldsymbol{M}, \boldsymbol{I}_p), \ \boldsymbol{M} \in \omega_m.$$

If $\boldsymbol{X} \in \mathcal{F}_1^+$, then $\boldsymbol{X} \stackrel{d}{=} \boldsymbol{UA}$ and $\boldsymbol{U} \stackrel{d}{=} X(X'X)^{-1/2}$ by Lemma 3.2.6. Hence, if a statistic $t(\boldsymbol{X})$ satisfies

(5.1.7) $$t(\boldsymbol{X}) \stackrel{d}{=} t(\boldsymbol{X}(\boldsymbol{X}'\boldsymbol{X})^{-1/2}), \text{ for each } \boldsymbol{X} \in \mathcal{F}_1^+,$$

then $t(\boldsymbol{X})$ is distribution-free on $\mathcal{F}_1^+$. Sometimes, it is easy to check the condition (5.1.7).

Kariya (1981) tried to prove that a necessary and sufficient condition for the distribution of $t(\boldsymbol{X})$ to remain the same for all $\boldsymbol{X} \in \mathcal{F}_1^+(\boldsymbol{M}, \Sigma)$, $\boldsymbol{M} \in \omega_m$ and $\Sigma \in \delta_p$, is that when $\boldsymbol{Z} \sim N_{n \times p}(\boldsymbol{M}, \boldsymbol{I}_n \otimes \Sigma)$, the following conditions hold:

(i) $t(\boldsymbol{Z} + \boldsymbol{M}) \stackrel{d}{=} t(\boldsymbol{Z})$ for each $\boldsymbol{M} \in \omega_m$ and $\Sigma \in \delta_p$, and

(ii) $t(\boldsymbol{Z}) \stackrel{d}{=} t(\boldsymbol{U})$ for $\boldsymbol{M} = \boldsymbol{O}$ and $\Sigma \in \delta_p$, where $\boldsymbol{U}$ is defined in Definiton 3.1.2.

Bian and Zhang (1984) pointed out that the Kariya's theorem is not always true. They gave a counterexample. They proved that if $\boldsymbol{X}$ has a density, then the Kariya's theorem is right (See Exercise 5.3).

Now we can use Theorem 5.1.1 and its corollaries to verify whether a statistic $t(\boldsymbol{X})$ is distribution-free.

Example 5.1.1. Suppose $\boldsymbol{X} \in \mathcal{F}_1^+$, $\boldsymbol{X} = \begin{pmatrix} \boldsymbol{X}_1 \\ \vdots \\ \boldsymbol{X}_{k+1} \end{pmatrix}$, $\boldsymbol{X}_i$: $n_i \times p$, $n_i \geqslant p$, $i = 1, \ldots, k$. Then $\boldsymbol{D}_i = (\boldsymbol{X}'\boldsymbol{X})^{-1/2}(\boldsymbol{X}_i'\boldsymbol{X}_i)(\boldsymbol{X}'\boldsymbol{X}_i)^{-1/2}$, $i = 1, \ldots, k$, has matrix Dirichlet distribution. In fact, for $\boldsymbol{U} = \boldsymbol{X}(\boldsymbol{X}'\boldsymbol{X})^{-1/2} = \begin{pmatrix} \boldsymbol{X}_1 \\ \vdots \\ \boldsymbol{X}_{k+1} \end{pmatrix} \left(\sum_{i=1}^{k+1} \boldsymbol{X}_i'\boldsymbol{X}_i \right)^{-1/2}$, we have $\boldsymbol{D}_i = \boldsymbol{U}_i'\boldsymbol{U}_i$, $i = 1, \ldots, k$, where $\boldsymbol{U}_i = \boldsymbol{X}_i(\boldsymbol{X}'\boldsymbol{X})^{-1/2}$, $i = 1, \ldots, k$. Thus $\boldsymbol{D}_1, \ldots, \boldsymbol{D}_k$ are functions of $\boldsymbol{U} = \boldsymbol{X}(\boldsymbol{X}'\boldsymbol{X})^{-1/2}$ only and by the above argument $\boldsymbol{D}_1, \ldots, \boldsymbol{D}_k$ have the same distribution as $\boldsymbol{X}$ which is normal (See Section 3.5.2). When $k = 1$, $\boldsymbol{D}_1$ has the matrix beta distribution.

Example 5.1.2. Suppose $\boldsymbol{X} = \begin{pmatrix} \boldsymbol{X}_1 \\ \vdots \\ \boldsymbol{X}_k \end{pmatrix} \in \mathcal{F}_1^+$, $\boldsymbol{X}_i$: $n_i \times p$, $i = 1, \ldots, k$. Then the roots of

$$|\boldsymbol{X}_i'\boldsymbol{A}_i\boldsymbol{X}_i - \boldsymbol{X}_j'\boldsymbol{A}_j\boldsymbol{X}_j| = 0, \ i \neq j, \ i, j = 1, \ldots, k,$$

for fixed symmetric $\boldsymbol{A}_1, \ldots, \boldsymbol{A}_k$, have the same distribution as $\boldsymbol{X}$ which is normal. In fact, now for any $\boldsymbol{T} \in UT(p)$ with positive diagonal elements which is a fixed constant matrix, we have

$$0 = |\boldsymbol{X}_i'\boldsymbol{A}_i\boldsymbol{X}_i - \boldsymbol{X}_j'\boldsymbol{A}_j\boldsymbol{X}_j|, \ i \neq j, \ i, j = 1, \ldots, k,$$

iff

$$0 = |\boldsymbol{T}'\boldsymbol{X}_i'\boldsymbol{A}_i\boldsymbol{X}_i\boldsymbol{T} - \boldsymbol{T}'\boldsymbol{X}_j'\boldsymbol{A}_j\boldsymbol{X}_j\boldsymbol{T}|, \ i = j, \ i, j = 1, \ldots, k,$$

By noting $\boldsymbol{XT} = (\boldsymbol{T}'\boldsymbol{X}_1', \ldots, \boldsymbol{T}'\boldsymbol{X}_k')'$ and using Theorem 5.1.1., the assertion follows.

In fact, the statistics of Example 5.1.1 and Example 5.1.2 remain to have the same distributions on $\mathcal{F}_1^+(\boldsymbol{O}, \Sigma)$ for each $\Sigma \in \delta_p$. There are some statistics which are distribution-free on $\mathcal{F}_3^+$, but are not on $\mathcal{F}_1^+$ (See Exercise 5.1).

5.2. Testing Hypotheses About Mean Vectors

5.2.1. Likelihood Ratio Criteria

Let $\boldsymbol{X} \sim LS_{n \times p}(\mu, \Sigma, f)$ with $n > p$, $\Sigma > 0$, $\mu \in \Omega_m$ and $\Sigma \in \Omega_c$, where $\Omega_m \subset R^p$ and $\Omega_c \in \delta_p$. If we are required to test the following hypothesis

$$H_0 : \mu \in \omega_m \text{ and } \Sigma \in \omega_c,$$

where $\omega_m \subset \Omega_m$ and $\omega_c \subset \Omega_c$, we always assume $\mathbf{0} \subset \omega_m$ and $\boldsymbol{I}_p \in \omega_c$. There are many methods to obtain a statistic testing the hypothesis H_0. Usually., one can use likelihood ratio criteria, because they, in general, have many good properties. The likelihood ratio criterion of the hypothesis H_0 is as follows:

$$\lambda = \frac{\max\limits_{\mu \in \omega_m, \ \Sigma \in \omega_c} L(\mu, \Sigma)}{\max\limits_{\mu \in \Omega_m, \ \Sigma > \Omega_c} L(\mu, \Sigma)},$$

where $L(\mu, \Sigma)$ is the likelihood function of $\boldsymbol{X}$, i.e., the density of $\boldsymbol{X}$, but regarded as the function of parameter (μ, Σ). The rejection region is $\lambda \leqslant \lambda_\alpha$, where λ_α is determined by

$$P(\lambda \leqslant \lambda_\alpha | H_0) = \alpha.$$

5.2.2. Testing that a Mean Vector Equals a Specified Vector

Let $\boldsymbol{X} \sim VS_{n \times p}(\mu, \Sigma, f)$ with $n > p$ and let f be nonincreasing and continuous. In most practical problems, μ, and Σ are all unknown. We are interested in whether μ equals a specific μ_0, that is, we are required to test the hypothesis

$$H_0 : \mu = \mu_0, \text{ against } H_1 : \mu \neq \mu_0. \tag{5.2.1}$$

Without loss of generality we assume $\mu_0 = \mathbf{0}$, otherwise, we may consider to replace $\boldsymbol{X}$ by $\boldsymbol{X} - \mu_0'$. Hence the above hypothesis becomes

$$H_0 : \mu = \mathbf{0}, \text{ against } H_1 : \mu \neq \mathbf{0}. \tag{5.2.2}$$

The likelihood function of the observation $\boldsymbol{X}$ is given by

$$\begin{aligned} L(\mu, \Sigma) &= |\Sigma|^{-n/2} f(\operatorname{tr}(\boldsymbol{X} - \mathbf{1}\mu')\Sigma^{-1}(\boldsymbol{X} - \mathbf{1}\mu)') \\ &= |\Sigma|^{-n/2} f(\operatorname{tr}\Sigma^{-1}\boldsymbol{S} + n(\bar{\boldsymbol{x}} - \mu)'\Sigma^{-1}(\bar{\boldsymbol{x}} - \mu)) \end{aligned} \tag{5.2.3}$$

where

$$\bar{\boldsymbol{x}} = \frac{1}{n}\sum_{i=1}^{n} \boldsymbol{x}_{(i)} = \frac{1}{n}\boldsymbol{X}'\mathbf{1} \text{ and } \boldsymbol{S} = \sum_{i=1}^{n} (\boldsymbol{x}_{(i)} - \bar{\boldsymbol{x}})(\boldsymbol{x}_{(i)} - \bar{\boldsymbol{x}})'.$$

Putting $\boldsymbol{P}_n = \frac{1}{n}\mathbf{1}\mathbf{1}$, we have $\boldsymbol{S} = \boldsymbol{X}'(\boldsymbol{I}_n - \boldsymbol{P}_n)\boldsymbol{X}$, $\boldsymbol{P}_n\mathbf{1} = \mathbf{1}$.

Lemma 5.2.1. *Under the above assumptions, we have*

$$\max_{\substack{\mu = \mathbf{0} \\ \Sigma > \mathbf{0}}} L(\mu, \Sigma) = |\lambda_{\max}(f)\boldsymbol{X}'\boldsymbol{X}|^{-n/2} f(p/\lambda_{\max}(f)) \tag{5.2.4}$$

where $\lambda_{\max}(f)$ *is defined by Theorem* 4.1.1.

Proof. By (5.2.3),

$$\max_{\substack{\mu = \mathbf{0} \\ \Sigma > \mathbf{0}}} L(\mu, \Sigma) = \max_{\Sigma > 0} L(\mathbf{0}, \Sigma) = \max_{\Sigma > \mathbf{0}} |\Sigma|^{-n/2} f(\operatorname{tr}\Sigma^{-1}\boldsymbol{X}'\boldsymbol{X}). \tag{5.2.5}$$

Let $\boldsymbol{X}'\boldsymbol{X} = (\boldsymbol{X}'\boldsymbol{X})^{1/2}(\boldsymbol{X}'\boldsymbol{X})^{1/2}$ and let $\tilde{\Sigma} = (\boldsymbol{X}'\boldsymbol{X})^{-1/2}\Sigma(\boldsymbol{X}'\boldsymbol{X})^{-1/2}$. Then

$$|\Sigma|^{-n/2} f(\mathrm{tr}\Sigma^{-1} X'X) = |X'X|^{-n/2} |\tilde{\Sigma}|^{-n/2} f(tr\tilde{\Sigma}^{-1})$$

$$= |X'X|^{-n/2}(\lambda_1 \cdots \lambda_p)^{-n/2} f(\lambda_1^{-1} + \cdots + \lambda_p^{-1}) \triangleq |X'X|^{-n/2} g(\lambda_1, \ldots, \lambda_p),$$

where $\lambda_1, \ldots, \lambda_p$ are the p eigenvalues of $\tilde{\Sigma}$. Just as in the proof of Theorem 4.1.1, g arrives at its maximum at $\lambda_1 = \cdots = \lambda_p = \lambda$, say. By Lemma 4.1.2., (5.2.5) becomes

$$\max_{\substack{\mu=0\\ \Sigma>0}} L(\mu, \Sigma) = \max_{\lambda>0} |X'X|^{-n/2} \lambda^{-np/2} f(p/\lambda)$$

$$= |X'X|^{-n/2} [\lambda_{\max}(f)]^{-np/2} f(p/\lambda_{\max}(f)),$$

which is desired. □

By Theorem 4.1.1 we get

$$\max_{\substack{\mu\in R^p\\ \Sigma>0}} L(\mu, \Sigma) = |\lambda_{\max}(f) S|^{-n/2} f(p/\lambda_{\max}(f)). \tag{5.2.6}$$

From (5.2.5) and (5.2.6), it follows that the likelihood ratio is

$$\lambda = \frac{\max_{\mu=0,\,\Sigma>0} L(\mu, \Sigma)}{\max_{\mu\in R^p,\,\Sigma>0} L(\mu, \Sigma)} = \frac{|X'X|^{-n/2}}{|S|^{-n/2}} = \frac{|S|^{n/2}}{|X'X|^{n/2}} \tag{5.2.7}$$

$$= \frac{|S|^{n/2}}{|S + n\bar{x}\bar{x}'|^{n/2}} = (1 + n\bar{x}'S^{-1}\bar{x})^{-n/2}.$$

The last step of (5.2.7) follows from Exercise 1.1. Noting that λ is a decreasing function of $n\bar{x}'S^{-1}\bar{x}$, we can use $n\bar{x}'S^{-1}\bar{x}$ as a statistic to test (5.2.2), or, equivalently, use $T^2 = n(n-1)\bar{x}'S^{-1}\bar{x}$ to test it, and the rejection region is

$$n(n-1)\bar{x}'S^{-1}\bar{x} \geqslant c_\alpha, \tag{5.2.8}$$

where c_α is a constant depending on the level of significance α of the test. By Theorem 5.1, we easily verify that the statistic T^2 is invariant in $VS_{n\times p}(f)$. Then, to determine the constant c_α for a given level α, we need only to derive the null distribution of T^2 in the normal population.

5.2.3. The Distribution of T^2

Let $X \sim N_{n\times p}(1\mu', I_p \otimes \Sigma)$ with $n>p$ and

$$T^2 = T^2(p, n-1, \mu) = n(n-1)\bar{x}'S^{-1}\bar{x},$$

where

$$\bar{x} = \frac{1}{n}\sum_{i=1}^{n} x_{(i)}, \qquad S = \sum_{i=1}^{n} (x_{(i)} - \bar{x})(x_{(i)} - \bar{x})'.$$

By the argument of Section 4.1.1, we have

$$\bar{x} = n^{1/2} z_{(n)} \text{ and } S = \sum_{i=1}^{n-1} z_{(i)} z_{(i)}', \tag{5.2.9}$$

where

$$Z = \begin{pmatrix} z'_{(1)} \\ \vdots \\ z'_{(n)} \end{pmatrix} = \Gamma X$$

and $\Gamma \in O(n)$ with thw last row $(n^{-1/2}, \ldots, n^{-1/2})$. Hence, $Z \sim N_{n\times p}(M, I_n \otimes \Sigma)$ where $M = \Gamma 1 \mu' = \begin{pmatrix} O \\ n^{-1/2}\mu' \end{pmatrix}$, i.e., $z_{(1)}, \ldots, z_{(n)}$ are independent. $z_{(i)} \sim N(0, \Sigma)$, $i = 1, \ldots, n-1$ and $z_{(n)} \sim N(n^{-1/2}\mu, \Sigma)$. Write

$$\begin{aligned} &S = (s_{ij}), \qquad S^{-1} = (s^{ij}), \qquad S_r = (s_{ij},\ i,j = 1, \ldots, r.), \qquad 0 < r < p, \\ &\Sigma = (\sigma_{ij}), \qquad \Sigma^{-1} = (\sigma^{ij}). \end{aligned}$$

First, we need the following lemma.

Lemma 5.2.2. *Using the above notations, we have*
(i) $\sigma^{pp}/S^{pp} \sim \chi^2_{n-p}$ *and is independent of* S_{p-1}*;*
(ii) *for each* $0 \neq b \in R^p$, $b'\Sigma^{-1}b \,/\, b'S^{-1}b \sim \chi^2_{n-p}$.
Proof. Partition Σ and S into

$$\Sigma = \begin{pmatrix} \Sigma_{11} & \sigma_{12} \\ \sigma_{21} & \sigma_{pp} \end{pmatrix} \quad \text{and} \quad S = \begin{pmatrix} S_{11} & s_{12} \\ s_{21} & s_{pp} \end{pmatrix},$$

where Σ_{11}: $(p-1) \times (p-1)$ and S_{11}: $(p-1) \times (p-1)$. We get

$$S^{pp} = \frac{|S_{11}|}{|S|} = \frac{|S_{11}|}{|S_{11}|(s_{pp} - s_{21}S_{11}^{-1}s_{12})} = (s_{pp} - s_{21}S_{11}^{-1}s_{12})^{-1}.$$

As $S = Z'_*Z_*$ where $Z_* = (z_{(1)}, \ldots, z_{(n-1)})'$, it is easy to see

$$S_{11} = Z'_1Z_1, \qquad s_{12} = Z'_1z_p = s'_{21}, \qquad s_{pp} = z'_pz_p,$$

where $Z^* = (Z_1\ z_p)$ with Z_1: $(n-1) \times (p-1)$. Hence

$$\begin{aligned} (s^{pp})^{-1} &= z'_pz_p - z'_pZ_1(Z'_1Z_1)^{-1}Z'_1z_p \\ &= z'_p(I - Z_1(Z'_1Z_1)^{-1}Z'_1)z_p \overset{\Delta}{=} z'_pCz_p. \end{aligned}$$

say. Clearly, $C' = C$ and $C^2 = C$. By Corollary 3 of Lemma 2.6.1., it can be shown that

$$\begin{aligned} z_p|Z_1 &\sim N_{n-1}(Z_1\Sigma_{11}^{-1}\sigma_{12}, (\sigma_{pp} - \sigma_{21}\Sigma_{11}^{-1}\sigma_{12})I_{n-1}) \\ &= N_{n-1}(Z_1\Sigma_{11}^{-1}\sigma_{12}, (\sigma^{pp})^{-1}I_{n-1}). \end{aligned}$$

Given Z_1, by Corollary 1 of Theorem 2.8.2., we have

$$\sigma^{pp}/s^{pp} = z'_pCz_p\sigma^{pp} \sim \chi^2_{n-p}, \tag{5.2.10}$$

because

$$\text{rk}C = \text{tr}C = \text{tr}(I - Z_1(Z'_1Z_1)^{-1}Z'_1) = (n-1) - (p-1) = n - p$$

with probability one and

$$CZ_1\Sigma_{11}^{-1}\sigma_{12} = (I - Z_1(Z'_1Z_1)^{-1}Z'_1)Z_1\Sigma_{11}^{-1}\sigma_{12} = 0.$$

As the conditional distribution of σ^{pp}/s^{pp} is independent of Z_1, the first assertion follows.

To prove (ii) we may assume $\|b\| = 1$. Make an $\Gamma \in O(p)$ with b' as its pth row. With $\Gamma S\Gamma'$ and $\Gamma\Sigma\Gamma'$in the first part instead of S and Σ, the second assertion follows because

$\boldsymbol{b}'\boldsymbol{S}^{-1}\boldsymbol{b}$ and $\boldsymbol{b}'\Sigma^{-1}\boldsymbol{b}$ are the last diagonal elements of $\Gamma\boldsymbol{S}\Gamma'$ and $\Gamma\Sigma\Gamma'$, respectively. □

Theorem 5.2.1. *Assume that* $\boldsymbol{X} \sim N_{n\times p}(\boldsymbol{1}\mu', \boldsymbol{I}_n \otimes \Sigma)$ *with* $n > p$ *and* T^2 *is defined as before. Then*

$$\frac{n-p}{(n-1)p}T^2 \sim F(p, n-p, \lambda),$$

where $\lambda = n\mu'\Sigma^{-1}\mu$.

Proof. Write

$$T^2 = (n-1)\frac{\bar{\boldsymbol{x}}'\boldsymbol{S}^{-1}\bar{\boldsymbol{x}}'}{\bar{\boldsymbol{x}}'\Sigma^{-1}\bar{\boldsymbol{x}}}(\bar{\boldsymbol{x}}'(\frac{1}{n}\Sigma)^{-1}\bar{\boldsymbol{x}}) \equiv (n-1)T_1T_2,$$

say. Hence

$$\frac{n-p}{(n-1)p}T^2 = \frac{n-p}{p}\frac{T_2}{1/T_1}.$$

As $\bar{\boldsymbol{x}}$ is independent of $\boldsymbol{S}$ (Section 4.2.3), given $\bar{\boldsymbol{x}}$, we have $1/T_1 \sim \chi^2_{n-p}$ by Lemma 5.2.2 which is independent of $\bar{\boldsymbol{x}}$, i.e., $1/T_1 \sim \chi^2_{n-p}$ is independent of $\bar{\boldsymbol{x}}$. On the other hand, $\bar{\boldsymbol{x}} \sim N(\mu, \frac{1}{n}\Sigma)$. Let $\boldsymbol{y} = (\frac{1}{n}\Sigma)^{-1/2}\bar{\boldsymbol{x}}$. Then $\boldsymbol{y} \sim N_p((\frac{1}{n}\Sigma)^{-1/2}\mu, \boldsymbol{I}_p)$ and

$$T_2 = \bar{\boldsymbol{x}}'(\frac{1}{n}\Sigma)^{-1/2}\bar{\boldsymbol{x}} = \boldsymbol{y}'\boldsymbol{y} \sim \chi^2_p(\lambda),$$

where $\lambda = \mu'(\frac{1}{n}\Sigma)^{-1/2}(\frac{1}{n}\Sigma)^{-1/2}\mu = n\mu'\Sigma^{-1}\mu$. □

5.2.4 T^2-Testing and Invariance of Tests

Let $\mathscr{X}$ be the sample space and $\mathscr{B}_x$ be the σ–algebra of subsets of $\mathscr{X}$. In practice, $\mathscr{X}$ is often R^n and $\mathscr{B}_x$ is the Borel field on R^n. Let $\Theta = \{\theta\}$ be the parametric space. Denote by $\mathscr{P}$ the family of probability distributions P_θ on $\mathscr{B}_x$, $\theta \in \Theta$. We are concerned here with the problem of testing the null hypothesis H_0: $\theta \in \Theta_0$ against the alternative H_1:$\theta \in \Theta_1(\Theta_1 \cap \Theta_0 = \phi$, $\Theta_1 \subset \Theta$, and $\Theta_0 \subset \Theta)$. The principle of invariance for testing problems involves transformations mainly on two spaces, the sample space $\mathscr{X}$ and the parametric space Θ. Consider a transformation group G (cf. Section 1.7) on $\mathscr{X}$. We assume that for each $g \in G$, it is (i) one-to-one from $\mathscr{X}$ onto $\mathscr{X}$; i.e., for every $x_1 \in \mathscr{X}$, there exists $x_2 \in \mathscr{X}$ such that $x_2 = gx_1$ and $gx_1 = gx_2$ implies $x_1 = x_2$, and (ii) bimeasurable to ensure that whenever ξ is a random variable with values in $\mathscr{X}$, $g\xi$ is also a random variable with values in $\mathscr{X}$, and for any $A \in \mathscr{B}_x$, gA and $g^{-1}A$ (the image and the transformed set) both belong to $\mathscr{B}_x$. The *induced transformation* $\bar{g}$ *corresponding to* g *on* $\mathscr{X}$ is defined as follows: Let X be a random variable (vector) with values in $\mathscr{X}$ and probability distribution $P_{\theta'}$, $\theta' \in \Theta$, such that we can define a transformation $\bar{g}$ on Θ: $\bar{g}\theta = \theta'$, denoting the set of all such $\bar{g}$ by $\bar{G} = \{\bar{g}\}$. Suppose that all P_θ, $\theta \in \Theta$, are distinct, i.e.,$\theta_1 \neq \theta_2, \theta_1 \in \Theta, \theta_2 \in \Theta$, implies $P_{\theta 1} \neq P_{\theta 2}$. Then g determines $\bar{g}$ uniquely. It is easy to see that $\bar{G}$ is a transformation group on Θ and $\overline{g_2g_1} = \bar{g}_2\bar{g}_1$, $\overline{g^{-1}} = \bar{g}^{-1}$. The corespondence between g and $\bar{g}$ is a homomorphism.

Definition 5.2.1. The problem of testing H_0: $\theta \in \Theta_0$ against H_1: $\theta \in \Theta_1$ is said to *remain invariant with respect to a group G if*

(ii) for $g \in G$, $A \in \mathscr{B}_x$, $P_{\bar{g}\theta}\{gA\} = P_\theta(A)$, and

(ii) for $\bar{g} \in \bar{G}$, $\bar{g}\Theta_0 = \Theta_0$, $\bar{g}\Theta_1 = \Theta_1$.

Example 5.2.1. Define a group $G = GL(p) = \{A|A: p \times p, |A| \neq 0\}$. $GL(p)$ is often called a *linear group*. Let $\mathscr{X} = R^{np}$ and group GL(p) on $\mathscr{X}$ act as

$$A: X \to XA, \ A \in GL(p).$$

Let $X \sim VS_{n\times p}(\mu, \Sigma, f)$, $(\mu, \Sigma) \in \Theta = R^p \times \delta_p$, where δ_p is the positive definite matrix group. It is easy to see that $XA \sim VS_{n\times p}(A'\mu, A'\Sigma A, f)$. Hence the induced group $\bar{G}$ $= G$ acts on Θ as

$$A: (\mu, \Sigma) - (A'\mu, A'\Sigma A), \quad A \in \bar{G} = GL(p).$$

Consider the hypothesis H_0: $\mu = \mathbf{0}$ against H_1: $\mu \neq \mathbf{0}$ (See (5.2.2)). Now $\Theta = \{0\} \times \delta_p$ and $\Theta_1 = \Theta - \Theta_0$. So the problem of testing (5.2.2) remains invariant with respect to the group $GL(p)$.

In Section 1.7, we have defined the invariant function and the maximal invariant and given many examples. The following theorem points out that if a family of distributions which is invariant under a group$G$ then the distribution of any invariant function (under G) depends only on a maximal invariant parameter (under $\bar{G}$).

Theorem 5.2.2. *Let $X \sim P_\theta$, $\theta \in \Theta$ and G be a group of transformations on sample space $\mathscr{X}$ and $T(x)$ be a measurable and invariant function under G. Let $\bar{G}$ be the induced group on Θ and $v(\theta)$ be a maximal invariant function under $\bar{G}$. Then the distribution of $T(X)$ depends on θ only through $v(\theta)$.*

Proof. Suppose $v(\theta_1) = v(\theta_2)$, $\theta_1, \theta_2 \in \Omega$. Then, since $v(\theta)$ is maximal invariant on Θ under $\bar{G}$, there exists a $\bar{g} \in \bar{G}$ such that $\theta_2 = \bar{g}\theta_1$. Now, for any Borel function $h \geqslant 0$, we have

$$(5.2.11) \quad E_{\theta_1}[h(T(X))] = E_{\theta_1}[h(T(gX))] = E_{\bar{g}\theta_1}[h(T(X))] = E_{\theta_2}[h(T(X))].$$

The proof is complete. □

By Theorem 5.2.2, the power function of any invariant test, $E_\theta(\phi)$ say, depends only on the maximal invariant in the parametric space. However, in general, the converse is not true.

To study the optimality of T^2-test, let X have a density

$$|\Sigma|^{-n/2} f(\mathrm{tr}\Sigma^{-1}S + n(\bar{x} - \mu)'\Sigma^{-1}(\bar{x} - \mu)).$$

Then when H_0: $\mu = \mathbf{0}$ is true, $\dfrac{np}{n-p}\bar{x}'S^{-1}\bar{x} = pT^2/(n-1)(n-p) \sim F(p, n-p)$. And when H_1: $\mu \neq \mathbf{0}$ is ture, $(np/(n-p))\bar{x}S^{-1}\bar{x}'$ is distributed as the generalized non-central F-distribution (cf. Section 2.9.3), i.e., $T^2 = \dfrac{np}{n-p}\bar{x}'S^{-1}\bar{x} \sim GF_{p,n-p}(\delta^2, f)$, δ^2 $= \mu\Sigma^{-1}\mu$ which is the maximal invariant in $\Theta = R^p \times \delta_p$ under group $\bar{G} = GL(p)$. The T^2 has a density

$$(5.2.12) \quad g(t^2|\delta^2) = c\left(\frac{p}{n-p}t^2\right)^{p/2-1}\left(1 + \frac{p}{n-p}t^2\right)^{-n/2} \int_0^\pi \int_0^\infty \int_0^\infty u^{n-1} r^{np-p-1}$$

$$\cdot f\left(u^2 - 2\left(\frac{pt^2}{n-p+pt^2}\right)^{1/2} u\delta\cos\beta + \delta^2 + r^2\right)\sin^{p-2}\beta\, du\, dr\, d\beta,$$

where c is a constant being independent of $\delta^2 = \mu'\Sigma^{-1}\mu$. When $\delta = 0$, it is obvious that $g(t^2|0)$ is the density of $F(p, n-p)$ (cf. Section 2.9.3).

Theorem 5.2.3 *Under the above assumptions, if $f''(u) \geqslant 0$, $u \geqslant 0$, then, among all tests $\phi(\boldsymbol{X})$ of level α for testing H_0: $\mu = \boldsymbol{0}$ against H_1: $\mu \neq \boldsymbol{0}$ which are invariant with respect to the group of nonsingular linear transformations, T^2-test or its equivalent* (5.2.11) *is a uniformly most powerful invariant test.*

Proof. Since $(\bar{\boldsymbol{x}}, \boldsymbol{S})$ is sufficient for (μ, Σ) (See Section 4.3.2), for any test $\phi(\boldsymbol{X})$, $\tilde{\phi} = E[\phi(\boldsymbol{X})|\bar{\boldsymbol{x}}, \boldsymbol{S}]$ is independent of *paramator* (μ, Σ) and depends only on $(\bar{\boldsymbol{x}}, \boldsymbol{S})$. When ϕ is invariant under $G = GL(p)$, i.e., $\phi(\boldsymbol{XA}) = \phi(\boldsymbol{X})$, for each $\boldsymbol{A} \in GL(p)$, it is easy to see that $\tilde{\phi}$ is invariant under $\boldsymbol{G}$. Since $E_\theta[\tilde{\phi}] = E_\theta[\phi]$, $\tilde{\phi}$ and ϕ have the same power function. Hence, we may confine our attention to invariant tests based only on $(\bar{\boldsymbol{x}}, \boldsymbol{S})$. Now, under the group of tansformations $\bar{\boldsymbol{x}} \to \boldsymbol{A}'\bar{\boldsymbol{x}}$, $\boldsymbol{S} \to \boldsymbol{A}'\boldsymbol{S}\boldsymbol{A}$ for each $\boldsymbol{A} \in GL(p)$, $\boldsymbol{T}^2$ is a maximal invariant function. Using Theorem 5.2.2, the power function of any invariant test depends only on $\delta^2 = \mu'\Sigma^{-1}\mu$. Thus, considering invariant tests, testing H_0: $\mu = \boldsymbol{0}$ against H_1: $\mu = \boldsymbol{0}$, is equivalent to testing H_0': $\delta = 0$ against H_1': $\delta \neq 0$. Applying the Neyman-Pearson fundamental Lemma to find the most powerful test, for testing $\delta^2 = 0$ against $\delta^2 > 0$, where δ is specified, we rejects H_0' whenever

$$\frac{g(t^2|\delta^2)}{g(t^2|0)} \geqslant c_\alpha', \tag{5.2.13}$$

where g is given by (5.2.12) and c_α' is chosen such that the test has level α. If we can prove that $g(t^2|\delta^2)/g(t^2|0)$ is a nondecreasing function of t^2 for any $\delta > 0$, there exists c such that (5.2.13) holds iff $t^2 \geqslant c$, for any $\delta > 0$, where c is independent of δ. Thus t^2-testing, equivalently T^2-testing, is uniformly most powerful invariant and the theorem follows. Now,

$$\frac{g(t^2|\delta^2)}{g(t^2|0)} = k\int_0^\pi\int_0^\infty\int_0^\infty u^{n-1}r^{np-p-1}f\left(u^2 - 2\sqrt{\frac{pt^2}{n-p+pt^2}}\,u\delta\,\cos\beta + \delta^2 + r^2\right)$$

$$\cdot(\sin\beta)^{p-2}\,du\,dr\,d\beta,$$

where k is a constant. So $(pt^2/(n-p+pt^2))^{1/2}$ is an increasing function of t^2. Therefore, if

$$\int_0^\pi f(u^2 - 2xu\delta\,\cos\beta + \delta^2 + r^2)\sin^{p-2}\beta d\beta$$

is a nondecreasing function of $x > 0$ for any $u > 0$, $\delta > 0$, $r > 0$, then $g(t^2|\delta^2)/g(t^2|0)$ is a nondecreasing function of t^2 for any $\delta^2 > 0$. Let

$$h(x) = \int_0^\pi f(u^2 - 2xu\delta\cos\beta + \delta^2 + r^2)\sin^{p-2}\beta d\beta.$$

Then, for $x \geqslant 0$,

$$h'(x) = -2u\delta\int_0^\pi f'(u^2 - 2xu\delta\cos\beta + \delta^2 + r^2)\cos\beta\,\sin^{p-2}\beta d\beta.$$

$$= -2u\delta\left[\int_0^1 f'(u^2 - 2xu\delta(1-y^2)^{1/2} + \delta^2 + r^2)y^{p-2}dy - \int_0^1 f'(u^2 + 2xu\delta(1-y^2)^{1/2} + \delta^2 + r^2)y^{p-2}dy\right]$$

$$= 4u^2\delta^2 x\int_0^1 (1-y^2)^{1/2} f''(\xi)y^{p-2}dy \geqslant 0,$$

where $u^2 - 2xu(1-y^2)^{1/2}\delta + \delta^2 + r^2 \leqslant \xi \leqslant u^2 + 2xu\delta(1-y^2)^{1/2} + \delta^2 + r^2$. This shows that $h(x)$ is nondecreasing. ∎

Theorem 5.2.3 is due to Quan (1986).

5.2.5. Testing Equality of Several Means with Equal and Unknown Covariance Matrices

Consider the case that the samples are from q populations each of which is p-dimensional with mean vectors $\mu_1, \ldots, \mu_q$ and covariance matrices $\Sigma_1, \ldots, \Sigma_q$, respectively. The i-th sample size is n_i, $n_i > p$, $i = 1, \ldots, q$ and $n = n_1 + \cdots + n_q$. Put $\bar{n}_0 = 0$ and $\bar{n}_i = n_1 + \cdots + n_i$, $i = 1, \ldots, q$. Assume that $\boldsymbol{x}_{(\bar{n}_{i-1}+1)}, \ldots, \boldsymbol{x}_{(\bar{n}_i)}$ are from the i-th population, $i = 1, \ldots, q$. Let $\boldsymbol{X}_i = (\boldsymbol{x}_{(\bar{n}_{i-1}+1)}, \ldots, \boldsymbol{x}_{(\bar{n}_i)})$ be the i-th sample observation, and assume that the joint density of the samples is

$$\left(\prod_{j=1}^{q} |\Sigma_j|^{-n_j/2}\right) f\left(\sum_{j=1}^{q} \operatorname{tr}\Sigma_j^{-1} G_j\right) \tag{5.2.14}$$

where $\boldsymbol{G}_j = \sum_{i=1}^{n_j} (\boldsymbol{x}_{(\bar{n}_{j-1}+i\cdot} - \mu_i)(\boldsymbol{x}_{(\bar{n}_{j-1}+i\cdot} - \mu_i)'$, $j = 1, \ldots, q$. Puting $\boldsymbol{X}' = (\boldsymbol{X}_1, \ldots, \boldsymbol{X}_q)$, we have $\boldsymbol{G}_j = (\boldsymbol{X}_j - \mu_j\boldsymbol{1}')(\boldsymbol{X}_j - \mu_j\boldsymbol{1})'$, $j = 1, \ldots, q$.

Let

$$\bar{\boldsymbol{x}}_j = \frac{1}{n}\sum_{i=1}^{n_j} \boldsymbol{x}_{(\bar{n}_{j-1}+i)} = \frac{1}{n_j}\boldsymbol{X}_j\boldsymbol{1}$$

and let

$$\boldsymbol{S}_j = \sum_{i=1}^{n_j} (\boldsymbol{x}_{(\bar{n}_{j-1}+i)} - \bar{\boldsymbol{x}}_j)(\boldsymbol{x}_{(\bar{n}_{j-1}+i)} - \bar{\boldsymbol{x}}_j)' = \boldsymbol{X}_j(\boldsymbol{I}_{n_j} - \boldsymbol{P}_j)\boldsymbol{X}_j', \tag{5.2.15}$$

where $\boldsymbol{P}_j = \frac{1}{n_j}\boldsymbol{1}\boldsymbol{1}'$, $j = 1, \ldots, q$. Using (5.2.15) we can rewrite (5.2.14) as follows:

$$\left(\prod_{j=1}^{q} |\Sigma_j|^{-n_j/2}\right) f\left(\sum_{j=1}^{q} \operatorname{tr}\Sigma_j^{-1}\boldsymbol{G}_j\right)$$

$$= \left(\prod_{j=1}^{q} |\Sigma_j|^{-n_j/2}\right) f\left(\sum_{j=1}^{q}\left[\operatorname{tr}\Sigma_j^{-1}\boldsymbol{S}_j + n_j(\bar{\boldsymbol{x}}_j - \mu_j)'\Sigma_j^{-1}(\bar{\boldsymbol{x}}_j - \mu_j)\right]\right). \tag{5.2.16}$$

By assuming that $\Sigma_1 = \cdots = \Sigma_q = \Sigma$, say, it is desired to test the hypothesis

$$H_0\colon \mu_1 = \cdots = \mu_q = \mu, \text{ (say)}. \tag{5.2.17}$$

The related likelihood function is

$$L(\mu_1, \ldots, \mu_q, \Sigma) = |\Sigma|^{-n/2} f\left(\sum_{j=1}^{q} [\operatorname{tr}\Sigma^{-1}\boldsymbol{S}_j + n_j(\bar{\boldsymbol{x}}_j - \mu_j)'\Sigma^{-1}(\bar{\boldsymbol{x}}_j - \mu_j)]\right). \tag{5.2.18}$$

The likelihood ratio statistic corresponding to (5.2.17) is as follows:

$$\lambda_1 = \frac{\max\limits_{\mu\in R^p,\ \Sigma>0} L(\mu, \ldots, \mu, \Sigma)}{\max\limits_{\substack{\mu_i\in R^p\ i=1,\ldots,q\\ \Sigma>0}} L(\mu_1, \ldots, \mu_q, \Sigma)}. \tag{5.2.19}$$

If f is nonincreasing and continuous, it is obvious that

$$\max_{\mu\in R^p,\ \Sigma>0} L(\mu, \ldots, \mu, \Sigma) = |\lambda_{\max}(f) S|^{-n/2} f(p/\lambda_{\max}(f)), \tag{5.2.20}$$

where

$$S = \sum_{i=1}^{n} (x_{(i)} - \bar{x})(x_{(i)} - \bar{x})' \text{ and } \bar{x} = \frac{1}{n}\sum_{i=1}^{n} x_{(i)} = \frac{1}{n}\sum_{j=1}^{q} n_j \bar{x}_j,$$

because the likelihood $L(\mu, \ldots, \mu, \Sigma)$ is equal to the likelihood (5.2.3).

Lemma 5.2.3. *Suppose that f is nonincreasing and continuous. Then*

(5.2.21)

$$\max_{\substack{\mu_i\in R^p,\ i=1,\ldots,q\\ \Sigma>0}} L(\mu_1, \ldots, \mu_q, \Sigma) = |\lambda_{\max}(f)(S_1 + \cdots + S_q)|^{-n/2} f(p/\lambda_{\max}(f)).$$

Proof. First, since f is nonincreasing, for any $\Sigma > 0$ we have

$$L(\mu_1, \ldots, \mu_q, \Sigma) \leqslant L(\bar{x}_1, \ldots, \bar{x}_q, \Sigma)$$

and the equality arises if $\mu_i = \bar{x}_i$, $i = 1, \ldots, q$. Then we have

$$L(\bar{x}_1, \ldots, \bar{x}_q, \Sigma) = |\Sigma|^{-n/2} f(tr\Sigma^{-1}(S_1 + \cdots + S_q)).$$

The rest of the proof of the lemma is the same as that of Theorem 4.1.1 by replacing S by $S_1 + \cdots + S_q$, which completes the proof of the lemma. □

Theorem 5.2.4. *Under the assumptions of Lemma 5.2.3, the likelihood ratio criterion for testing the hypothesis (5.2.17) is*

$$\lambda_1 = \frac{\max\limits_{\mu\in R^p,\ \Sigma>0} L(\mu, \ldots, \mu, \Sigma)}{\max\limits_{\substack{\mu_i\in R^p\ i=1,\ldots,q\\ \Sigma>0}} L(\mu_1, \ldots, \mu_q, \Sigma)} = \frac{|S|^{-n/2}}{|S_1 + \cdots + S_q|^{-n/2}},$$

which has the same distribution as in the normal case.

Proof. This follows from Lemma 5.2.3, Theorem 5.1.1 and its corollaries. □

In order to obtain the distribution of λ_2, we need the following definition:

Definition 5.2.2. Assume that $A \sim W_p(n, \Sigma)$ is independent of $B \sim W_p(m, \Sigma)$, where $W_p(\cdot, \cdot)$ is the Wishart distribution defined in Section 3.4.1. The distribution of

$$\Lambda = \frac{|A|}{|A + B|}$$

is called the *Wilks* Λ-*distribution* and we write $\Lambda \sim \Lambda(p, n, m)$.

Beacause the statistic $\lambda_1^{n/2}$ has the same distribution as in the normal case, it can be shown that $\lambda_1^{n/2} \sim \Lambda(p, n-q, q)$ in the following theorem.

Theorem 5.2.5. *Using the above notations, when H_0 is true we have*

$$\lambda_1^{n/2} = \frac{|\boldsymbol{S}_1 + \cdots + \boldsymbol{S}_q|}{|\boldsymbol{S}|} \sim \Lambda(p, n-q, q).$$

Proof. By invariance of λ_1 we can assume that the sample is from the normal population, i.e., $\boldsymbol{X} \sim N_{n \times p}(\boldsymbol{M}, \boldsymbol{I}_n \otimes \Sigma)$ where $\boldsymbol{M} = (\mu_1 \boldsymbol{1}_{n_1}, \ldots, \mu_q \boldsymbol{1}_{n_q})'$. So $\boldsymbol{X}_j'$. $\sim N_{n \times p}(\boldsymbol{1}_{n_1} \mu_j', \Sigma,)$, $j = 1, \ldots, q$. Noting that

$$\boldsymbol{S} = \boldsymbol{X}'\boldsymbol{D}_n\boldsymbol{X} \text{ and } \boldsymbol{S}_j = \boldsymbol{X}_j\boldsymbol{D}_{n_j}\boldsymbol{X}_j',$$

where $\boldsymbol{D}_r = \boldsymbol{I}_r - \frac{1}{r}\boldsymbol{1}_r\boldsymbol{1}_r'$, we have

$$\boldsymbol{E} = \boldsymbol{S}_1 + \cdots + \boldsymbol{S}_q = \sum_{j=1}^{q} \boldsymbol{X}_j\boldsymbol{D}_{n_j}\boldsymbol{X}_j' = \boldsymbol{X}'\boldsymbol{C}_1\boldsymbol{X},$$

where $\boldsymbol{C}_1 = \operatorname{diag}(\boldsymbol{D}_{n_1}, \ldots, \boldsymbol{D}_{n_q})$. It is obvious that $\boldsymbol{C}_1^2 = \boldsymbol{C}_1$, $\boldsymbol{C}_1' = \boldsymbol{C}_1$, $\operatorname{rk}\boldsymbol{C}_1 = n_1 - 1 + \cdots + n_q - 1 = n - q$ and $\boldsymbol{C}_1\boldsymbol{M} = \boldsymbol{O}$. By the theory of Section 2.8.2 and 3.4.2 we get $\boldsymbol{E} \sim W_p(n-q, \Sigma)$. Let $\boldsymbol{C}_2 = \boldsymbol{D}_n - \boldsymbol{C}_1$ and $\boldsymbol{B} = \boldsymbol{S} - \boldsymbol{E}$. Then $\boldsymbol{B} = \boldsymbol{X}'\boldsymbol{C}_2\boldsymbol{X}$. It can be verified that $\boldsymbol{C}_2' = \boldsymbol{C}_2$, $\boldsymbol{C}_2^2 = \boldsymbol{C}_2$ and $\operatorname{rk}\boldsymbol{C}_2 = q$. When H_0 is true we have $\boldsymbol{C}_2\boldsymbol{M} = \boldsymbol{0}$ so that $\boldsymbol{B} \sim W_p(q, \Sigma)$. Then the assertion follows from Definition 5.2.2. □

The Λ-distributions are useful in multivariate analysis. There are detailed discussions in many textbooks, for example, Anderson (1984), Muirhead (1982) and Zhang & Fang (1982). Its exact distribution has been calculated by Schatzoff (1966). For practice, the following results are furnished without proof:

(1) If $\Lambda \sim \Lambda(p, n, 1)$, then

$$\frac{1-\Lambda}{\Lambda}\frac{n-p+1}{p} \sim F(p, n-p+1)$$

and

$$n\frac{1-\Lambda}{\Lambda} \sim T^2(p, n)$$

(2) If $\Lambda \sim \Lambda(p, n, 2)$, then

$$\frac{1-(\Lambda)^{1/2}}{(\Lambda)^{1/2}}\frac{n-p}{p} \sim F(2p, 2(n-p)).$$

(3) If $\Lambda \sim \Lambda(1, n, m)$, then

$$\frac{1-\Lambda}{\Lambda}\frac{n}{m} \sim F(m, n).$$

(4) If $\Lambda \sim \Lambda(2, n, m)$, then

$$\frac{1-(\Lambda)^{1/2}}{(\Lambda)^{1/2}}\frac{n-1}{m} \sim F(2m, 2(n-1)).$$

(5) If $\Lambda \sim \Lambda(p, n, m)$, then $\Lambda \overset{d}{=} \beta_1 \cdots \beta_p$, where $\beta_1, \ldots, \beta_p$ are independent and $\beta_j \sim B((n-p+j)/2, m/2)$, (beta-distribution), $j = 1, \ldots, p$. The reader can find the table of $100\alpha\%$ point of $\Lambda(p, m, n)$ for $p=1(1)8$, $n=1 \sim \infty$ and $m=1 \sim 120$ in Zhang and Fang (1982), or in Kres (1975). Further discussion on testing the hypothesis (5.2.17) can be found in Section 6.4.1.

In the above we have studied the test for several means with equal covariance matrices. Concerning tests for several means with unequal and unknown covariance matrices, it has not well solved even in the case of both univariate and normal.

5.3. Tests for Covariance Matrices

5.3.1. The Spherical Test

Let $\boldsymbol{X}$ have a density

(5.3.1)
$$|\Sigma|^{-n/2} f(\mathrm{tr}(\boldsymbol{X} - \mathbf{1}\mu')\Sigma^{-1}(\boldsymbol{X} - \mathbf{1}\mu')') = |\Sigma|^{-n/2} f(\mathrm{tr}\Sigma^{-1}\boldsymbol{S} + n(\bar{\boldsymbol{x}} - \mu)'\Sigma^{-1}(\bar{\boldsymbol{x}} - \mu)).$$

We want to test

(5.3.2)
$$H_0\colon \Sigma = \alpha\Sigma_0,\ \Sigma_0 > 0 \text{ known and } \alpha > 0 \text{ unknown}.$$

First, we assume $\Sigma_0 = \boldsymbol{I}_p$. By (5.3.1) we have the likelihood function

$$L(\mu, \Sigma) = |\Sigma|^{-n/2} f(\mathrm{tr}\Sigma^{-1}\boldsymbol{S} + n(\bar{\boldsymbol{x}} - \mu)'\Sigma^{-1}(\bar{\boldsymbol{x}} - \mu)).$$

Because the MLE of (μ, Σ) is $(\bar{\boldsymbol{x}}, \lambda_{\max}(f)\boldsymbol{S})$ when f is nonincreasing and continuous, we have

(5.3.3)
$$\max_{\mu \in R^p,\ \Sigma > 0} L(\mu, \Sigma) = |\lambda_{\max}(f)\boldsymbol{S}|^{-n/2} f(p/\lambda_{\max}(f)),$$

and

(5.3.4)
$$\begin{aligned}\max_{\mu \in R^p,\ \alpha > 0} L(\mu, \alpha\boldsymbol{I}_p) &= \max_{\alpha > 0} L(\bar{\boldsymbol{x}}, \alpha\boldsymbol{I}_p) = \max_{\alpha > 0} \alpha^{-np/2} f(\mathrm{tr}\boldsymbol{S}/\alpha) \\ &= \max_{\alpha > 0} (\mathrm{tr}\boldsymbol{S}/\alpha p)^{+np/2} (\mathrm{tr}\boldsymbol{S}/p)^{-np/2} f(p(tr\boldsymbol{S}/p\alpha)) \\ &= \max_{\lambda > 0} (\mathrm{tr}\boldsymbol{S}/p)^{-np/2} \lambda^{-np/2} f(p/\lambda) \\ &= (\mathrm{tr}\boldsymbol{S}/p)^{-np/2} (\lambda_{\max}(f))^{-np/2} f(p/\lambda_{\max}(f)).\end{aligned}$$

Theorem 5.3.1. *Let* $\boldsymbol{X} \sim VS_{n \times p}(\mu, \Sigma, f)$ *with* $n > p$ *and* f *being noninceasing and continuous. Then, for testing* $H_0\colon \Sigma = \alpha\boldsymbol{I}_p$, *the likelihood ratio criterion is*

$$\lambda_2 = \frac{\max_{\mu \in R^p,\ \alpha > 0} L(\mu, \alpha\boldsymbol{I}_p)}{\max_{\mu \in R^p,\ \Sigma > 0} L(\mu, \Sigma)} = \frac{|\boldsymbol{S}|^{n/2}}{(\mathrm{tr}\boldsymbol{S}/p)^{np/2}},$$

which has the same distribution as in the normal case.

Corollary. *Under the assumptions of Theorem* 5.3.1, *for testing* $H_0\colon \Sigma = \alpha\Sigma_0$, Σ_0 *being known, the likelihood ratio criterion is*

$$\frac{|\Sigma_0^{-1} \boldsymbol{S}|^{n/2}}{(\mathrm{tr}\Sigma_0^{-1}\boldsymbol{S}/p)^{np/2}}.$$

For unbiasedness of λ_2, someone revised λ_2 as (when $\Sigma_0 = \boldsymbol{I}$)

$$\lambda_2^* = \frac{|\boldsymbol{S}|^{(n-1)/2}}{(\mathrm{tr}\boldsymbol{S}/p)^{(n-1)p/2}},$$

replacing n by $n-1$. Davis (1971) got an asymptotic distribution of λ_2^* and showed that the asymptotic null distribution as $n \to \infty$ of

$$-2\,\frac{6p(n-1)-(2p^2+p+2)}{6p(n-1)}\log\lambda_2^*$$

is $\chi^2_{p(p+1)/2-1}$. Mauchy (1940) derived the moments of $\lambda_2^{2/n}$ and, Khatri and Srivastava (1971) got the exact distribution of λ_2^*.

5.3.2 Equality of Several Covariance Matrices

Let $\boldsymbol{X}$ have a density (5.2.14). We want to test

(5.3.5) $$\mathrm{H}_0\colon \Sigma_1 = \cdots = \Sigma_q \equiv \Sigma \text{ (say)}$$

when μ_i, $i=1, \ldots, q$, are unknown. The likelihood function is

$$L(\mu_1, \ldots, \mu_q, \Sigma_1, \ldots, \Sigma_q)$$
$$= \prod_{i=1}^{q} |\Sigma_i|^{-n_i/2} f\left(\sum_{i=1}^{q} [\mathrm{tr}\Sigma_i^{-1}\boldsymbol{S}_i + n_i(\bar{\boldsymbol{x}}_i - \mu_i)'\Sigma_i^{-1}(\bar{\boldsymbol{x}}_i - \mu_i)\right),$$

where $\boldsymbol{S}_1, \ldots, \boldsymbol{S}_q$ are given by (5.2.15). By Lemma 5.2.3, when f is nonincreasing and continuous, we have

(5.2.21) $$\max_{\substack{\mu_i \in R^p,\, i=1, \ldots, \\ q, \Sigma>0}} L(\mu_1, \ldots, \mu_q, \Sigma, \ldots, \Sigma)$$
$$= |\lambda_{\max}(f)(\boldsymbol{S}_1 + \cdots + \boldsymbol{S}_q)|^{-n/2} f(p/\lambda_{\max}(f)).$$

To obtain the likelihood ratio criteria we need the following lemma.

Lemma 5.3.1. *Let f be noninceasing and differentiable. Then*

(5.3.5) $$\max_{\substack{\mu_i \in R^p,\, \Sigma_i>0 \\ i=1, \ldots, q}} L(\mu_1, \ldots, \mu_q, \Sigma_1, \ldots, \Sigma_q)$$
$$= (\lambda_{\max}(f))^{-np/2}\left[\prod_{i=1}^{q}\left(\frac{n_i}{n}\right)^{n_i p/2} |\boldsymbol{S}_i|^{-n_i/2}\right] f(p/\lambda_{\max}(f)),$$

where $\boldsymbol{S}_1, \ldots, \boldsymbol{S}_q$ are given by (5.2.15).

Proof. By a technique similar to the proof of Lemma 5.2.3, we get

$$\max_{\substack{\mu_i \in R^p,\, \Sigma_i>0 \\ i=1, \ldots, q}} L(\mu_1, \ldots, \mu_q, \Sigma_1, \ldots, \Sigma_q)$$

$$= \max_{\substack{\Sigma_i>0 \\ i=1, \dots, q}} L(\bar{x}_1, \dots, \bar{x}_q, \Sigma_1, \dots, \Sigma_q)$$

$$= \max_{\substack{\Sigma_i>0 \\ i=1, \dots, q}} \prod_{i=1}^{q} |\Sigma_i|^{-n_i/2} f\left(\sum_{i=1}^{q} \operatorname{tr} \Sigma_i S_i\right)$$

$$= \max_{\substack{\lambda_{ij}>0 \\ i=1, \dots, q \\ j=1, \dots, p}} \left\{\prod_{j=1}^{q} |S_j|^{-n_j/2} \prod_{i=1}^{q} (\lambda_{i1} \cdots \lambda_{ip})^{-n_i/2} f\left(\sum_{i=1}^{q} \sum_{j=1}^{p} \lambda_{ij}^{-1}\right)\right\},$$

where $\lambda_{i1}, \dots, \lambda_{ip}$ are the p eigenvalues of $\tilde{\Sigma}_i = S_i^{-1/2} \Sigma_i S_i^{-1/2}$. Because $n_i > p$, $S_i > 0$ with probability one by Lemma 4.1.1, $(i=1, \dots, q)$, just as in the proof of Theorem 4.1.1 the function in the above braces arrives its maximum at $\lambda_{i1} = \cdots = \lambda_{ip} \equiv \lambda_i$ (say), $i=1, \dots, q$, and the quantity in the braces becomes

$$(5.3.6) \qquad \left(\prod_{i=1}^{q} |S_i|^{-n_i/2}\right)\left(\prod_{i=1}^{q} \lambda_i^{-n_i p/2}\right) f\left(p \sum_{i=1}^{q} \lambda_i^{-1}\right) \hat{=} L^*,$$

say. Let $\frac{\partial L^*}{\partial \lambda_i} = 0$, $i=1, \dots, q$. We have

$$\lambda_i = -(2/n_i) \frac{f'\left(p \sum_{i=1}^{q} \lambda_i^{-1}\right)}{f\left(p \sum_{i=1}^{q} \lambda_i^{-1}\right)}, \quad i=1, \dots, q.$$

Hence $\lambda_i = (n_1/n_i)\lambda_1$, $i=1, \dots, q$. Put $\lambda = (n_1/n)\lambda_1$. Then $\lambda_i = (n/n_i)\lambda$, $i=1, \dots, q$ and (5.3.6) becomes

$$\left(\prod_{i=1}^{q} |S_i|^{-n_i/2}\right)\left(\prod_{i=1}^{q} \left(\frac{n_i}{n}\right)^{pn_i/2}\right) \lambda^{-np/2} f(p/\lambda),$$

which completes the proof by recalling that $\lambda^{-np/2} f(p/\lambda)$ takes its maximum value at $\lambda_{\max}(f)$. □

Theorem 5.3.2. *Under the assumptions of Lemma* 5.3.1, *for testing the hypothesis* (5.3.5), *the likelihood ratio criterion is*

$$\lambda_3 = \frac{\max_{\Sigma>0,\, \mu_i \in R^p,\, i=1, \dots, q} L(\mu_1, \dots, \mu_q, \Sigma, \dots, \Sigma)}{\max_{\mu_i \in R^p,\, \Sigma_i>0,\, i=1, \dots, q} L(\mu_1, \dots, \mu_q, \Sigma_1, \dots, \Sigma_q)}$$

$$= \frac{\prod_{i=1}^{q} |S_i|^{n_i/2}}{|S_1 + \cdots + S_q|^{n/2}} \prod_{i=1}^{q} \left(\frac{n_i}{n}\right)^{-n_i p/2},$$

which has the same distribution as in the normal case.

Bartlett suggested to replace n_i by $n_i - 1$ and n by $n-q$ in λ_3, denoting the revised λ_3 by λ_3^*. Box gave the asymptotical null distribution of $-2\log\lambda_3^*$ which is $(1/(1-d_1))\chi^2_{f_1}$ where $f_1 = (1/2)p(p+1(q-1))$ and

$$d_1 = \begin{cases} \dfrac{(2p^2+3p-1)(q+1)}{6(p+1)(q-1)}, \text{ if } n_1 = \cdots = n_q, \\ \dfrac{2p^2+3p-1}{6(p+1)(q-1)}\left(\sum_{i=1}^{q}\dfrac{1}{n_i-1} - \dfrac{1}{n-q}\right), \text{ otherwise.} \end{cases}$$

The null distribution of $-2\log\lambda_3^*$ can also be approximated by $bF(f_1, f_2)$, where

$b = f_1/(1-d_1 - f_1 f_2^{-1})$, $f_2 = (f_1+2)/(d_2 - d_1^2)$ and

$$d_2 = \begin{cases} \dfrac{(p-1)(p+2)(q^2+q+1)}{6q^2(n-1)^2}, \text{ if } n_1 = \cdots = n_q, \\ \dfrac{(p-1)(p+2)}{6(q-1)}\left(\sum_{i=1}^{q}\dfrac{1}{(n_i-1)^2} - \dfrac{1}{(n-q)^2}\right), \text{ otherwise.} \end{cases}$$

When $q=2$, Khatri and Srivatava (1971) derived the exact distribution of λ_3^*. Using $\theta_1 \geqslant \cdots \geqslant \theta_p$ to indicate the eigenvalues of $\Sigma_1\Sigma_2^{-1}$, testing $\Sigma_1 = \Sigma_2$ is equivalent to testing $\boldsymbol{H}_0{:}\theta_1 = \cdots = \theta_p = 1$. The problem of testing the hypothesis remains invariant under the group $\boldsymbol{G} = (\mathrm{GL}(p), R^p)$ which acts on $\boldsymbol{X}$ as

$$\boldsymbol{X} \to \boldsymbol{X}\boldsymbol{A} + \mathbf{1}\boldsymbol{a}', \ \boldsymbol{A} \in \mathrm{GL}(p), \ \boldsymbol{a} \in R^p.$$

The set of eigenvalues of $\boldsymbol{S}_1\boldsymbol{S}_2^{-1}$, $(\xi_1, \ldots, \xi_p)$, is a maximal invariant statistic under the group $\boldsymbol{G}$. Hence we may easily establish many invariant testing statistics, e.g., (a) $|\boldsymbol{S}_1\boldsymbol{S}_1^{-1}| = \xi_1 \cdots \xi_p$; (b) $\mathrm{tr}(\boldsymbol{S}_1\boldsymbol{S}_2^{-1}) = \xi_1 + \cdots + \xi_p$; (c) $(\max(\xi_1, \ldots, \xi_p), \min(\xi_1, \ldots, \xi_p))$ and (d) $|(\boldsymbol{S}_1 + \boldsymbol{S}_2)\boldsymbol{S}_2^{-1}| = (\xi_1+1)\cdots(\xi_p+1)$. These statistics have many attractive properties.

5.3.3 Simultaneously Testing Equality of Several Means and Covariance Matrices

Let $\boldsymbol{X}$ have a density (5.2.16) or (5.2.14). We are required to test

(5.3.7) $$\mathrm{H}_0{:}\mu_1 = \cdots = \mu_q \equiv \mu(\text{say}) \text{ and } \Sigma_1 = \cdots = \Sigma_q \equiv \Sigma \text{ (say)}$$

Now we have the likelihood

$$L(\mu_1, \ldots, \mu_q, \Sigma_1, \ldots, \Sigma_q) = \prod_{i=1}^{q} |\Sigma_i|^{-n_i/2} f\left(\sum_{i=1}^{q} [\mathrm{tr}\Sigma_i \boldsymbol{S}_i + n_i(\bar{\boldsymbol{x}}_i - \mu_i)'\Sigma_i^{-1}(\bar{\boldsymbol{x}}_i - \mu_i)]\right)$$

where $n_i > p$ and $\boldsymbol{S}_i$ is given by (5.2.15), $i=1, \ldots, q$. Assume that f is nonincreasing and continuous and further, differentiable. By Lemma 5.3.1, we immediately get

$$\lambda_4 = \frac{\max\limits_{\mu_i \in R^p, \Sigma>0} L(\mu, \ldots, \mu, \Sigma, \ldots, \Sigma)}{\max\limits_{\substack{\mu_i \in R^p, \Sigma_i>0 \\ i=1,\ldots,q}} L(\mu_1, \ldots, \mu_q, \Sigma_1, \ldots, \Sigma_q)} = \frac{|\boldsymbol{S}|^{-n/2}}{\prod_{i=1}^{q}\left[\left(\dfrac{n_i}{n}\right)^{n_i p/2} |\boldsymbol{S}_i|^{-n_i/2}\right]}$$

$$= \frac{\prod_{i=1}^{q} \left| \frac{1}{n_i} S_i \right|^{n_i/2}}{\left| \frac{1}{n} S \right|^{n/2}}.$$

As in the previous discussion, we replace n by $n-1$, and denote the revised λ_4 by λ_4^*. Box showed that

$$P(-2\rho\log\lambda_4^* \leqslant x|H_0) = P(\chi_f^2 \leqslant x) + \omega\,[P(\chi_{f+4}^2 \leqslant x) - P(\chi_f^2 \leqslant x)] + o(n^{-3}),$$

where

$$f = \frac{1}{2}(q-1)(p+1)p,$$

$$\rho = 1 - \left(\sum_{i=1}^{q} \frac{1}{n_i-1} - \frac{1}{n-q}\right)\left(\frac{2p^2+3p-1}{6(q-1)(p+3)}\right) + \frac{p-q+2}{(p-q)(p+3)},$$

and

$$\omega = \frac{q}{288\rho^2}\left[6\left(\sum_{i=1}^{q} \frac{1}{(n_i-1)^2} - \frac{1}{(n-q)^2}\right)(p^2-1)(p+2)\right.$$

$$- \sum_{i=1}^{q}\left(\frac{1}{n_i-1} - \frac{1}{n-q}\right)^2 \frac{(2p^2+3p-1)^2}{(q-1)(p+3)} - 12\left(\sum_{i=1}^{q} \frac{1}{n_i-1} - \frac{1}{n-q}\right)$$

$$\cdot \frac{(2p^2+3p-1)(p-q+2)}{(n-q)(p+3)} - 36\frac{(q-1)(p-q+2)^2}{(n-q)^2(p+3)}$$

$$\left. - \frac{12(q-1)}{(n-q)^2}(7q-2q^2+3pq-2p^2-6p-4)\right],$$

so that the asymptotic null distribution of $-2\rho\log\lambda_4^*$ is $\chi^2_{(q-1)(p+1)p/2}$.

The testing problem is invariant under the $G = (\mathrm{GL}(p), R^p)$ which acts on $\boldsymbol{X}$ as

$$\boldsymbol{X} \to \boldsymbol{X}\boldsymbol{A} + \mathbf{1}\boldsymbol{a}', \ \boldsymbol{A} \in \mathrm{GL}(p), \ \boldsymbol{a} \in R^p.$$

Clearly, the likelihood ratio criterion λ_4 is an invariant statistic under G.

5.3.4. Testing Lack of Correlation Between Sets of Variates

Let $\boldsymbol{X}$ have a density function

$$(5.3.8) \qquad |\Sigma|^{-n/2} f(\mathrm{tr}\Sigma^{-1}(\boldsymbol{X} - \mathbf{1}\mu')'(\boldsymbol{X} - \mathbf{1}\mu'))$$
$$= |\Sigma|^{-n/2} f(\mathrm{tr}\Sigma^{-1}\boldsymbol{S} + n(\bar{\boldsymbol{x}} - \mu)'\Sigma^{-1}(\bar{\boldsymbol{x}} - \mu))$$

with $n > p$, where $\boldsymbol{X} = (\boldsymbol{x}_{(1)}, \ldots, \boldsymbol{x}_{(n)})'$, $\bar{\boldsymbol{x}} = \frac{1}{n}\sum_{i=1}^{n} \boldsymbol{x}_{(i)}$, $\boldsymbol{S} = \boldsymbol{X}'\left(\boldsymbol{I}_n - \frac{1}{n}\mathbf{1}\mathbf{1}'\right)\boldsymbol{X}$. Partition $\bar{\boldsymbol{x}}$, μ, $\boldsymbol{S}$, Σ as

$$\mu = \begin{pmatrix} \mu_{(1)} \\ \vdots \\ \mu_{(k)} \end{pmatrix} \begin{pmatrix} p_1 \\ \vdots \\ p_k \end{pmatrix} \quad \bar{\boldsymbol{x}} = \begin{pmatrix} \bar{\boldsymbol{x}}_{(1)} \\ \vdots \\ \bar{\boldsymbol{x}}_{(k)} \end{pmatrix} \quad \boldsymbol{x}_{(1)} = \begin{pmatrix} \boldsymbol{x}_{(11)} \\ \\ \boldsymbol{x}_{(1k)} \end{pmatrix}, \quad \Sigma = \begin{pmatrix} \Sigma_{11}, \cdots, \Sigma_{1k} \\ \cdots \\ \Sigma_{k1}, \cdots, \Sigma_{kk} \end{pmatrix}$$

$$S = \begin{pmatrix} S_{11}, \cdots, S_{1k} \\ \cdots \\ S_{k1}, \cdots, S_{kk} \end{pmatrix} = (S_{ij}).$$

We are interested in testing the null hypothesis

$$\text{(5.3.9)} \qquad \mathrm{H}_0: \quad \Sigma_{ij} = \boldsymbol{O} \quad \text{for all } i \neq j.$$

Since, when $\boldsymbol{x}_{(1)}$ has the second order finite moment,

$$\mathcal{E}[\boldsymbol{x}_{(1)}\boldsymbol{x}'_{(1)}] = \delta^2\Sigma,$$

that H_0 holds means that the subvectors $\boldsymbol{x}_{(11)}, \ldots, \boldsymbol{x}_{(1k)}$ are mutually uncorrelated and in particular, independent if $\boldsymbol{x}_{(1)} \sim N(\mu, \Sigma)$.

Let $\Omega = \{(\mu, \Sigma) | \Sigma > 0, \mu \in R^p\}$ and $\Omega_0 = \{(\mu, \Sigma) | \Sigma = \mathrm{diag}(\Sigma_{11}, \ldots, \Sigma_{kk}) > 0, \mu \in R^p\}$.

Lemma 5.3.2. *Let f be nonincreasing and differentiable. Then*

$$\max_{(\mu, \Sigma) \in \Omega_0} L(\mu, \Sigma) = \lambda_{\max}(f)^{-np/2} \left(\prod_{i=1}^{k} |\boldsymbol{S}_{ii}|^{-n/2} \right) f(p/\lambda_{\max}(f)),$$

where $L(\mu, \Sigma)$ *is the likelihood function defined in* (5.3.6).

Proof. As in the previous discussion, we have

$$\begin{aligned} &\max_{(\mu, \Sigma) \in \Omega_0} L(\mu, \Sigma) \\ &= \max_{\substack{\Sigma_{ii} > 0 \\ i=1,\cdots,k}} L(\bar{\boldsymbol{x}}, \mathrm{diag}(\Sigma_{11}, \cdots, \&_{kk}), \\ &= \max_{\substack{\Sigma_{ii} > 0 \\ i=1,\ldots,k}} \prod_{i=1}^{k} |\Sigma_{ii}|^{-n/2} f\left(\mathrm{tr} \sum_{i=1}^{k} \Sigma_{ii}^{-1} \boldsymbol{S}_{ii} \right) \\ &= \max_{\substack{\lambda_{ij} > 0 \\ j=1,\ldots,p_i \\ i=1,\ldots,k}} \left\{ \left(\prod_{i=1}^{k} |\boldsymbol{S}_{ii}| \right)^{-n/2} \prod_{i=1}^{k} \prod_{j=1}^{p_i} \lambda_{ij}^{-n/2} f\left(\sum_{i=1}^{k} \sum_{j=1}^{p_i} \lambda_{ij}^{-1} \right) \right\} \end{aligned}$$

where $\lambda_{i1}, \ldots, \lambda_{ip_i}$ are the eigenvalues of $\boldsymbol{S}_{ii}^{-1/2} \Sigma_{ii} \boldsymbol{S}_{ii}^{-1/2}$, $i = 1, \ldots, k$.

Just as in the proof of Theorem 4.1.1, the $\{\lambda_{ij}\}$ in the above braces must be equal. We have

$$\begin{aligned} \max_{(\mu, \Sigma) \in \Omega_0} L(\mu, \Sigma) &= \left(\prod_{i=1}^{k} |\boldsymbol{S}_{ii}| \right)^{-n/2} \max_{\lambda > 0} \lambda^{-np/2} f(p/\lambda) \\ &= \left(\prod_{i=1}^{k} |\boldsymbol{S}_{ii}| \right)^{-n/2} \lambda_{\max}(f)^{-np/2} f(p/\lambda_{\max}(f)), \end{aligned}$$

which completes the proof. □

By Lemma 5.3.2, we obtain immediately the following theorem.

Theorem 5.3.3. *Under the assumptions of Lemma* 5.3.2, *the likelihood ratio criterion for testing* (5.3.9) *is*

$$\lambda_5 = \frac{\max_{(\mu, \Sigma)\in\Omega_0} L(\mu, \Sigma)}{\max_{(\mu, \Sigma)\in\Omega} L(\mu, \Sigma)} = \frac{|\boldsymbol{S}|^{n/2}}{\left(\prod_{i=1}^{k} \boldsymbol{S}_{ii}\right)^{n/2}},$$

which has the same distribution as in the normal case.

By writing $\hat{\rho}_{ij} = S_{ij}/(S_{ii}S_{jj})^{1/2}$, the matrix $\hat{\boldsymbol{R}}$ of sample correlation coefficients $\hat{\rho}_{ij}$ is

$$\hat{\boldsymbol{R}} = \begin{pmatrix} 1 & \hat{\rho}_{12} & \cdots & \hat{\rho}_{1p} \\ \hat{\rho}_{21} & 1 & & \\ \vdots & \vdots & \ddots & \vdots \\ \hat{\rho}_{p1} & \hat{\rho}_{p2} & \cdots & 1 \end{pmatrix}$$

Obviously $|\boldsymbol{S}| = \left(\prod_{i=1}^{p} S_{ii}\right)|\hat{\boldsymbol{R}}|$. Partition $\hat{\boldsymbol{R}}$ after the fashion of $\boldsymbol{S}$ as

$$\hat{\boldsymbol{R}} = \begin{pmatrix} \hat{\boldsymbol{R}}_{11} & \cdots & \hat{\boldsymbol{R}}_{1k} \\ & \cdots & \\ \hat{\boldsymbol{R}}_{k1} & \cdots & \hat{\boldsymbol{R}}_{kk} \end{pmatrix}.$$

Then

$$\lambda_5 = \frac{|\hat{\boldsymbol{R}}|^{n/2}}{\left(\prod_{i=1}^{k} |\hat{\boldsymbol{R}}_{ii}|\right)^{n/2}}$$

gives a representation of λ_5 in terms of the sample correlation coefficients.

Let G_{BD} be the group of $p\times p$ *nonsingular block diagonal matrices* $\boldsymbol{D}=\mathrm{diag}(\boldsymbol{D}_{11}, \ldots, \boldsymbol{D}_{kk})$, where $\boldsymbol{D}_{ii}$:$p_i \times p_i$, $\sum_{i=1}^{k} p_i = p$. The problem of testing (5.3.9) remains invariant under the group G_{BD} of affine transformations $(\boldsymbol{D}, \boldsymbol{a})$, $\boldsymbol{D}\in G_{BD}$ and $\boldsymbol{a}\in R^p$, transforming each $\boldsymbol{X}$ to $\boldsymbol{XD}+\mathbf{1}\boldsymbol{a}'$. Obviously, the λ_5 is an invariant statistic under the group G_{BD}.

Let $\lambda_5^{2/n}=v$. By noting $0\leqslant v\leqslant 1$, the distribution of v is determined by its moments. Anderson (1958) gave

$$E(v^h|H_0) = \frac{\prod_{i=1}^{p}\Gamma\left(\frac{n-i}{2}+h\right)\prod_{i=1}^{k}\prod_{j=1}^{p_i}\Gamma\left(\frac{n-j}{2}\right)}{\prod_{i=1}^{p}\Gamma\left(\frac{n-i}{2}\right)\prod_{i=1}^{k}\prod_{j=1}^{p_i}\Gamma\left(\frac{n-j}{2}+h\right)}, \quad h=0, 1, \ldots. \tag{5.3.10}$$

Since these moments are independent of Σ_{ii}, $i=1, \ldots, k$, it follows (See Anderson, 1958, Section 9.4) that when H_0 is true v is distributed as $\prod_{i=1}^{k}\prod_{j=1}^{p_i} Y_{ij}$ where the $\{Y_{ij}\}$ are independently distributed central beta random variables with parameters $((n-\delta_{i-1}-j)/2, \delta_{i-1}/2)$ with $\delta_j = \sum_{i=1}^{j} p_i$, $\delta_0=0$. In practice, we use the asymptotical distribution of v. Let

$$f=\frac{1}{2}\left[p(p+1) - \sum_{i=1}^{k} p_i(p_i+1)\right],$$

$$\rho=1-\frac{2\left(p^3-\sum_{i=1}^{k} p_i^3\right)+9\left(p^2-\sum_{i=1}^{k} p_i^2\right)}{6n\left(p^2-\sum_{i=1}^{k} p_i^2\right)},$$

$$r=\frac{p^4-\sum_{i=1}^{k} p_i^4}{48}-\frac{5\left(p^2-\sum_{i=1}^{k} p_i^2\right)}{96}-\frac{\left(p^3-\sum_{i=1}^{k} p_i^3\right)^2}{72\left(p^2-\sum_{i=1}^{k} p_i^2\right)},$$

$$a=n\rho.$$

Box (1949) showed that

$$P(-a\log v\leqslant x)=P(\chi_f^2\leqslant x)+(r/a^2)\left[P(\chi_{f+4}^2\leqslant x)-P(\chi_f^2\leqslant x)\right]+O(a^{-3}).$$

Thus for large n the distribution of $-a\log v$ approximates to that of χ_f^2.

Take $k=2$, $p_1=1$ and $p_2=p-1$. Then testing $\Sigma_{12}=\boldsymbol{O}$ is equal to testing that the multiple correlation coefficient (cf. Section 4.1.4) is zero. Now

$$v=\frac{|\boldsymbol{S}|}{s_{11}|\boldsymbol{S}_{22}|}=\frac{s_{11}-\boldsymbol{s}_{12}\boldsymbol{S}_{22}^{-1}\boldsymbol{s}_{21}}{s_{11}}=1-\frac{\boldsymbol{s}_{12}\boldsymbol{S}_{22}^{-1}\boldsymbol{s}_{21}}{s_{11}}=1-R^2\ \text{(say)},$$

It can be shown that

$$v\sim\Lambda(1,\ n-p,\ p-1),$$

and

(5.3.11) $$F=\frac{R^2}{1-R^2}\,\frac{n-p}{p-1}\sim F(p-1,\ n-p)$$

which gives the statistic of testing that the multiple correlation coefficient is zero. In particular, using (5.3.11), we can test that the correlation coefficient between two univariables, x and y, (say) is zero.

Regarding the optimal properties of R^2-testing, the reader may refer to Giri (1977).

5.4 A Note on Likelihood Ratio Method

In the previous several sections by means of a unified technique suggested by Anderson and Fang (1982c) we have derived the likelihood ratio criteria for testing means and covariance matrices respectively by the maximum likelihood procedure. The forms of those likelihood ratio criteria and their distributions remain the same in the whole class including the normal case. The following is a theorem establishing a relationship between the likelihood ratio criterion of the normal population and that of elliptically contoured population, due to Anderson Fang, and Hsu (1986). They worked with $\boldsymbol{x}=\text{vec}\boldsymbol{X}$ instead of $\boldsymbol{X}$ itself.

Let $\boldsymbol{x}\sim EC_N(\mu,\ \Sigma,\ \phi)$ have a density

(5.4.1) $$|\Sigma|^{-1/2}f((\boldsymbol{x}-\mu)'\Sigma^{-1}(\boldsymbol{x}-\mu)),$$

where $f(t)$ is decreasing and continuous on $[0,\ \infty]$. Put $\omega=\omega_m\times\omega_c$, where $\omega_m\subset R^N$ and ω_c is a subset of $N\times N$ positive definition matrices such that $\Sigma\in\omega_c$ implies $\alpha\Sigma\in\omega_c$ for each $\alpha>0$.

Theorem 5.4.1. *Under the above assumptions, if the MLE's under normality,*

$\tilde{\mu} \in \omega_m$ and $\tilde{\Sigma} \in \omega_c$, exist and are unique, and $\tilde{\Sigma} > 0$ with probability one, then $\hat{\mu} = \tilde{\mu}$ and $\hat{\Sigma} = N\lambda_{\max}(f)\tilde{\Sigma}$ are the MLE's under $f(\cdot)$, and the maximum of the likelihood function is

$$[\tilde{c}\lambda_{\max}(f)]^{-N/2} f(1/\lambda_{\max}(f)), \tag{5.4.2}$$

where

$$\tilde{c} = \min_{\substack{\mu\in\omega_m, B\in\omega_c \\ |B|=1}} (\boldsymbol{x}-\mu)'\boldsymbol{B}^{-1}(\boldsymbol{x}-\mu). \tag{5.4.3}$$

Proof. Let $a = |\Sigma|^{1/N}$ and $\boldsymbol{B} = (1/a)\Sigma$. Then $|\boldsymbol{B}| = 1$, $\boldsymbol{B} > 0$ and $\boldsymbol{B} \in \omega_c$ iff $\Sigma \in \omega_c$. The normal function is $(2\pi)^{-N/2}$ times

$$a^{-N/2} \exp\left\{-\frac{1}{2a}(\boldsymbol{x}-\mu)'\boldsymbol{B}^{-1}(\boldsymbol{x}-\mu)\right\}.$$

Now the maximum of $a^{-N/2}\exp\left\{-\dfrac{c}{2a}\right\}$, $(0 \leqslant a < \infty)$, is attained at $a_c = c/N$ for any $c > 0$. Hence we have

$$\max_{\mu\in\omega_m,\, \Sigma\in\omega_c} |\Sigma|^{-1/2}\exp\left\{-\frac{1}{2}(\boldsymbol{x}-\mu)'\Sigma^{-1}(\boldsymbol{x}-\mu)\right\}$$

$$= \max_{\substack{\mu\in\omega_m \\ B\in\omega_c,\, |B|=1}} \max_{a>0} a^{-N/2}\exp\left\{-\frac{1}{2a}(\boldsymbol{x}-\mu)'\boldsymbol{B}^{-1}(\boldsymbol{x}-\mu)\right\}$$

$$= \max_{\substack{\mu\in\omega_m \\ B\in\omega_c,\, |B|=1}} \left(\frac{N}{(\boldsymbol{x}-\mu)'\boldsymbol{B}^{-1}(\boldsymbol{x}-\mu)}\right)^{N/2} e^{-N/2}.$$

Thus by the assumption that the maximum likelihood estimators under normality, $\tilde{\mu}\in\omega_m$ and $\tilde{\Sigma}\in\omega_c$, exist and are unique, and $\tilde{\Sigma} > 0$, a.e., there exist $\mu^* = \tilde{\mu}\in\omega_m$ and $\tilde{\boldsymbol{B}}\in\omega_c$ such that

$$(\boldsymbol{x}-\tilde{\mu})'\tilde{\boldsymbol{B}}^{-1}(\boldsymbol{x}-\tilde{\mu}) = \min_{\substack{\mu\in\omega_m \\ B\in\omega_c,\, |B|=1}} (\boldsymbol{x}-\mu)'\boldsymbol{B}^{-1}(\boldsymbol{x}-\mu) = \tilde{c}.$$

Now consider the likelihood $L(\mu, a\boldsymbol{B}) = a^{-N/2} f((\boldsymbol{x}-\mu)'\boldsymbol{B}^{-1}(\boldsymbol{x}-\mu)/a)$.

$$\max_{\substack{\mu\in\omega_m,\, B\in\omega_c \\ |B|=1,\, a>0}} L(\mu, a\boldsymbol{B}) = \max_{a>0} \max_{\substack{\mu\in\omega_m \\ B\in\omega_c,\, |B|=1}} a^{-N/2} f((\boldsymbol{x}-\mu)'\boldsymbol{B}^{-1}(\boldsymbol{x}-\mu)/a)$$

$$= \max_{a>0} a^{-N/2} f(\tilde{c}/a), \text{ for } f \text{ is decreasing,}$$

$$= \max_{a>0} \tilde{c}^{-N/2} a^{-N/2} f(1/a) = \tilde{c}^{-N/2}(\lambda_{\max}(f))^{-N/2} f(1/\lambda_{\max}(f)),$$

which completes (5.4.2) and the MLE's of $\mu\in\omega_m$ and $\Sigma\in\omega_c$ are obviously $\hat{\mu} = \tilde{\mu}$ and $\hat{\Sigma} = N\lambda_{\max}(f)\tilde{c}\tilde{\boldsymbol{B}} = N\lambda_{\max}(f)\tilde{\Sigma}$. □

Corollary 1. *Let the conditions of Theorem 5.4.1 hold for $f(\cdot)$, ω_m, Ω_m, ω_c and Ω_c, and let the null hypothesis be $\mu \in \omega_m \subset \Omega_m$ and $\Sigma \in \omega_c \subset \Omega_c$ against alternatives $\mu \in \Omega_m - \omega_m$ and $\Sigma \in \Omega_c - \omega_c$. If the likelihood ratio criterion exists and is unique when $f(\cdot)$ is the normal density, the criterion is*

$$(\tilde{c}_\omega / \tilde{c}_\Omega)^{N/2}$$

where

$$\tilde{c}_\omega = \min_{\substack{\mu\in\omega_m \\ B\in\omega_c,\ |B| = 1}} (x-\mu)'B^{-1}(x-\mu)$$

and

$$\tilde{c}_\Omega = \min_{\substack{\mu\in\Omega_m \\ B\in\Omega_c,\ |B| = 1}} (x-\mu)'B^{-1}(x-\mu).$$

Corollary 2. *Let the conditions of Theorem 5.4.1 hold for $f(\cdot)$, ω_m, Ω_m, ω_c and Ω_c. Suppose that $\mu \in \omega_m$ and $\eta \in \Omega_m$ imply $\alpha\mu \in \omega_m$ and $\alpha\eta \in \Omega_m$, respectively, for all $\alpha \geqslant 0$. If the distribution of the likelihood ratio criterion is independent of $\mu\in\omega_m$ and $\Sigma\in\omega_c$, then it is independent of (μ, Σ) and $f(\cdot)$.*

Proof. It follows from Theorem 5.1.1 and

$$t(x) = \frac{\tilde{c}_\omega}{\tilde{c}_\Omega} = \frac{\min_{\mu\in\omega_m,\ B\in\omega_c,\ |B|=1} (\alpha x - \alpha\mu)'B^{-1}(\alpha x - \alpha\mu)}{\min_{\mu\in\Omega_m,\ B\in\omega_c,\ |B|=1} (\alpha x - \alpha\mu)'B^{-1}(\alpha x - \alpha\mu)}$$

$$= t(\alpha x), \qquad \text{for all } \alpha > 0. \qquad \square$$

Using Theorem 5.4.1 and its corollaries, we can obtain many likelihood criteria by means of a unified technique, while in the previous discussion, we got them one by one. Although the methods as used there would also be important. This is the reason why we conclude this section with this significant Theorem 5.4.1. The interested reader may verify the criteria again by Theorem 5.4.1.

So far the discussion is essentially based on $\mathscr{F}_3^+$, we are required develop theories on MLE and likelihood ratio criteria in bigger classes, for instance, $\mathscr{F}_2^+$, $\mathscr{F}_1^+$ or $\mathscr{F}_s^+$. Fang and Xu (1986) have given some of the related results.

5.5. Robust Tests with Invariance

5.5.1. Robust Tests for Spherical Symmetry

Let $\mathscr{F}^n$ be the class of all pdf's (probability density functions) with respect to the Lebesgue measure on R^n. Let $\mathscr{F}_0^n$ denote the class of spherical symmetric pdf's, i.e.,

$$\mathscr{F}_0^n = \{ f\in\mathscr{F}^n | f \text{ depends on } x \text{ only through } x'x \},$$

and let

$$\mathscr{F}_0^n(\Sigma) = \{ f\in\mathscr{F}^n | f(x) = |\Sigma|^{-1/2} q(x'\Sigma^{-1}x) \}.$$

Suppose a sample x is from a population with pdf h in $\mathscr{F}^n$. We are expected to test

$$(5.5.1) \qquad H_0: h\in\mathscr{F}_0^n \qquad \text{against } H_1: h\in\mathscr{F}_0^n(\Sigma),$$

where Σ is known and $\Sigma \neq \sigma^2 \boldsymbol{I}_n$ for any $\sigma > 0$.

$G = (0, \infty)$ is a transformation group, $\alpha \in G$, acting on R^n as

$$\boldsymbol{x} \to \alpha \boldsymbol{x}. \tag{5.5.2}$$

Then the problem of testing (5.5.1) remains invariant under the group G.

Lemma 5.5.1. *Under the above assumption, a maximal invariance statistic is*

$$w(\boldsymbol{x}) = \boldsymbol{x} / (\boldsymbol{x}'\boldsymbol{x})^{1/2}.$$

Proof. It is obvious that $w(\boldsymbol{x})$ is invariant under G, i.e., $w(\alpha\boldsymbol{x}) = w(\boldsymbol{x})$, for any $\boldsymbol{x} \in \boldsymbol{R}^n, \alpha > 0$. Now, if $w(\boldsymbol{x}_1) = w(\boldsymbol{x}_2)$, i.e., $\boldsymbol{x}_1 / (\boldsymbol{x}'_1 \boldsymbol{x}_1)^{1/2} = \boldsymbol{x}_2 / (\boldsymbol{x}'_2 \boldsymbol{x}_2)^{1/2}$, then $\boldsymbol{x}_1 = \alpha \boldsymbol{x}_2$ where $\alpha = (\boldsymbol{x}'_1 \boldsymbol{x}_1)^{1/2} / (\boldsymbol{x}'_2 \boldsymbol{x}_2)^{1/2} > 0$, which completes the proof. □

Lemma 5.5.2. *Let x be a sample from a population F_θ with pdf $f_\theta(x)$.* Consider testing H_0: $\theta \in \omega$ *against* H_1: $\theta = \theta_1$. *Let λ be a finite measure on ω. Then the test*

$$\varphi(x) = \begin{cases} 1, & \text{if } f_{\theta 1}(x) \geqslant k \int_\omega f_\theta(x)\, d\lambda(\theta) \\ 0, & \text{if } f_{\theta 1}(x) < k \int_\omega f_\theta(x)\, d\lambda(\theta), \end{cases} \tag{5.5.3}$$

is a UMP (*uniformly most powerful*) *test with level* $\alpha \in [0, 1]$, *where constant k is chosen such that $E_\theta[\varphi(x)] \leqslant \alpha$, for each $\theta \in \omega$.*

Proof. Let

$$h(x) = c \int_\omega f_\theta(x)\, d\lambda(\theta)$$

where $c^{-1} = \lambda(\omega)$. Then $h(x)$ is a density. Applying Neyman-Pearson's Lemma to testing $H_0 : x \sim$ pdf $h(x)$ against H_1: $x \sim f(x)$, we get a test defined by (5.5.3). Then it can be verfied that the test φ is also UMP for testing $H_0 : \theta \in \omega$ against $H_1 : \theta = \theta_1$. For details of the proof the reader is refered to Lehman and Stein (1948). □

Lemma 5.5.3. *Let $\boldsymbol{x}$ be an n-dimensional random vector. Consider testing* (5.5.4) $H_0 : \boldsymbol{x} \sim N(\boldsymbol{0}, \sigma^2 \boldsymbol{I}_n)$, *where σ^2 is unknown, against* $H_1 : \boldsymbol{x} \sim N(\boldsymbol{0}, \Sigma)$, *where Σ is known and $\Sigma \neq \sigma^2 \boldsymbol{I}_n$. Then the test defined by*

$$\varphi(\boldsymbol{x}) = \begin{cases} 1, & \text{if } \boldsymbol{x}'\Sigma^{-1}\boldsymbol{x} / \boldsymbol{x}'\boldsymbol{x} \leqslant k \\ 0, & \text{if } \boldsymbol{x}'\Sigma^{-1}\boldsymbol{x} / \boldsymbol{x}'\boldsymbol{x} > k \end{cases} \tag{5.5.5}$$

is a UMP test for (5.5.4) *with level α. For a given α, k in* (5.5.5) *is determined by*

$$\int_A p_n(t_1, \ldots, t_{n-1})\, dt_1 \cdots dt_{n-1} = \alpha, \tag{5.5.6}$$

where p_n is the pdf of Dirichlet's distribution $D_n(1/2, \ldots, 1/2; 1/2)$,

$$A = \left\{ (t_1, \ldots, t_{n-1}) \,\middle|\, \sum_{i=1}^n d_i t_i \leqslant k,\ t_n = 1 - \sum_{i=1}^{n-1} t_i \right\}$$

and the d_i's are the eigenvalues of Σ^{-1}.

Proof. We can apply Lemma 5.5.2 to prove it. Using the notations in Lemma 5.5.2, denote $\omega = \{\sigma^2 | \sigma > 0\}$. Take λ to be a finite measure on ω. **The proof is left to the reader.**

Lemmas 5.5.2 and 5.5.3 are due to Lehmann and Stein (1948) but Lemma 5.5.3 as given here is a little different from the original. The following theorem is due to King (1980).

Theorem 5.5.1. *Suppose* $\boldsymbol{x}$ *is an* $n\times 1$ *random vector with pdf* h. *Then for a fixed* $\Sigma>0$, *when testing*

$$H_0: h\in \mathscr{F}_0^n \text{ against } H_1: h\in \mathscr{F}_0^n(\Sigma),\ \Sigma\neq\sigma^2\boldsymbol{I}_n,$$

the test defined by (5.5.5) *is an UMP invariant test under group* $G=(0,\infty)$ *acting on* R^n *as in* (5.5.2), *having level* α.

Proof. Since $w(\boldsymbol{x})=\boldsymbol{x}/(\boldsymbol{x}'\boldsymbol{x})^{1/2}$ is a maximal invariant statistic under the group G by Lemma 5.5.1, any invariant test is necessarily a function of $w(\boldsymbol{x})$. When H_0 is true, as $\boldsymbol{x}\sim S_n(\phi)$, we have $w(\boldsymbol{x})\overset{d}{=}\boldsymbol{u}^{(n)}$; when H_1 is true, as $\boldsymbol{x}\sim \mathrm{EC}_n(\boldsymbol{0},\Sigma,\phi)$, we have $w(\boldsymbol{x})\overset{d}{=}\Sigma^{1/2}\boldsymbol{u}^{(n)}/(\boldsymbol{u}^{(n)\prime}\Sigma\boldsymbol{u}^{(n)})^{1/2}$. Hence when H_0 is true,

$$w(\boldsymbol{x})\overset{d}{=}w(\boldsymbol{z}),$$

where $\boldsymbol{z}\sim N(\boldsymbol{0},\sigma^2\boldsymbol{I}_n)$, $\sigma^2>0$ arbitrary; when H_1 is true, $w(\boldsymbol{x})\overset{d}{=}w(\boldsymbol{z})$,

where $\boldsymbol{z}\sim N(\boldsymbol{0},\Sigma)$. The theorem then follows from Lemma 5.5.3. □

Theorem 5.5.1 can be extended to cover the following situations when Σ is partially unknown:

(i) $\Sigma=\sigma^2\Sigma_0$, Σ_0 known, σ^2 unknown.

(ii) $\Sigma=\lambda_1(\boldsymbol{I}_n-\boldsymbol{M})+\lambda_2\boldsymbol{M}$, $\boldsymbol{M}'=\boldsymbol{M}$, $\boldsymbol{M}^2=\boldsymbol{M}$, $\boldsymbol{M}$ known, and $\lambda_1>\lambda_2>0$ (or $\lambda_2>\lambda_1>0$), λ_1, λ_2 unknown.

(iii) $\Sigma^{-1}=\lambda_1\boldsymbol{I}_n+\lambda_2\boldsymbol{A}$, $\boldsymbol{A}$ known, λ_1, λ_2 unknown with $\lambda_1>0$, $\lambda_2>0$ such that Σ is positive definite.

The case (i) is left to the reader as an exercise. For (ii), since $\Sigma^{-1}=\lambda_1^{-1}(\boldsymbol{I}_n-\boldsymbol{M})+\lambda_2^{-1}\boldsymbol{M}$ (why), $\boldsymbol{x}'\Sigma^{-1}\boldsymbol{x}/\boldsymbol{x}'\boldsymbol{x}=\lambda_1^{-1}+(\lambda_2^{-1}-\lambda_1^{-1})\boldsymbol{x}'\boldsymbol{M}\boldsymbol{x}/\boldsymbol{x}'\boldsymbol{x}$, thus, firstly, given $\lambda_1>\lambda_2>0$, the test with critical region $\boldsymbol{x}'\boldsymbol{M}\boldsymbol{x}/\boldsymbol{x}'\boldsymbol{x}\leqslant k$ is a UMP invariant test for testing H_0: $h\in\mathscr{F}_0^n$ against H_1: $h\in\mathscr{F}_0^n(\Sigma)$, $\Sigma=\lambda_1(\boldsymbol{I}_n-\boldsymbol{M})+\lambda_2\boldsymbol{M}$ known. Since the distribution of $\boldsymbol{x}'\boldsymbol{M}\boldsymbol{x}/\boldsymbol{x}'\boldsymbol{x}$ has nothing to do with λ_1 and λ_2, the rest with critical region $\boldsymbol{x}'\boldsymbol{M}\boldsymbol{x}/\boldsymbol{x}'\boldsymbol{x}\leqslant k$ is also a UMP invariant test for testing H_0: $h\in\mathscr{F}_0^n$ against H_1: $h\in\{\mathscr{F}_0^n(\Sigma)|\Sigma=\lambda_1(\boldsymbol{I}_n-\boldsymbol{M})+\lambda_2\boldsymbol{M},\ \lambda_1>\lambda_2>0\}$, where $\boldsymbol{M}^2=\boldsymbol{M}$, $\boldsymbol{M}'=\boldsymbol{M}$, $\boldsymbol{M}$ known. For (iii), it is similar to (ii), the test with critical region $\boldsymbol{x}'\boldsymbol{A}\boldsymbol{x}/\boldsymbol{x}'\boldsymbol{x}\leqslant k$ is UMP.

As a special case of (iii), consider Σ of the form

$$\Sigma^{-1}=\tau(1+\rho^2)\boldsymbol{I}_n-\tau\rho\begin{Bmatrix} 0 & 1 & & 0\\ 1 & 0 & \ddots & \vdots\\ & \ddots & \ddots & \\ 0 & & 1 & 0\end{Bmatrix},\quad |\rho|<1.$$

This form for Σ^{-1} arises in serial correlation problems. In this case, to reject $\sum_{i=2}^{n} x_i x_{i-1}\geqslant k$ is UMP for $\rho^2>0$.

5.5.2. A Multivariate Test

Let $\overline{\delta}_p = \{V | V \geqslant 0,\ V: p \times p\}$ and $\mathcal{Q}$ be the class of functions from the set of $p \times p$ matrices into $[0, \infty)$ such that for $q \in \mathcal{Q}$, q is convex on $\overline{\delta}_p$,

$$\int_{R^{np}} q(X'X)\, dX = 1$$

and

$$q(BV) = q(VB) \tag{5.5.7}$$

for all $V \in \overline{\delta}_p$ and $B \in \mathrm{GL}(p)$. Further, let

$$\mathcal{F}(M, \Sigma) = \{f | f(X|M, \Sigma) = |\Sigma|^{-n/2}\, q(\Sigma^{-1}(X-M)'(X-M)),\ q \in \mathcal{Q}\}, \tag{5.5.8}$$

where $M \in R^{np}$ and $\Sigma \in \delta_p$ which is the set of $p \times p$ positive definite matrices. Obviously, if $X \sim SS_{n \times p}(f)$, i.e., $X \overset{d}{=} \Gamma X Q$, for each $\Gamma \in O(n)$ and $Q \in O(p)$, having a density f convex on $\overline{\delta}_p$ and $f \in \mathcal{Q}$. If $p = 1$, $\mathcal{F}(\mu, \Sigma)$ coincides with the class of elliptical densities of the form $f((x-\mu)'\Sigma^{-1}(x-\mu))$ with f convex.

Example 5.5.1. $\mathcal{F}(M, \Sigma)$ contains the density of $N_{n \times p}(M, I_n \otimes \Sigma)$. It follows that $\exp\left\{-\frac{1}{2}W\right\}$ is a convex function of W on $\overline{\delta}_p$, or equivalently, $e^{-t/2}$ is convex on $[0, \infty)$.

Example 5.5.2. Let $f(X|M, \Sigma) = c|\Sigma|^{-n/2}|I_n + \Sigma^{-1}(X-M)'(X-M)|^{-(n+p)/2}$. Then $f \in \mathcal{F}(M, \Sigma)$. To see that $q(V) = c|I_n + V|^{-(n+p)/2}$ is convex on $\overline{\delta}_p$, it suffices to note that $q(X'X)$ is the density of a convex mixture of $N_{n \times p}(O, I_n \otimes \Sigma)$ and that the density of $N_{n \times p}(O, I_n \otimes \Sigma)$ is convex in $X'X$. In fact,

$$k \int_{\delta_p} |S|^{(n-1)/2} \exp\left\{-\frac{1}{2}(I_p + X'X)S\right\} dS = c|I_p + X'X|^{-(n+p)/2} = q(X'X).$$

Let h be the density of an $n \times p$ random matrix X and consider the testing problem

$$\begin{aligned} &H_0\colon h \in \mathcal{F}(M, \Sigma),\ M_2 = O,\ \Sigma \in \delta_p \text{ against} \\ &H_1\colon h \in \mathcal{F}(M, \Sigma),\ M_2 \neq O,\ \Sigma \in \delta_p, \end{aligned} \tag{5.5.9}$$

where $M = (M_1', M_2', O)'$ is partitioned as $M_i\colon n_i \times p$, $i = 1, 2$, and $n_3 = n - n_1 - n_2 \geqslant p$ is assumed. This is the well-known MANOVA (multivariate analysis of variance) problem, when X has a density $h \in \mathcal{F}(M, \Sigma)$. If $p = 1$, it is the ANOVA (analysis of variance) problem. In Chapter 6, we shall have a chance to meet them.

Partition $X = (X_1', X_2', X_3')'$ as $X_i\colon n_i \times p$, $i = 1, 2, 3$. Then the problem (5.5.9) remains invariant under $G = O(n_1) \times O(n_2) \times O(n_3) \times \mathrm{GL}(p) \times R^{n_1 p}1$ acting on X as

$$X \to gX = (Q_1 X_1 A' + F,\ Q_2 X_2 A',\ Q_3 X_3 A'), \tag{5.5.10}$$

where $g = (Q_1, Q_2, Q_3, A, F) \in G$. The induced group $\overline{G} = O(n_1) \times O(n_2) \times GL(p) \times R^{n_1 p}1$ acts on the parametric space as

$$(M_1, M_2, \Sigma) \to \overline{g}(M_1, M_2, \Sigma) = (Q_1 M_1 A' + F,\ Q_2 M_2 A',\ A \Sigma A') \tag{5.5.11}$$

where $\overline{g} = (Q_1, Q_2, A, F) \in \overline{G}$.

Lemma 5.5.4. *Under the above assumptions, when $\min(n_2, p)=1$, maximal invariant functions under the group G and $\bar{G}$ are*

$$T = \operatorname{tr} \boldsymbol{X}_2 (\boldsymbol{X}_3' \boldsymbol{X}_3)^{-1} \boldsymbol{X}_2' \quad \text{and} \quad \delta = \operatorname{tr} \boldsymbol{M}_2 \Sigma^{-1} \boldsymbol{M}_2', \tag{5.5.12}$$

respectively.

Proof. Obviously, T and δ are invariant under the group G and $\bar{G}$, respectively. To see that they are maximal invariant, firstly, we assume $n_2=1$. If

$$\operatorname{tr}\boldsymbol{X}_2 (\boldsymbol{X}_3' \boldsymbol{X}_3)^{-1}\boldsymbol{X}_2 = \operatorname{tr}\boldsymbol{Y}_2(\boldsymbol{Y}_3'\boldsymbol{Y}_3)^{-1}\boldsymbol{Y}_2$$

and

$$\operatorname{tr}\boldsymbol{M}_2 \Sigma^{-1}\boldsymbol{M}_2' = \operatorname{tr}\boldsymbol{V}_2\boldsymbol{W}^{-1}\boldsymbol{V}_2', \ (\Sigma>0,\ \boldsymbol{W}>0),$$

it is required to show that there exist $g = (\boldsymbol{Q}_1, \boldsymbol{Q}_2, \boldsymbol{Q}_3, \boldsymbol{A}, \boldsymbol{F})\in G$ and $\bar{g} = (\Gamma_1, \Gamma_2, \boldsymbol{B}, \boldsymbol{E})\in\bar{G}$ such that, $\boldsymbol{X} = g\boldsymbol{Y}$ and $(\boldsymbol{M}_1, \boldsymbol{M}_2\Sigma) = \bar{g}(\boldsymbol{V}_1, \boldsymbol{V}_2, \boldsymbol{W})$ respectively. In fact, since $n_2 =1$, $\operatorname{tr}\boldsymbol{X}_2(\boldsymbol{X}_3' \boldsymbol{X}_3)^{-1}\boldsymbol{X}_2'=\boldsymbol{X}_2(\boldsymbol{X}_3'\boldsymbol{X}_3)^{-1} \boldsymbol{X}_2' = \operatorname{tr}\boldsymbol{Y}_2 (\boldsymbol{Y}_3'\boldsymbol{Y}_3)^{-1}\boldsymbol{Y}_2' = \boldsymbol{Y}_2(\boldsymbol{Y}_3'\boldsymbol{Y}_3)^{-1}\boldsymbol{Y}_2'$, by (6) of Section 1.6, there exists $\Gamma\in O(p)$ such that

$$(\boldsymbol{X}_3' \boldsymbol{X}_3)^{-1/2} \boldsymbol{X}_2' = \Gamma (\boldsymbol{Y}_3'\boldsymbol{Y}_3)^{-1/2}\boldsymbol{Y}_2',$$

or equivalently,

$$\boldsymbol{X}_2 = \boldsymbol{Y}_2 (\boldsymbol{Y}_3'\boldsymbol{Y}_3)^{-1/2} \Gamma' (\boldsymbol{X}_3'\boldsymbol{X}_3)^{-1/2}.$$

Now,

$$\begin{aligned}\boldsymbol{X}_3' \boldsymbol{X}_3 &= (\boldsymbol{X}_3'\boldsymbol{X}_3)^{1/2}\Gamma\Gamma'(\boldsymbol{X}_3'\boldsymbol{X}_3)^{1/2}\\ &= (\boldsymbol{X}_3'\boldsymbol{X}_3)^{1/2}\ \Gamma (\boldsymbol{Y}_3' \boldsymbol{Y}_3)^{-1/2}\\ &\quad\cdot(\boldsymbol{Y}_3'\boldsymbol{Y}_3)(\boldsymbol{Y}_3'\boldsymbol{Y}_3)^{-1/2}\Gamma' (\boldsymbol{X}_3'\boldsymbol{X}_3)^{1/2}.\end{aligned}$$

By (6) of Section 1.2 again, there exists $\boldsymbol{Q}_3\in O(n_3)$ such that

$$\boldsymbol{X}_3 = \boldsymbol{Q}_3\boldsymbol{Y}_3[(\boldsymbol{X}_3'\boldsymbol{X}_3)^{1/2}\Gamma (\boldsymbol{Y}_3' \boldsymbol{Y}_3)^{-1/2}]'.$$

Put $\boldsymbol{A} = (\boldsymbol{X}_3' \boldsymbol{X}_3)^{1/2}\Gamma (\boldsymbol{Y}_3'\boldsymbol{Y}_3)^{-1/2}$, $\boldsymbol{Q}_2=1$, $\boldsymbol{Q}_1=\boldsymbol{I}_{n_1}$ and $\boldsymbol{F} = -\boldsymbol{Y}_1\boldsymbol{A}' + \boldsymbol{X}_1$. Then $\boldsymbol{X}_1 = \boldsymbol{Q}_1\boldsymbol{Y}_1\boldsymbol{A}' + \boldsymbol{F}$, $\boldsymbol{X}_2 = \boldsymbol{Q}_2\boldsymbol{Y}_2\boldsymbol{A}'$ and $\boldsymbol{X}_3 = \boldsymbol{Q}_3\boldsymbol{Y}_3\boldsymbol{A}'$, i.e., $\boldsymbol{X} = g\boldsymbol{Y}$ where $g=(\boldsymbol{Q}_1, \boldsymbol{Q}_2, \boldsymbol{Q}_3, \boldsymbol{A}, \boldsymbol{F})\in G$. Similarly, it can be shown that $\operatorname{tr} \boldsymbol{M}_2\Sigma^{-1}\boldsymbol{M}_2' = \operatorname{tr}\boldsymbol{V}_2\boldsymbol{W}^{-1}\boldsymbol{V}_2'$ implies that there exists $\bar{g}\in\bar{G}$ such that $(\boldsymbol{M}_1, \boldsymbol{M}_2, \Sigma) = \bar{g}(\boldsymbol{V}_1, \boldsymbol{V}_2, \boldsymbol{W})$.

Next, if $p=1$, noting that $\operatorname{tr}\boldsymbol{X}_2 (\boldsymbol{X}_3'\boldsymbol{X}_3)^{-1}\boldsymbol{X}_2' = \boldsymbol{X}_2'\boldsymbol{X}_2 / \boldsymbol{X}_3'\boldsymbol{X}_3 = [(\boldsymbol{X}_3'\boldsymbol{X}_3)^{-1/2} \cdot\boldsymbol{X}_2]'\ [(\boldsymbol{X}_3'\boldsymbol{X}_3)^{-1/2}\boldsymbol{X}_2]$, we may prove that what we want is a way similar to the above proof. □

To obtain a formal expression of the distribution of T, we first consider the marginal distribution of $\tilde{\boldsymbol{X}} = (\boldsymbol{X}_2', \boldsymbol{X}_3')'$.

Lemma 5.5.5. *Let $h\in\mathfrak{F}(\boldsymbol{M}, \Sigma)$, where $\boldsymbol{M} = (\boldsymbol{M}_1', \boldsymbol{M}_2', \boldsymbol{O})'$ and $\Sigma>0$, have a form*

$$h(\boldsymbol{X}|\boldsymbol{M}, \Sigma) = |\Sigma|^{-n/2}\, q(\Sigma^{-1}\{(\boldsymbol{X}_1 - \boldsymbol{M}_1)'(\boldsymbol{X} - \boldsymbol{M}_1) + (\boldsymbol{X}_2 - \boldsymbol{M}_2)'(\boldsymbol{X}_2 - \boldsymbol{M}_2) + \boldsymbol{X}_3'\boldsymbol{X}_3\}), \tag{5.5.13}$$

where $q\in\mathfrak{Q}$. Then the marginal density of $\tilde{\boldsymbol{X}} = (\boldsymbol{X}_2', \boldsymbol{X}_3')'$ is

$$\tilde{h}(\tilde{\boldsymbol{X}}|\boldsymbol{M}_2, \Sigma) = |\Sigma|^{-(n_2+n_3)/2}\,\tilde{q}(\Sigma^{-1/2}\{(\boldsymbol{X}_2 - \boldsymbol{M}_2)'(\boldsymbol{X}_2 - \boldsymbol{M}_2) + \boldsymbol{X}_3'\boldsymbol{X}_3\}\Sigma^{-1/2}), \tag{5.5.14}$$

where $\Sigma^{-1/2}\in\delta_p$, and $\tilde{q}(\boldsymbol{V})=\int_{R^{n_1p}} q(\boldsymbol{Y}'\boldsymbol{Y} + \boldsymbol{V})\,d\boldsymbol{Y}$. Further, $\tilde{q}$ is convex on $\bar{\delta}_p$, and independent of $(\boldsymbol{M}_1, \boldsymbol{M}_2, \Sigma)$, and it satisfies

(5.5.15) $$\tilde{q}(\boldsymbol{Q}'\boldsymbol{V}\boldsymbol{Q}) = \tilde{q}(\boldsymbol{V}) \quad \textit{for any } \boldsymbol{V} \in \bar{\delta}_p \textit{ and } \boldsymbol{Q} \in O(p).$$

Proof. From (5.5.13), using (5.5.7), transforming $\boldsymbol{X}_1$ into $\boldsymbol{Y} = (\boldsymbol{X}_1 - \boldsymbol{M}_1)\Sigma^{-1/2}$ and integrating (5.5.13) with respect to $\boldsymbol{Y}$ yields (5.5.14). Now $\tilde{q}$ turns out to be independent of $(\boldsymbol{M}_1, \boldsymbol{M}_2, \Sigma)$ and the convexity of $\tilde{q}$ follows from the convexity of q. Further, by (5.5.7),

$$\tilde{q}(\boldsymbol{Q}'\boldsymbol{V}\boldsymbol{Q}) = \int q(\boldsymbol{Q}'(\boldsymbol{Q}\boldsymbol{Y}'\boldsymbol{Y}\boldsymbol{Q}' + \boldsymbol{V})\boldsymbol{Q})\, d\boldsymbol{Y} = \int q(\boldsymbol{Q}\boldsymbol{Y}'\boldsymbol{Y}\boldsymbol{Q}' + \boldsymbol{V})\, d\boldsymbol{Y} = \tilde{q}(\boldsymbol{V})$$

because the Jacobian of transformation $\boldsymbol{Y} \to \boldsymbol{Y}\boldsymbol{Q}'$ is 1. □

Applying a theorem by Wijsman (1967), Kariya (1981) obtained the distribution of T.

Lemma 5.5.6. *Let P_δ^T be the distribution of T under parameter δ. Then when $\min(n_2, p) = 1$, the density h_T of T with respect to P_0^T evaluated at $T = \operatorname{tr} \boldsymbol{X}_2(\boldsymbol{X}_3'\boldsymbol{X}_3)^{-1}\boldsymbol{X}_2'$ is given by*

$$h_T(\operatorname{tr}\boldsymbol{X}_2(\boldsymbol{X}_3'\boldsymbol{X}_3)^{-1}\boldsymbol{X}_2' | \delta)$$

(5.5.16)

$$= \frac{\displaystyle\int_{\mathrm{GL}(p)\times O(n_2)} \tilde{q}(\boldsymbol{A}\boldsymbol{A}' + \delta \boldsymbol{e}\boldsymbol{e}' - v\delta^{1/2} r_{11}(\boldsymbol{a}_1\boldsymbol{e}' + \boldsymbol{e}\boldsymbol{a}_1'))|\boldsymbol{A}\boldsymbol{A}'|^{k/2}\, d\boldsymbol{A}\, dU(\boldsymbol{R})}{\displaystyle\int_{GL(p)} \tilde{q}(\boldsymbol{A}\boldsymbol{A}')|\boldsymbol{A}\boldsymbol{A}'|^{k/2}\, d\boldsymbol{A}}$$

where $v = (T/(1+T))^{1/2} = [\operatorname{tr}\boldsymbol{X}_2(\boldsymbol{X}_3'\boldsymbol{X}_3)^{-1}\boldsymbol{X}_2'/(1 + \operatorname{tr}\boldsymbol{X}_2(\boldsymbol{X}_3'\boldsymbol{X}_3)^{-1}\boldsymbol{X}_2)]^{1/2}$, $k = n_2 + n_3 - p$, $\boldsymbol{e} = (1, 0, \ldots, 0)' \in R^p$, *$\boldsymbol{a}_i$ is the ith column of $\boldsymbol{A}$, r_{ij} is the (i, j) element of $\boldsymbol{R}$ and $dU(\boldsymbol{R})$ is the unique invariant probability measure over $O(n_2)$.*

Proof. Applying Wijsman's (1967) Theorem 4, h_T is given by N_δ / N_0, where

$$N_\delta = \int_{O(n_2)\times GL(p)} \tilde{q}\{\Sigma^{-1/2}(\boldsymbol{R}\boldsymbol{X}_2\boldsymbol{A}' - \boldsymbol{M}_2)'(\boldsymbol{R}\boldsymbol{X}_2\boldsymbol{A}' - \boldsymbol{M}_2)$$

$$\Sigma^{-1/2} \cdot \Sigma^{-1/2}\boldsymbol{A}\boldsymbol{X}_3'\boldsymbol{X}_3\boldsymbol{A}'\Sigma^{-1/2}\} \cdot |\boldsymbol{A}\boldsymbol{A}'|^{(n_2+n_3)/2}\, d\mu(\boldsymbol{A})\, dU(\boldsymbol{R})$$

in which $d\mu(\boldsymbol{A}) = d\boldsymbol{A}/|\boldsymbol{A}\boldsymbol{A}'|^{p/2}$. The measure $\mu(\cdot)$ is left and right invariant under the group GL(p), i.e., $\mu(\boldsymbol{A}E\boldsymbol{B}) = \mu(E)$, for all $\boldsymbol{A} \in \mathrm{GL}(p)$, $\boldsymbol{B} \in \mathrm{GL}(p)$ and the Borel set E, where $\boldsymbol{A}E\boldsymbol{B} \,\hat{=}\, \{\boldsymbol{X} | \boldsymbol{X} = \boldsymbol{A}\boldsymbol{Y}\boldsymbol{B},\ \boldsymbol{Y} \in E\}$. Hence replacing $\boldsymbol{A}$ by $\Sigma^{1/2}\boldsymbol{A}(\boldsymbol{X}_3'\boldsymbol{X}_3)^{-1/2}$ leaves the integral unchanged; therefore

(5.5.17) $$N_\delta = \int_{O(n_2)\times \mathrm{GL}(p)} c_1\tilde{q}(\boldsymbol{A}\boldsymbol{u}'\boldsymbol{u}\boldsymbol{A}' - \boldsymbol{A}\boldsymbol{u}'\boldsymbol{R}'\eta' - \eta\boldsymbol{R}\boldsymbol{u}\boldsymbol{A}' + \eta\eta' + \boldsymbol{A}\boldsymbol{A}')|\boldsymbol{A}\boldsymbol{A}'|^{k/2}\, d\boldsymbol{A}\, dU(\boldsymbol{R}),$$

where $c_1^{-1} = |\boldsymbol{X}_3'\boldsymbol{X}_3|^{(n_2+n_3)/2}$, $\boldsymbol{u} = \boldsymbol{X}_2(\boldsymbol{X}_3'\boldsymbol{X}_3)^{-1/2}$ and $\eta = \Sigma^{-1/2}\boldsymbol{M}_2'$. Suppose $n_2 = 1$ and let $\boldsymbol{U}_1$ and $\boldsymbol{U}_2$ be orthogonal matrices with $\eta'(\eta\eta')^{-1/2}$ and $\boldsymbol{u}(\boldsymbol{u}'\boldsymbol{u})^{-1/2}$ as their first rows, respectively. Then replacing $\boldsymbol{A}$ by $\boldsymbol{U}_1'\boldsymbol{A}\boldsymbol{U}_2$ and using (5.5.15) yields

$$N_\delta = \int c_1\tilde{q}(\boldsymbol{U}_1\boldsymbol{a}_1\boldsymbol{a}_1'\boldsymbol{U}_1 T - \boldsymbol{U}_1^1\boldsymbol{a}_1\boldsymbol{R}'\eta' T^{1/2} - \eta\boldsymbol{R}\boldsymbol{a}_1'\boldsymbol{U}_1 T^{1/2} + \eta\eta' + \boldsymbol{U}_1'\boldsymbol{A}\boldsymbol{A}'\boldsymbol{U}_1)|\boldsymbol{A}\boldsymbol{A}'|^{k/2}\, d\boldsymbol{A}\, dU(\boldsymbol{R})$$

$$= \int c_1 \tilde{q} \left\{ (1+T)\boldsymbol{a}_1\boldsymbol{a}_1' + \sum_{i=1}^{p} \boldsymbol{a}_i\boldsymbol{a}_i' + \delta \boldsymbol{e}\boldsymbol{e}' - T^{1/2}\delta^{1/2} \boldsymbol{R}(\boldsymbol{a}_1\boldsymbol{e}' + \boldsymbol{e}\boldsymbol{a}_1') \right\} |\boldsymbol{A}\boldsymbol{A}'|^{k/2}\, d\boldsymbol{A}\, dU(\boldsymbol{R})$$

where $T = \boldsymbol{u}\boldsymbol{u}' = \boldsymbol{X}_2(\boldsymbol{X}_3'\boldsymbol{X}_3)^{-1}\boldsymbol{X}_2'$ and $\delta = \eta'\eta = \boldsymbol{M}_2\Sigma^{-1}\boldsymbol{M}_2'$. Finally transforming $\boldsymbol{a}_1$ into $\boldsymbol{a}_1/(1+T)^{1/2}$ and taking the ratio of N_δ and N_0 in gives (5.5.16). Next, suppose p $=1$ and let $\boldsymbol{V}_1$ and $\boldsymbol{V}_2$ be $n_2 \times n_2$ orthogonal matrices with $\eta'(\eta\eta')^{-1/2}$ and $\boldsymbol{u}(\boldsymbol{u}'\boldsymbol{u})^{-1/2}$ as their first columns respectively. Then replacing $\boldsymbol{R}$ in (5.5.17) by $\boldsymbol{V}_1\boldsymbol{R}\boldsymbol{V}_2'$ yields

$$N_\delta = \int c_1\tilde{q}(\boldsymbol{A}^2T + \boldsymbol{A}^2 + \delta\boldsymbol{e}\boldsymbol{e}' - 2\boldsymbol{A}r_{11}\delta^{1/2}T^{1/2})\, |\boldsymbol{A}\boldsymbol{A}'|^{k/2} d\boldsymbol{A}\, dU(\boldsymbol{R}),$$

where $T = \boldsymbol{u}'\boldsymbol{u} = \boldsymbol{X}_2'\boldsymbol{X}_2/\boldsymbol{X}_3'\boldsymbol{X}_3$. Hence, transforming $\boldsymbol{A}$ into $\boldsymbol{A}/(1+T)^{1/2}$ gives (5.5.16). □

Theorem 5.5.2. *When* $\min(n_2, p) = 1$, *the test with critical region*

$$T = \operatorname{tr} \boldsymbol{X}_2(\boldsymbol{X}_3'\boldsymbol{X}_3)^{-1}\boldsymbol{X}_2' > c$$

is UMP invariant for problem (5.5.9), *and the null distribution of* T *under* $h \in \mathfrak{F}(\boldsymbol{M}, \Sigma)$ *with* $\boldsymbol{M}_2 = \boldsymbol{0}$ *is the same as that under* $N_{n\times p}(\boldsymbol{O}, \boldsymbol{I}_n \otimes \boldsymbol{I}_p)$. *That is, when* H_0 *is true,* $((n_2+n_3-p)/p)T \sim F(p, n_2+n_3-p)$ *if* $n_2 = 1$ *and* $(n_3/n_2)T \sim F(n_2, n_3)$ *if* $p = 1$, *where* $F(i, j)$ *denotes the F-distribution with degrees of freedom i and j.*

Proof. The last part follows from Theorem 5.1.1. To show the first part, let $H(v)$ be the numerator of (5.5.16). Then transforming $\boldsymbol{A}$ into $-\boldsymbol{A}$ leaves $H(v)$ unchanged and so $H(-v) = H(v)$. Hence, using the convexity of $\tilde{q}$ on $\bar{\delta}_p$, we obtain for $1/2 < \alpha < 1$

$$H(v) = \alpha H(v) + (1-\alpha)H(-v) \geqslant H((2\alpha-1)v), \quad v \in [0, 1].$$

This implies that $H(v)$ is a nondecreasing function of $v \in [0, 1]$. Therefore by applying the Neyman-Pearson Lemma to h_T in (5.5.16), a most powerful test is given by critical region

$$v = (T/(1+T))^{1/2} > c,$$

or equivalently,

$$T > c'.$$

Since this region does not depend on δ and h, the test is UMP invariant by noting that as for invariant tests, H_0 is true iff $\delta = 0$. □

Most of material in Section 5.5.1 and Section 5.5.2 is partly adopted from the paper by King (1980) and Kariya (1981).

5.6. Goodness of Fit Test for Elliptical Symmetry

In the previous discussion, we assumed that a sample is from the population of elliptically symmetric distribution. At present, we would like to test that a population has elliptically symmetric distribution, which is a goodness of fit problem. As in non-parametric univariate statistic some characteristic of interested distribution is used to establish a testing procedure. In this section, we always assume that the random vector involved has all finite moments. The material in the section is mainly adopted from Deng (1984).

5.6.1. A characteristic of spherical symmetry

A random variable X is said to be uniquely determined by its moment sequence $\{\alpha_i = E(X^i), i \geqslant 1\}$, if $E(X^i) = E(Y^i)$, $i = 1, 2, \ldots$, which then implies $X \stackrel{d}{=} Y$. Similarly, we can define the conception for a random vector $\boldsymbol{x}$ to be uniquely determined by its moment sequence

$$\{\Gamma_{2k} = \varepsilon(\underbrace{\boldsymbol{x} \otimes \boldsymbol{x}' \otimes \boldsymbol{x} \otimes \cdots \otimes \boldsymbol{x}'}_{2k \text{ factors}}), \Gamma_{2k-1} = \varepsilon(\underbrace{\boldsymbol{x} \otimes \boldsymbol{x}' \otimes \boldsymbol{x} \otimes \cdots \otimes \boldsymbol{x}}_{2k-1 \text{ factors}}), k \geqslant 1\},$$

(cf. Section 2.2). It is well-known that the distribution of a random vector $\boldsymbol{x}$ of order n is given by all distributions of $\boldsymbol{a}'\boldsymbol{x}$, $\boldsymbol{a} \in R^n$, i.e., $\boldsymbol{a}'\boldsymbol{x} \stackrel{d}{=} \boldsymbol{a}'\boldsymbol{y}$, for all $\boldsymbol{a} \in R^n$ iff $\boldsymbol{x} \stackrel{d}{=} \boldsymbol{y}$.

Lemma 5.6.1. *Suppose $\boldsymbol{x}$ is a random vector. If $\boldsymbol{a}'\boldsymbol{x}$ is uniquely determined by its moment sequence for all $\boldsymbol{a}$, then $\boldsymbol{x}$ is uniquely determined by the moment sequence of $\boldsymbol{x}$.*

Proof. Assume that $\boldsymbol{x}$ and $\boldsymbol{y}$ have the same moment sequence. Then $\boldsymbol{a}'\boldsymbol{x}$ and $\boldsymbol{a}'\boldsymbol{y}$ have the same moment sequence for any given $\boldsymbol{a}$. Thus $\boldsymbol{a}'\boldsymbol{y} \stackrel{d}{=} \boldsymbol{a}'\boldsymbol{x}$ for any given $\boldsymbol{a}$ which shows that $\boldsymbol{x} \stackrel{d}{=} \boldsymbol{y}$. □

Using Lemma 5.6.1, to verify a random vector $\boldsymbol{x}$ to be uniquely determined by its moment sequence, it is sufficient to show that $\boldsymbol{a}'\boldsymbol{x}$ does for any given $\boldsymbol{a}$. In the univariate case, we have

(i) A random variable X is uniquely determined by its moment sequence $\{\alpha_n = E(X^n), n \geqslant 1\}$ if the series $\sum_{n=0}^{\infty} (\alpha_n / n!) t^n$ has a positive radius of convergence (See Loève, 1977, p.230).

(ii) A random variable X is uniquely determined by its moment sequence $\{\alpha_n = E(X^n), n \geqslant 1\}$ if either

$$\overline{\lim_{n}} \frac{(\alpha_{2n})^{1/2n}}{2n} < \infty,$$

or

$$\sum_{n=1}^{\infty} \alpha_{2n}^{-1/2n} = \infty$$

(See Chow, 1978, P.280).

From (i) or (ii), it immediately follows that a bounded random variable is uniquely determined by its moment sequence.

Example 5.6.1. Let $x \sim N(0, 1)$. Then

$$\alpha_n = \begin{cases} 0, & \text{if } n = 2k+1 \\ (2\pi)^{-1/2} \int_{-\infty}^{\infty} t^n e^{-t^2/2} \, \mathrm{dt} = 2^{-1/2} (n-1)!!, & \text{if } n = 2k. \end{cases}$$

By (i), X is uniquely determined by its moment sequence. Now if $\boldsymbol{x} \sim N(\mathbf{0}, \mathbf{I}_p)$, then for any given $\boldsymbol{a} \in R^p$, $\boldsymbol{a}'\boldsymbol{x} \sim N(\mathbf{0}, \boldsymbol{a}'\boldsymbol{a})$ and $\boldsymbol{a}'\boldsymbol{x} / (\boldsymbol{a}'\boldsymbol{a})^{1/2} \sim N(0, 1)$, and $\boldsymbol{a}'\boldsymbol{x} / (\boldsymbol{a}'\boldsymbol{a})^{1/2}$, or equivalently $\boldsymbol{a}'\boldsymbol{x}$, is uniquely determined by its moment sequence. Thus $\boldsymbol{x} \sim N(\mathbf{0}, \boldsymbol{I}_p)$ is uniquely determined by its moment sequence.

Eaxmple 5.6.2. The $\boldsymbol{u}^{(n)}$ is uniquely determined by its moment sequence. This follows from $|\boldsymbol{a}'\boldsymbol{u}^{(n)}| \leqslant \|\boldsymbol{a}\|$, i.e., $\boldsymbol{a}'\boldsymbol{u}^{(n)}$ being bounded, for each $\boldsymbol{a} \in R^n$.

Lemma 5.6.2. *Let* $\boldsymbol{x}$ *be a random vector and let* $\boldsymbol{a}'\boldsymbol{x}$ *be uniquely determined by its moment sequence for each* $\boldsymbol{a}$. *Then any affine transformation of* $\boldsymbol{x}$, $\boldsymbol{A}\boldsymbol{x}+\boldsymbol{b}$ *say, is also uniquely determined by its moment sequence.*

Proof. By the assumption and Lemma 5.6.1, it suffices to show the assertion that if a univariate X is uniquely determined by its moment sequence, then $X+a$ does, for any $a \in (-\infty, \infty)$. Let Y be a random variable having

$$E[(X+a)^n] = E(Y^n) = E[((Y-a)+a)^n], \qquad n \geqslant 1.$$

Then, for $n=1$, $E(X+a) = E[(Y-a)+a]$ implies $E(X) = E(Y-a)$. By $E(X) = E(Y-a)$, $E(X+a)^2 = E[(Y-a)+a]^2$ implies $E(X^2) = E[(Y-a)^2]$. In general, if $E(X^k) = E[(Y-a)^k]$, $k = 1, \dots, n-1$, then $E[(X+a)^n] = E[(Y-a)^n]$. Thus $X \overset{\mathrm{d}}{=} Y - a$, or equivalently, $X + a \overset{\mathrm{d}}{=} Y$. □

Theorem 5.6.1. *Let a random vector* $\boldsymbol{x}$ *of order p have the moment sequence* $\{\Gamma_k, k \geqslant 1\}$ *and let* $\{\tilde{\Gamma}_k, k \geqslant 1\}$ *be the moment sequence of* $\boldsymbol{y} \sim N(\boldsymbol{0}, \boldsymbol{I}_p)$. *Then*

(1) *if* $\boldsymbol{x} \sim S_p(\phi)$, $\Gamma_k = c_k \tilde{\Gamma}_k$, *where* $c_k = E((\boldsymbol{x}'\boldsymbol{x})^{k/2}) / E(R)^{k/2}$, $k = 1, \dots, R^* \sim \chi_p^2$ *and*

(2) *if* $\Gamma_k = a_k \tilde{\Gamma}_k$, $k \geqslant 1$, *there is* $\boldsymbol{x}_0 \sim S_p(\phi_0)$ *such that* $\boldsymbol{x}_0$ *and* $\boldsymbol{x}$ *have the same moment sequence; further, if* $\boldsymbol{x}$ *is uniquely determined by its moment sequence* $\{\Gamma_k, k \geqslant 1\}$, *then* $\boldsymbol{x} \overset{\mathrm{d}}{=} \boldsymbol{x}_0$.

Proof. To show (1), suppose $\boldsymbol{x} \overset{\mathrm{d}}{=} R\boldsymbol{u}^{(p)}$ where R and $\boldsymbol{u}^{(p)}$ are independent, $R^2 \overset{\mathrm{d}}{=} \boldsymbol{x}'\boldsymbol{x}$, $\boldsymbol{y} \overset{\mathrm{d}}{=} \sqrt{R^*}\boldsymbol{u}^{(p)}$, and R^* is independent of $\boldsymbol{u}^{(p)}$. By (2.2.18), we get

$$\begin{aligned}
\Gamma_{2k} &= \mathcal{E}(\boldsymbol{x} \otimes \boldsymbol{x}' \otimes \boldsymbol{x} \otimes \cdots \otimes \boldsymbol{x}') \\
&= \mathcal{E}(R^{2k}(\boldsymbol{u}^{(p)} \otimes \boldsymbol{u}^{(p)\prime} \otimes \cdots \otimes \boldsymbol{u}^{(p)\prime})) \\
&= E(R^{2k}) \cdot \mathcal{E}(\boldsymbol{u}^{(p)} \otimes \boldsymbol{u}^{(p)\prime} \otimes \cdots \otimes \boldsymbol{u}^{(p)\prime}) \\
&= c_{2k}\tilde{\Gamma}_{2k}, \qquad k = 1, 2, \dots .
\end{aligned}$$

Similarly, $\Gamma_{2k+1} = c_{2k+1}\tilde{\Gamma}_{2k+1}$, $k = 0, 1, \dots$, which proves (1).

To show (2), let $\Gamma_{2k} = c_{2k}\tilde{\Gamma}_{2k} = c_{2k}\mathcal{E}(\boldsymbol{u}^{(p)} \otimes \boldsymbol{u}^{(p)\prime} \otimes \cdots \otimes \boldsymbol{u}^{(p)\prime})$. Then

$$\begin{aligned}
E[(\boldsymbol{x}'\boldsymbol{x})^n] &= \mathrm{tr}\Gamma_{2n} = c_{2n}\mathrm{tr}\mathcal{E}(\boldsymbol{u}^{(p)} \otimes \boldsymbol{u}^{(p)\prime} \otimes \cdots \otimes \boldsymbol{u}^{(p)\prime}) \\
&= c_{2n}E(\boldsymbol{u}^{(p)\prime}\boldsymbol{u}^{(p)})^n = c_{2n}
\end{aligned}$$

via Theorem 2.2.1. Take $R \overset{\mathrm{d}}{=} (\boldsymbol{x}'\boldsymbol{x})^{1/2}$ independent of $\boldsymbol{u}^{(p)}$, and put $\boldsymbol{x}_0 = R\boldsymbol{u}^{(p)}$ so that $\boldsymbol{x}_0 \sim S_p(\psi)$, and $\boldsymbol{x}_0$ and $\boldsymbol{x}$ have the same moment sequence. Finally, if $\boldsymbol{x}$ is uniquely determined by its moment sequence, then $\boldsymbol{x} \overset{\mathrm{d}}{=} \boldsymbol{x}_0$, i.e., $\boldsymbol{x}$ is spherically symmetric. □

Corollary. *Suppose* $\boldsymbol{x}$ *is uniquely determined by its moment sequence. Then* $\boldsymbol{x} \sim S_p(\phi)$ *iff the moment sequence* $\{\Gamma_k, k \geqslant 1\}$ *is correspondingly proportional to* $\{\tilde{\Gamma}_k, k \geqslant 1\}$ *of* $\boldsymbol{y} \sim N(\boldsymbol{0}, \boldsymbol{I}_p)$, *i.e., there exists a sequence* $\{a_k, k \geqslant 1\}$ *such that* $\Gamma_k = a_k\tilde{\Gamma}_k$, $k = 1, 2, \dots$.

Theorem 5.6.1, or its corollary, is called the *moment characteristic theorem of spherically symmetric distributions.* By this corollary, a procedure can be designed to test spherical symmetry. Let $\boldsymbol{x}_1, ..., \boldsymbol{x}_n$ be i.i.d. such that $\boldsymbol{x}_1 \stackrel{d}{=} \boldsymbol{x}$. If the ratios of the same order sample moments, $\hat{r}^n_{ij} / \hat{r}^n_{kl}$ say, where $\hat{\Gamma}_n = (\hat{r}^n_{ij})$, are quite different from the $\tilde{r}^n_{ij} / \tilde{r}^n_{kl}$, where $\tilde{r}^n_{ij}$ is the (i, j) element of the n-th moment matrix $\tilde{\Gamma}_n$ of $N(\mathbf{0}, \boldsymbol{I}_p)$, then we reject the hypothesis that $\boldsymbol{x}$ is spherical symmetry. The sequel of the chapter is devoted to discuss the details.

5.6.2. Significance Tests for Spherical Symmetry (I)

Let $\boldsymbol{x}_{(1)}, ..., \boldsymbol{x}_{(n)}$ i.i.d. and $\boldsymbol{x}_{(1)} \stackrel{d}{=} \boldsymbol{x}$. Consider a significance test problem

$$H_0: \boldsymbol{x} \sim S_p(\phi), \text{ for some } \phi. \tag{5.6.1}$$

It is known that H_0 is true iff $\boldsymbol{x} \stackrel{d}{=} R\boldsymbol{u}^{(p)}$ where $R \geqslant 0$ and $\boldsymbol{u}^{(p)}$ are independent. Hence when $P(\boldsymbol{x}'\boldsymbol{x} = 0) = 0$, H_0 can be decomposed into two parts:

$$H_{01}: \boldsymbol{x} / (\boldsymbol{x}'\boldsymbol{x})^{1/2} \stackrel{d}{=} \boldsymbol{u}^{(p)} \tag{5.6.2}$$

and

$$H_{02}: \boldsymbol{x}'\boldsymbol{x} \text{ and } \boldsymbol{x} / (\boldsymbol{x}'\boldsymbol{x})^{1/2} \text{ are independent.} \tag{5.6.3}$$

Firstly we would like to test H_{01}. By Example 5.6.2, $\boldsymbol{u}^{(p)}$ is uniquely determined by its moment sequence, $\{\Gamma^u_k, k \geqslant 1\}$ say. Hence we can compare the sample moments with $\{\Gamma^u_k, k \geqslant 1\}$ to test H_{01}. Let

$$\boldsymbol{u}_{(i)} = \boldsymbol{x}_{(i)} / \|\boldsymbol{x}_{(i)}\|, \qquad i = 1, 2, ..., n. \tag{5.6.4}$$

Since $\boldsymbol{x}_{(1)}, ..., \boldsymbol{x}_{(n)}$ are i.i.d. and $\boldsymbol{x}_{(1)} \stackrel{d}{=} \boldsymbol{x}$, we have $\boldsymbol{u}_{(1)}, ..., \boldsymbol{u}_{(n)}$ i.i.d. and $\boldsymbol{u}_{(1)} \stackrel{d}{=} \boldsymbol{x} / \|\boldsymbol{x}\|$, and $\boldsymbol{u}_{(1)}, ..., \boldsymbol{u}_{(n)}$ are regarded as n samples from $\boldsymbol{x} / \|\boldsymbol{x}\|$.

Lemma 5.6.3. *Let the k-th moment matrix of $\boldsymbol{u}^{(p)} = (u_1, ..., u_p)'$ be $\Gamma^u_k = (r^{(k)}_{ij})$. If $r^{(k)}_{ij} = E(u_1^{ip} \cdots u_p^{ip}) \equiv \mu^u_{i_1, ..., i_p}$, $i_1 + \cdots + i_p = k$, then*

$$r^{(k)}_{ij} = \begin{cases} 0 & \textit{if some } i \textit{ is odd,} \\ \dfrac{\prod_{\alpha=1}^{p} \left(\frac{1}{2}\right)^{[i_\alpha/2]}}{\left(\frac{p}{2}\right)^{[k/2]}} & \textit{if all of } i_1, ..., i_p \textit{ are even,} \end{cases} \tag{5.6.5}$$

where $t^{[k]} \stackrel{\Lambda}{=} t(t+1) \ldots (t+k-1)$. Further, let $m_i = h_1 + k_i$, $1 \leqslant i \leqslant p$. Then

$$\operatorname{cov}(u_1^{h_1} \cdots u_p^{h_p}, u_1^{k_1} \cdots u_p^{k_p}) = \begin{cases} 0, & \textit{if some } m_i \textit{ is odd,} \\ \mu^u_{m_1, ..., m_p}, & \textit{if all of } m_1, ..., m_p \textit{ are even and some } h_i \textit{ is odd,} \\ \mu^u_{m_1, ..., m_p} - \mu^u_{h_1, ..., h_p} \mu^u_{k_1, ..., k_p}, & \textit{otherwise.} \end{cases} \tag{5.6.6}$$

Proof. The result (5.6.5) follows from $\boldsymbol{u}^{(p)} \overset{\mathrm{d}}{=} -\boldsymbol{u}^{(p)}$ and Theorem 2.4.2. The result (5.6.6) follows from equality

$$\operatorname{cov}(u_1^{h_1} \cdots u_p^{h_p}, u_1^{k_1} \cdots u_p^{k_p}) = \mu^u_{m_1, \dots, m_p} - \mu^u_{h_1, \dots, h_p} \mu^u_{k_1, \dots, k_p}. \qquad \square$$

Thus, by (5.6.5) and (5.6.6), we can calculate Γ_k^u and $\Gamma_{2k}^u \equiv \boldsymbol{V}_k$ (say). Let $r_{ij}^{(k)} = E(u_1^{i} \cdots u_p^{i})$, $i_1 + \cdots + i_p = k$. Define the sample moment of $r_{ij}^{(k)}$ as follows:

$$\hat{r}_{ij}^{(k)} = \frac{1}{n} \sum_{t=1}^{n} (u_{t1}^{i_1} \cdots u_{tp}^{i_p}) \tag{5.6.7}$$

where $\boldsymbol{U} = (\boldsymbol{u}_{(1)}, \dots, \boldsymbol{u}_{(n)})' = (u_{ij}), \boldsymbol{u}_{(1)}, \dots, \boldsymbol{u}_{(n)}$ are given by (5.6.4.). Put $\hat{\Gamma}_k^u = (\hat{r}_{ij}^{(k)})$. Then, for a larger n, vec $\hat{\Gamma}_k^u$ is asymptotically distributed as $N(\operatorname{vec}\Gamma_k^u, \frac{1}{n}\boldsymbol{V}_k)$ by Kolmogorov's strong large number theorem. Therefore we reject H_{01}, whenever

$$n[\operatorname{vec}(\hat{\Gamma}_k^u - \Gamma_k^u)' \boldsymbol{V}_k^{-1} \operatorname{vec}(\hat{\Gamma}_k^u - \Gamma_k^u)] \geqslant \chi^2_{f,\alpha},$$

for some $k \geqslant 1$, where $f=p^k$ and $\chi^2_{f,\alpha}$ is the critical point of χ^2_f at significance level α.

If there is no significant difference between $\boldsymbol{x} / \|\boldsymbol{x}\|$ and $\boldsymbol{u}^{(p)}$, we start with testing H_{02}. H_{02} is true iff the conditional distribution of $\boldsymbol{x}$, given $\|\boldsymbol{x}\|$, is the uniform distribution on spherical face in R^p with radius $\|\boldsymbol{x}\|$, $S_{\|x\|} = \{\boldsymbol{y} \in R^p | \|\boldsymbol{y}\| = \|\boldsymbol{x}\|\}$ say. We partition $[0, \infty)$ into $k+1$ segments, $[0, R_1)$, $[R_1, R_2)$, ..., $[R_k, \infty)$. Then, given i, $1 \leqslant i \leqslant k+1$, for samples $\boldsymbol{x}_{(1)}, \dots, \boldsymbol{x}_{(k)}$, examine $\boldsymbol{x}_{(i_1)}, \dots, \boldsymbol{x}_{(i_t)}$ that drop into the annular region $C_i = \{\boldsymbol{y} \in R^p | R_{i-1} \leqslant \|\boldsymbol{y}\| < R_i\}$ $(R_0 \equiv 0, R_{k+1} \equiv \infty)$, and whether they are distributed uniformly in C_i. Of course, how to choose $\{R_i, 1 \leqslant i \leqslant k\}$ and how large k is taken are worth studying further. In general, the solutions of these problems are connected with practical models.

5.6.3. Significance Tests for Spherical Symmetry (II)

Let $\boldsymbol{x}$ be a random vector and let $E(\|\boldsymbol{x}\|^k)$, $k = 1, 2, \dots$, be known. Let $\boldsymbol{x}_{(1)}, \dots, \boldsymbol{x}_{(n)}$ be i.i.d. and $\boldsymbol{x}_{(1)} \overset{d}{=} \boldsymbol{x}$. We are required to test

$$H_0\colon \boldsymbol{x} \sim S_p(\phi), \text{ for some } \phi.$$

Here, it is not necessary that the procedure used in Section 5.6.2 is repeated. Let $\{\boldsymbol{T}_k, k \geqslant 1\}$ be the moment sequence of $\boldsymbol{x}$ and $\boldsymbol{T}_k = (t_{ij}^{(k)})$. If $t_{ij}^{(k)} = E(x_1^{h_1} \cdots x_p^{h_p})$, $h_1 + \cdots + h_p = k$, $\boldsymbol{x} = (x_1, \dots, x_p)'$, then when H_0 is true,

$$t_{ij}^{(k)} = E(x_1^{h_1} \cdots x_p^{h_p}) = \mu^u_{h_1, \dots, h_p} \cdot E(\|\boldsymbol{x}\|^k), \tag{5.6.8}$$

where $\mu^u_{h_1, \dots, h_p}$ is defined by Lemma 5.6.3. Similarly, when H_0 is true,

$$\begin{aligned}\operatorname{cov}(x_1^{h_1} \cdots x_p^{h_p}, x_1^{i_1} \cdots x_p^{i_p}) = E(\|\boldsymbol{x}\|^{2k})\mu^u_{m_1, \dots, m_p} - \\ [E(\|\boldsymbol{x}\|^k)]^2 \mu^u_{h_1, \dots, h_p}\mu^u_{i_1, \dots, i_p},\end{aligned} \tag{5.6.9}$$

where $h_1 + \cdots + h_p = i_1 + \cdots + i_p = k$, $m_j = h_j + i_j$, $j = 1, \dots, p$. Using (5.6.5), we can also rewrite (5.6.8) and (5.6.9). Note that $E(\|\boldsymbol{x}\|^m)$, $m \geqslant 1$, are known. So $\boldsymbol{T}_k$ and $\boldsymbol{T}_{2k}$ are known, $k \geqslant 1$. For samples $\boldsymbol{x}_{(1)}, \dots, \boldsymbol{x}_{(n)}$, define the sample moment of $t_{ij}^{(k)} = E(x_1^{h_1} \cdots x_p^{h_p})$ as follows:

$$\hat{t}_{ij}^{(k)} = \frac{1}{n} \sum_{t=1}^{n} (x_{t1}^{h_1} \cdots x_{tp}^{h_p}) \tag{5.6.10}$$

where $\boldsymbol{X} = (\boldsymbol{x}_{(1)}, \ldots, \boldsymbol{x}_{(n)})' = (x_{ij})$. Put $\hat{\boldsymbol{T}}_k = (\hat{t}_{ij}^{(k)})$. Then, for a large n, vec $\hat{\boldsymbol{T}}_k \sim N(\text{vec}\hat{\boldsymbol{T}}_k, \frac{1}{n}\boldsymbol{T}_{2k})$. We think that $\boldsymbol{x}$ is not from spherically symmetric distribution whenever

$$n[\text{vec}(\hat{\boldsymbol{T}}_k - \boldsymbol{T}_k)'\boldsymbol{T}_{2k}^{-1} \text{ vec}(\hat{\boldsymbol{T}}_k - \boldsymbol{T}_k)] \geqslant \chi^2_{f,\alpha},$$

for some $k \geqslant 1$, where $f = p^k$.

5.6.4. Significance Tests for Elliptical Symmetry

Let $\boldsymbol{x}$ be a random vector having

$$\mathcal{E}(\boldsymbol{x}) = \mu$$

and

$$\mathcal{D}(\boldsymbol{x}) = \sigma^2\Sigma$$

where μ and $\Sigma > 0$ are known. Let $\boldsymbol{x}_{(1)}, \ldots, \boldsymbol{x}_{(n)}$ be i.i.d. and $\boldsymbol{x}_{(1)} \overset{d}{=} \boldsymbol{x}$. Consider testing

$$H_0\colon \boldsymbol{x} \sim EC_p(\mu, \Sigma, \phi), \text{ for some } \phi. \tag{5.6.11}$$

Since μ and Σ are known, we only need to test $\boldsymbol{y} = \Sigma^{-1/2}(\boldsymbol{x} - \mu)$ to be of spherical symmetry. Through the samples $\boldsymbol{x}_{(1)}, \ldots, \boldsymbol{x}_{(n)}$ from $\boldsymbol{x}$ we produce the samples $\boldsymbol{y}_{(1)} = \Sigma^{-1/2}(\boldsymbol{x}_{(1)} - \mu), \ldots, \boldsymbol{y}_{(n)} = \Sigma^{-1/2}(\boldsymbol{x}_{(n)} - \mu)$ from $\boldsymbol{y}$. Using the procedure before, we can test hypothesis (5.6.11).

If μ and Σ are unknown, we may use $\bar{\boldsymbol{x}} = \frac{1}{n}\sum_{i=1}^{n} \boldsymbol{x}_{(i)}$ and $\frac{1}{n}\boldsymbol{S} = \frac{1}{n}\sum_{i=1}^{n} (\boldsymbol{x}_{(i)} - \bar{\boldsymbol{x}})(\boldsymbol{x}_{(i)} - \bar{\boldsymbol{x}})'$ to replace μ and Σ in the above. Put $\hat{\mu} = \bar{\boldsymbol{x}}$ and $\hat{\Sigma} = \frac{1}{n}\boldsymbol{S}$. Now the data are transformed as

$$\boldsymbol{x}_{(i)} \to \boldsymbol{z}_{(i)} = n^{-1/2}\boldsymbol{S}^{-1/2}(\boldsymbol{x}_{(i)} - \bar{\boldsymbol{x}}),\ i = 1, \ldots, n.$$

The $\boldsymbol{z}_{(1)}, \ldots, \boldsymbol{z}_{(n)}$ relpace $\boldsymbol{y}_{(1)}, \ldots, \boldsymbol{y}_{(n)}$ to test $\boldsymbol{y} = \Sigma^{-1/2}(\boldsymbol{x} - \mu)$ to be of spherical symmetry. Here there are some problems to be solved. Deng (1984) has investigated the asymptotic distribution of the moment statistics. Readers may refer to his paper.

Concerning the goodness of fit test for spherical symmetry, Beran (1979) and Koziol (1982) discussed it and the interested reader may further refer to their papers

References

Anderson and Fang (1982c), Anderson, Fang and Hsu (1986), Beran (1979), Bian, Wang and Zhang (1984), Blackwell (1956), Box (1949), Chow (1978), Davis (1971), Deng (1984), Fang and Chen (1984), Fang and Xu (1986), Giri (1977), Hsu (1940), Kariya (1981), Khatri and Srivastava (1971), King (1980), Kres (1975), Lehmann and Stein (1948), Loéve (1977), Mauchly (1940), Quan (1987), Schatzoff (1966) Wijsman (1967), Wilks (1932), Zhang and Fang (1982), Zhang, Fang and Chen (1985).

Exercises 5

5.1. Let $\boldsymbol{X}$ $n \times p$ be a random matrix and $\boldsymbol{S} = \boldsymbol{X}'(\boldsymbol{I}_n - \frac{1}{n}\boldsymbol{11}')\boldsymbol{X}$. Partition $\boldsymbol{S}$ into

$$S = \begin{pmatrix} S_{11} & S_{12} \\ S_{21} & S_{22} \end{pmatrix}, \ S_{11} : q \times q, \ S_{22} : (p - q) \times (p - q).$$

(i) The canonical correlation coefficients, $T(X)$ say, are the eigenvalues of $S_{12}S_{22}^{-1}S_{21}S_{11}^{-1}$. Show that $T(X)$ is distribution-free on $\mathcal{F}_2^+$ (hence on $\mathcal{F}_3^+$), but it is not on $\mathcal{F}_1^+$.

(ii) Show that $\lambda_3(X) = |S|^{n/2}/(\mathrm{tr} S/p)^{np/2}$ is distribution-free on $\mathcal{F}_3^+$ but it is not on $\mathcal{F}_2^+$ (hence on $\mathcal{F}_1^+$). (See Theorem 5.3.1.)

5.2. Let $X \sim LS_{n \times p}(\mu, \Sigma, f)$. Show that the distribution of $T^2 = n(n-1)\bar{x}'S^{-1}\bar{x}$ depends on (μ, Σ) only through $\delta^2 = \mu'\Sigma^{-1}\mu$.

5.3. Let $\mathcal{F} = \{X | \Gamma X \overset{d}{=} X, \ \Gamma \in O(n)\}$ and, for $X \in \mathcal{F}$, let $\mathcal{F}(X) = \{Y | Y \ll X\}$ where "$Y \ll X$" means that the distribution law of Y is absolutely continuous with respect to that of X. Show that if $t(X) \overset{d}{=} t(X(X'X)^{-1/2})$ for some $X \in \mathcal{F}$ with $P(X'X > 0) = 1$, then $t(X)$ is distribution-free on $\mathcal{F}(X)$. (Due to Bian, Wang and Zhang, 1984.)

5.4. Show Theorem 5.1.2 (Hint: use Exercise 5.3).

5.5. Let $X \sim VS_{n \times p}(\mu, \Sigma, f)$ and let f be nonincreasing and continuous, $n > p$. Establish an appropriate testing procedure to test H_0: $\mu_{(1)} = \mu_{(2)}$, where $\mu = (\mu_{(1)}', \mu_{(2)}')'$.

5.6. (Union–intersection principle) Let $X \sim VS_{n \times p}(\mu, \Sigma, f)$ with f being nonincreasing and continuous, and $n > p$. The hypothesis H_0: $\mu = 0$ is true iff H_a: $a'\mu = 0$, for any nonnull vector $a \in R^p$, is ture. Thus H_0 will be rejected if at least one of the hypotheses H_a, $a \in R^p - \{0\}$, is rejectedand hence $H_0 = \bigcap_a H_a$. Let ω_a denote the rejection region of hypothesis H_a. Obviously, the rejection region of H_0 is $\omega = \bigcup_a \omega_a$. The size of ω_a should be such that ω is rejection region of H_0 of size α. This is known as the *union-intersection principle* of Roy. Show that the union-intersection principle for testing H_0 leads to the T^2-test, too.

5.7. Suppose x has a density h. Establish a critical region for testing H_0: $h \in \mathcal{F}_0^n$ against H_1: $h \in \mathcal{F}_0^n(\Sigma)$, $\Sigma = \sigma^2\Sigma_0$, σ^2 unknown, $\Sigma_0 \neq \lambda I_n$ known, which is UMP invariant under the group $G = (0, \infty)$ acting on R^n as in (5.5.2) (See Section 5.5.).

5.8. Extend the results in Section 5.6 to the matrix case.

CHAPTER VI
LINEAR MODELS

Many useful models, such as regression models, variance analysis models and discriminant analysis models, can be expressed in terms of linear models. In this chapter, a general theory of the estimation and hypothesis testing of linear model are given. Then this general theory is applied to the study of above models (regression, variance analysis and discriminant analysis). A double screening stepwise regression procedure is suggested and an example using this procedure is given in Section 6.5.

6.1. Definition and Examples

6.1.1. Definition

Let $\boldsymbol{Y}$ be an $n \times p$ random matrix. If, for a given $n \times k$ matrix $\mathbf{X}$ and an unknown matrix $\boldsymbol{B}$,

$$\boldsymbol{Y} = \boldsymbol{XB} + \boldsymbol{E} \tag{6.1.1}$$

where $\boldsymbol{E}$ is a random matrix with $\mathcal{E}(\boldsymbol{E}) = \boldsymbol{O}$ and $\mathcal{D}(\text{vec}\boldsymbol{E}) = \boldsymbol{V} \in \Theta$ ——a specific set of $np \times np$ matrices, then $\boldsymbol{Y}$ is said to be a *generalized linear model* and denoted by $L(\boldsymbol{Y}; \mathbf{X}, \Theta)$.

If $\Theta = \{\Sigma \otimes \boldsymbol{I}_n | \Sigma > 0\}$, then $L(\boldsymbol{Y}; \mathbf{X}, \Theta)$ is the linear model in the ordinary sense. If $p = 1$, and $\Theta = \left\{ \sum_{i=1}^{t} \sigma_i \boldsymbol{V}_i \right\}$ where $\boldsymbol{V}_1, \ldots, \boldsymbol{V}_t$ are given symmetric matrices, the linear model $L(\boldsymbol{Y}; \mathbf{X}, \Theta)$ is the components of variance model or the mixed linear model. Many examples will be given in the following subsections.

6.1.2. Regression Model

Let $L(\boldsymbol{Y}; \mathbf{X}, \Theta)$ be a linear model. If $\mathbf{X} = (\mathbf{1}\ \boldsymbol{Z})$ then $L(\boldsymbol{Y}; \mathbf{X}, \Theta)$ is said to be a *regression model*. The matrix $\boldsymbol{Y}$ is the n observation values of the p response variables $y_1, \ldots, y_p$ corresponding to the n observation values of the m explanatory variables $z_1, \ldots, z_m$, and equation (6.1.1) expresses the relation between $\{y_i\}$ and $\{z_i\}$. By rewriting (6.1.1) as

$$\boldsymbol{Y} = (\mathbf{1}\ \boldsymbol{Z})\binom{\boldsymbol{b}'}{\boldsymbol{B}} + \boldsymbol{E} \tag{6.1.2}$$

the vector $\boldsymbol{b}$ is called the *regression constant* and $\boldsymbol{B}$ is called the *regression coefficient matrix of the* $\{z_i\}$.

6.1.3. Variance Analysis Model

Let $L(\boldsymbol{Y}; \mathbf{X}, \Theta)$ be a linear model. If the entries of $\mathbf{X}$ take values 0 and 1 only, then $L(\boldsymbol{Y}; \mathbf{X}, \Theta)$ is called to be a *variance analysis model.*

Example 6.1.1. Suppose $y_1^{(i)}, \ldots, y_{n_i}^{(i)} \sim N(\mu_i, \Sigma)$, $i = 1, 2, \ldots, k$. Consider the problem —— how to test the hypothesis

$$H_0 \colon \mu_1 = \mu_2 = \cdots = \mu_k.$$

Let $\boldsymbol{Y} = (\boldsymbol{Y}_1', \ldots, \boldsymbol{Y}_k')'$, where $\boldsymbol{Y}_i' = (y_1^{(i)}, y_2^{(i)}, \cdots, y_{n_i}^{(i)})$, $i = 1, 2, \ldots, k$. Then

$$\mathcal{E}(\boldsymbol{Y}) = \begin{pmatrix} \mathbf{1} & 0 & 0 & \cdots & 0 \\ \mathbf{1} & \mathbf{1} & 0 & \cdots & 0 \\ \mathbf{1} & 0 & \mathbf{1} & \cdots & 0 \\ \vdots & \vdots & \vdots & & \vdots \\ \mathbf{1} & 0 & 0 & \cdots & \mathbf{1} \end{pmatrix} \begin{pmatrix} \mu_1' \\ \delta_2' \\ \vdots \\ \delta_k' \end{pmatrix}$$

and

$$\mathcal{D}(\operatorname{vec}\boldsymbol{Y}) = \Sigma \otimes \boldsymbol{I}_n, \quad n = \sum_{i=1}^{k} n_i$$

where $\delta_i = \mu_i - \mu_1$, $i = 2, \ldots, k$, and H_0 can be rewritten as

$$H_0' \colon \delta_2 = \cdots = \delta_k = \mathbf{0}.$$

It is obvious that the entries of $\boldsymbol{X}$ are 0's or 1's.

Example 6.1.2. (*Factorial design*) There are k factors and s_i is the number of levels of the i-th factor, $i = 1, 2, \ldots, k$. Consider p response variables for each experiment. Let $\mu(i, j)$ be the main effect vector of the ith factor at level j, $1 \leqslant j \leqslant s_i$, $i = 1, 2, \ldots, k$. Put

$$\boldsymbol{M}_i = (\mu(i, 1)\ \mu(i, 2) \cdots \mu(i, s_i))', \qquad i = 1, 2, \ldots, k.$$

Then $\boldsymbol{M}_i'\mathbf{1}_{s_i} = \mathbf{0}$, $i = 1, 2, \ldots, k$, because the summation of the effects must be zero. If $\boldsymbol{a}_{\alpha i}$ denotes the level of the $\mathbf{i}$-th factor at the α-th experimental unit and the response variables at the α-th experiment are $(y_{\alpha 1}, y_{\alpha 2}, \cdots, y_{\alpha p}) = \boldsymbol{y}_\alpha'$, say, then

$$\mathcal{E}(\boldsymbol{y}_\alpha) = \mu_0 + \mu(1, a_{\alpha 1}) + \cdots + \mu(k, a_{\alpha k}), \qquad \alpha = 1, 2, \ldots, n,$$

where μ_0 is a constant vector. Let

$$x_{\alpha j}^{(i)} = \begin{cases} 1, & a_{\alpha i} = j \\ 0, & a_{\alpha i} \neq j \end{cases} \quad 1 \leqslant j \leqslant s_i, \qquad i = 1, 2, \ldots, k, \qquad \alpha = 1, 2, \ldots, n,$$

and $\boldsymbol{X}_i = (x_{\alpha j}^{(i)})$: $n \times s_i$, $i = 1, 2, \ldots, k$. Then

$$\mathcal{E}(\boldsymbol{Y}) = \mathcal{E}\begin{pmatrix} \boldsymbol{y}_1' \\ \vdots \\ \boldsymbol{y}_n' \end{pmatrix} = (\mathbf{1X}_1 \cdots \boldsymbol{X}_k)\begin{pmatrix} \mu_0' \\ \mu_1' \\ \vdots \\ \mu_k' \end{pmatrix}$$

is an $n \times p$ matrix and

$$\mathcal{D}(\operatorname{vec}\ \boldsymbol{Y}) = \Sigma \otimes \boldsymbol{I}_n.$$

Hence it is a linear model and the entries of $\boldsymbol{X}$ are 0 or 1 only.

In the variance analysis model $L(\boldsymbol{Y}; \boldsymbol{X}, \Theta)$ the matrix $\boldsymbol{X}$ is said to be the *design matrix*.

6.1.4. Discriminant Analysis

Let $L(\boldsymbol{Y}; \boldsymbol{X}, \Theta)$ be a linear model. If the entries of $\boldsymbol{Y}$ take values 0 and 1 only, then

$L(\boldsymbol{Y}; \boldsymbol{X}, \Theta)$ is called to be a *discriminant analysis model.* Sometimes the entries of $\boldsymbol{Y}$ may be real. (cf. Section 6.5.3.)

Suppose that there are k m-dimensional populations and the samples. $\boldsymbol{X}_{1i}, \boldsymbol{X}_{2i}, \ldots, \boldsymbol{X}_{n_i i}$ are the samples taken from the i-th population and

$$(6.1.3) \qquad \begin{cases} \mathcal{E}(\boldsymbol{X}_{ri}) = \mu_i, \\ \mathcal{D}(\boldsymbol{X}_{ri}) = \Sigma, \end{cases} \qquad r = 1, \ldots, n_i, \qquad i = 1, 2, \ldots, k,$$

where μ_i: $m \times 1$ and Σ: $m \times m$, $i = 1, \ldots, k$.

Let $\boldsymbol{y}_i' = (\underset{i-1}{0\ 0\ \ldots\ 0}\ 1\ \underset{k-1}{0\ \ldots\ 0})$, $\boldsymbol{Y}_i = \boldsymbol{1}_{n_i}\boldsymbol{y}_i'$ and $\boldsymbol{X}_i = (\boldsymbol{X}_{1i} \cdots \boldsymbol{X}_{n_i i})'$: $n_i \times m$. Consider the relationship between the y_i's and x_i's. Write

$$\boldsymbol{Y} \triangleq \begin{pmatrix} \boldsymbol{Y}_1 \\ \boldsymbol{Y}_2 \\ \vdots \\ \boldsymbol{Y}_k \end{pmatrix} = \begin{pmatrix} \boldsymbol{1} & \boldsymbol{X}_1 \\ \boldsymbol{1} & \boldsymbol{X}_2 \\ \vdots & \vdots \\ \boldsymbol{1} & \boldsymbol{X}_k \end{pmatrix} \begin{pmatrix} \boldsymbol{B}_0 \\ \boldsymbol{B} \end{pmatrix} + \boldsymbol{E} \triangleq (\boldsymbol{1X}) \begin{pmatrix} \boldsymbol{B}_0 \\ \boldsymbol{B} \end{pmatrix} + \boldsymbol{E}$$

and assume $\mathcal{E}(\boldsymbol{E}) = \boldsymbol{O}$, $\mathcal{D}(\mathrm{vec}\boldsymbol{E}) = \Sigma \otimes \boldsymbol{I}_n$. Then $\boldsymbol{Y}$ is a linear model, and $\boldsymbol{B}_0$, $\boldsymbol{B}$ are the coefficients of a linear discriminant function. Thus a discriminant problem can be regarded as a regression problem. Further results will be discussed at Section 6.5.3.

6.2. BLUE

6.2.1. Least Square Estimates

Let $L(\boldsymbol{Y}; \boldsymbol{X}, \Theta)$ be a linear model. If $\boldsymbol{B}$ satisfies

$$(6.2.1) \qquad \mathrm{tr}\, (\boldsymbol{Y} - \boldsymbol{X}\hat{\boldsymbol{B}})'(\boldsymbol{Y} - \boldsymbol{X}\hat{\boldsymbol{B}}) = \min_{\boldsymbol{B}}\ \mathrm{tr}\, (\boldsymbol{Y} - \boldsymbol{X}\boldsymbol{B})'(\boldsymbol{Y} - \boldsymbol{X}\boldsymbol{B})$$

then $\hat{\boldsymbol{B}}$ is said to be the *least square estimate of* $\boldsymbol{B}$.

Theorem 6.2.1. *The least square estimate* $\hat{\boldsymbol{B}}$ *is the solution of the normal equation*

$$(6.2.2) \qquad \boldsymbol{X}'\boldsymbol{X}\hat{\boldsymbol{B}} = \boldsymbol{X}'\boldsymbol{Y}$$

and hence $\hat{\boldsymbol{B}} = (\boldsymbol{X}'\boldsymbol{X})^{-}\boldsymbol{X}'\boldsymbol{Y}$, *where* $(\boldsymbol{X}'\boldsymbol{X})^{-}$ *is a generalized inverse of* $\boldsymbol{X}'\boldsymbol{X}$.

Proof. By (6.2.2) it is easy to prove

$$\mathrm{tr}(\boldsymbol{Y} - \boldsymbol{X}\boldsymbol{B})'(\boldsymbol{Y} - \boldsymbol{X}\boldsymbol{B}) = \mathrm{tr}(\boldsymbol{Y} - \boldsymbol{X}\hat{\boldsymbol{B}})'(\boldsymbol{Y} - \boldsymbol{X}\hat{\boldsymbol{B}}) + \mathrm{tr}(\hat{\boldsymbol{B}} - \boldsymbol{B})'\boldsymbol{X}'\boldsymbol{X}(\hat{\boldsymbol{B}} - \boldsymbol{B}).$$

Hence $\hat{\boldsymbol{B}}$ is the least square estimate of $\boldsymbol{B}$. □

Notice that, for $L(\boldsymbol{Y}; \boldsymbol{X}, \Theta)$,

$$(6.2.3) \qquad \mathcal{E}(\boldsymbol{AY}) = \boldsymbol{AXB}, \quad \mathcal{D}(\mathrm{vec}\boldsymbol{AY}) = (\boldsymbol{I} \otimes \boldsymbol{A})\boldsymbol{V}(\boldsymbol{I} \otimes \boldsymbol{A}'), \text{ for each } \boldsymbol{V} \in \Theta.$$

Thus we obtain that

$$(6.2.4) \qquad \begin{aligned} &\mathcal{E}(\hat{\boldsymbol{B}}) = (\boldsymbol{X}'\boldsymbol{X})^{-}\boldsymbol{X}'\boldsymbol{X}\boldsymbol{B} \\ &\mathcal{D}(\mathrm{vec}\hat{\boldsymbol{B}}) = (\boldsymbol{I} \otimes (\boldsymbol{X}'\boldsymbol{X})^{-}\boldsymbol{X}')\boldsymbol{V}(\boldsymbol{I} \otimes \boldsymbol{X}(\boldsymbol{X}'\boldsymbol{X})^{-}), \qquad \boldsymbol{V} \in \Theta. \end{aligned}$$

6.2.2. BLUE

Definition 6.2.1. Let $L(\boldsymbol{Y}; \boldsymbol{X}, \Theta)$ be a linear model. A parameter tr $\boldsymbol{A}'\boldsymbol{B}$ is said to be *estimable*, if there exists an $\boldsymbol{H}$ such that $E(\mathrm{tr}\ \boldsymbol{H}'\boldsymbol{Y}) = \mathrm{tr}\ \boldsymbol{A}'\boldsymbol{B}$, and tr $\boldsymbol{H}'\boldsymbol{Y}$ is a LUE (linear unbiased estimate) of tr $\boldsymbol{A}'\boldsymbol{B}$. If tr $\boldsymbol{H}'\boldsymbol{Y}$ is a LUE of tr $\boldsymbol{A}'\boldsymbol{B}$ and

$$\operatorname{Var}(\operatorname{tr}\ \boldsymbol{H}'\boldsymbol{Y}) = \min_{E(\operatorname{tr}\ \boldsymbol{G}'\boldsymbol{Y}) = \operatorname{tr}\ \boldsymbol{A}'\boldsymbol{B}} \operatorname{Var}(\operatorname{tr}\ \boldsymbol{G}'\boldsymbol{Y})$$

then tr $\boldsymbol{H}'\boldsymbol{Y}$ is said to be the *best linear unbiased estimate* ——BLUE—— of tr $A'B$.

Theorem 6.2.2. tr $\boldsymbol{A}'\boldsymbol{B}$ *is estimable iff*

$$\mathcal{L}(\boldsymbol{A}) \subset \mathcal{L}(\boldsymbol{X}'), \tag{6.2.5}$$

where $\mathcal{L}(\boldsymbol{A})$ *denotes the spanning space of columns of* $\boldsymbol{A}$.

Proof. $E(\operatorname{tr}\ \boldsymbol{H}'\boldsymbol{Y}) = \operatorname{tr}\ \boldsymbol{A}'\boldsymbol{B}$ iff $\operatorname{tr}\ \boldsymbol{A}'\boldsymbol{B} = \operatorname{tr}\ \boldsymbol{H}'\boldsymbol{X}\boldsymbol{B}$ for all $\boldsymbol{B}$. It is equivalent to $\boldsymbol{X}'\boldsymbol{H} = \boldsymbol{A}$, that is, $\mathcal{L}(\boldsymbol{A}) \subset \mathcal{L}(\boldsymbol{X}')$. □

From (6.2.5) all the solutions of

$$\boldsymbol{X}'\boldsymbol{H} = \boldsymbol{A} \tag{6.2.6}$$

correspond to all the LUE tr $\boldsymbol{H}'\boldsymbol{Y}$ of the estimable parameter tr $\boldsymbol{A}'\boldsymbol{B}$. Hence

$$\{\text{LUE of tr } \boldsymbol{A}'\boldsymbol{B}\} = \{\operatorname{tr}\ \boldsymbol{H}'\boldsymbol{Y} | \boldsymbol{H} = (\boldsymbol{X}')^{+}\boldsymbol{A} + (\boldsymbol{I} - \boldsymbol{X}\boldsymbol{X}^{+})\boldsymbol{U} \quad \text{for all } \boldsymbol{U}\} \tag{6.2.7}$$

if tr $\boldsymbol{A}'\boldsymbol{B}$ is estimable.

By (6.2.3), the variance of the LUE tr $\boldsymbol{H}'\boldsymbol{Y}$ of the estimable tr $\boldsymbol{A}'\boldsymbol{B}$ is

$$\operatorname{Var}(\operatorname{tr}\ \boldsymbol{H}'\boldsymbol{Y}) = \operatorname{Var}((\operatorname{vec}\boldsymbol{H})'(\operatorname{vec}\boldsymbol{Y})) = (\operatorname{vec}\boldsymbol{H})'\boldsymbol{V}(\operatorname{vec}\boldsymbol{H}), \tag{6.2.8}$$

for each $\boldsymbol{V} \in \Theta$.

Theorem 6.2.3. (Gauss–Markow) Let $L(\boldsymbol{Y}; \boldsymbol{X}, \Theta)$, $\Theta = \{\Sigma \otimes \boldsymbol{I}_n | \Sigma \geqslant 0\}$, *be a linear model. For each estimable parameter* tr$\boldsymbol{A}'\boldsymbol{B}$ *there exists a unique BLUE* tr $\boldsymbol{A}'\hat{\boldsymbol{B}}$, *where* $\hat{\boldsymbol{B}}$ *is the least square estimate, if there is a* $\Sigma_0 > 0$ *such that* $\Sigma_0 \otimes \boldsymbol{I} \in \Theta$.

Proof. By (6.2.8), for the LUE of tr $\boldsymbol{A}'\boldsymbol{B}$ we obtain

$$\begin{aligned}
\operatorname{Var}(\operatorname{tr}\ \boldsymbol{H}'\boldsymbol{Y}) &= (\operatorname{vec}\ \boldsymbol{H})'(\Sigma \otimes \boldsymbol{I})(\operatorname{vec}\ \boldsymbol{H}) \\
&= \operatorname{tr}\ \Sigma\boldsymbol{H}'\boldsymbol{H} \\
&= \operatorname{tr}\ \Sigma[\boldsymbol{U}'(\boldsymbol{I} - \boldsymbol{X}\boldsymbol{X}^{+}) + \boldsymbol{A}'\boldsymbol{X}^{+}][(\boldsymbol{X}^{+})'\boldsymbol{A} + (\boldsymbol{I} - \boldsymbol{X}\boldsymbol{X}^{+})\boldsymbol{U}] \\
&= \operatorname{tr}\ \Sigma[\boldsymbol{A}'(\boldsymbol{X}'\boldsymbol{X})^{+}\boldsymbol{A} + \boldsymbol{U}'(\boldsymbol{I} - \boldsymbol{X}\boldsymbol{X}^{+})\boldsymbol{U}] \\
&\geqslant \operatorname{tr}\ \Sigma\boldsymbol{A}'(\boldsymbol{X}'\boldsymbol{X})^{+}\boldsymbol{A}.
\end{aligned}$$

Hence $\operatorname{Var}(\operatorname{tr}\ \boldsymbol{H}'\boldsymbol{Y}) = \operatorname{tr}\ \Sigma\boldsymbol{A}'(\boldsymbol{X}'\boldsymbol{X})^{+}\boldsymbol{A}$ iff $(\boldsymbol{I} - \boldsymbol{X}\boldsymbol{X}^{+})\boldsymbol{U}\Sigma = \boldsymbol{O}$ for all $\Sigma \otimes \boldsymbol{I} \in \Theta$. For $\Sigma > 0$, it is equivalent to $(\boldsymbol{I} - \boldsymbol{X}\boldsymbol{X}^{+})\boldsymbol{U} = \boldsymbol{O}$, and then $\boldsymbol{H} = (\boldsymbol{X}^{+})'\boldsymbol{A}$, which is unique.

By (6.2.5) or (6.2.6) $\boldsymbol{A}' = \boldsymbol{C}'\boldsymbol{X}$ for the estimable tr$\boldsymbol{A}'\boldsymbol{B}$. Then using the least sqare estimate $\hat{\boldsymbol{B}} = (\boldsymbol{X}'\boldsymbol{X})^{-}\boldsymbol{X}'\boldsymbol{Y}$ we obtain

$$\begin{aligned}
\operatorname{tr}\ \boldsymbol{A}'\hat{\boldsymbol{B}} &= \operatorname{tr}\ \boldsymbol{C}'\boldsymbol{X}(\boldsymbol{X}'\boldsymbol{X})^{-}\boldsymbol{X}'\boldsymbol{Y} = \operatorname{tr}\ \boldsymbol{C}'\boldsymbol{X}(\boldsymbol{X}'\boldsymbol{X})^{+}\boldsymbol{X}'\boldsymbol{Y} \\
&= \operatorname{tr}\ \boldsymbol{A}'\boldsymbol{X}^{+}\boldsymbol{Y} = \operatorname{tr}\ ((\boldsymbol{X}^{+})'\boldsymbol{A})'\boldsymbol{Y}.
\end{aligned}$$ □

Corollary 1. *For the BLUE* tr $\boldsymbol{A}'\hat{\boldsymbol{B}}$ *of the estimable* tr$\boldsymbol{A}'\boldsymbol{B}$,

$$\begin{cases} E(\operatorname{tr}\boldsymbol{A}'\hat{\boldsymbol{B}}) = \operatorname{tr}\boldsymbol{A}'\boldsymbol{B} \\ \operatorname{Var}(\operatorname{tr}\ \boldsymbol{A}'\hat{\boldsymbol{B}}) = \operatorname{tr}\ \boldsymbol{A}'(\boldsymbol{X}'\boldsymbol{X})^{+}\boldsymbol{A}\Sigma \end{cases}. \tag{6.2.9}$$

Remark. In general, the LSE

$$\hat{\boldsymbol{B}} = (\boldsymbol{X}'\boldsymbol{X})^{-}\boldsymbol{X}'\boldsymbol{Y}$$

of $\boldsymbol{B}$ is not unique unless $\boldsymbol{X}$ has the full rank. But for each LUE of tr$\boldsymbol{A}'\boldsymbol{B}$, Theorem 6.2.3

points out that the BLEU $\mathrm{tr}\boldsymbol{A}'\hat{\boldsymbol{B}}$ of $\mathrm{tr}\ \boldsymbol{A}'\boldsymbol{B}$ is unique which is independent of what is taken for $(\boldsymbol{X}'\boldsymbol{X})^-$.

6.2.3. Regularity

In general it is more convenient to consider the linear model $L(\boldsymbol{Y};\ \boldsymbol{X},\Theta)$ when $\boldsymbol{Y}$ is an $n\times 1$ vector, i.e., $p=1$, because we can use $\boldsymbol{y}$ instead of matrix $\boldsymbol{Y}$ and rewrite $L(\boldsymbol{Y};\ \boldsymbol{X},\Theta)$ as

$$\mathcal{E}(y) = (\boldsymbol{I}\otimes \boldsymbol{X})\beta, \qquad \mathcal{D}(y) = \boldsymbol{V}\in\Theta.$$

Definition 6.2.2. $L(y;\ \boldsymbol{X},\Theta)$ is said to be *regular* if there exists a BLUE for each estimable parameter $\boldsymbol{a}'\boldsymbol{\beta}$.

It is obvious that Theorem 6.2.3 states that $L(\boldsymbol{Y};\ \boldsymbol{X},\Theta)$ is *regular* if $\Theta = \{\Sigma\otimes\boldsymbol{I}|\quad \Sigma\geqslant 0$ and inequality at least holds for a $\Sigma_0\}$. In this section we derive the conditions for regularity.

Lemma 6.2.1. *The parameter* $\boldsymbol{a}'\boldsymbol{\beta}$ *is estimable iff there is* $\boldsymbol{c}$ *such that* $\boldsymbol{a}'\boldsymbol{\beta} = \boldsymbol{c}'\mathcal{E}(y)$.

This is obvious from the definition of estimability. It means that each estimable parameter is a linear function of the expectation $\mathcal{E}(y)$. Let $\mu = \mathcal{E}(y)$. Then $\{\boldsymbol{c}'\mu|\boldsymbol{c}\in R^n\}$ is the collection of all estimable parameters.

Corollary 1. $\boldsymbol{c}_1'\mu = \boldsymbol{c}_2'\mu$ *iff* $\boldsymbol{c}_1 = \boldsymbol{c}_2 + (\boldsymbol{I}-\boldsymbol{X}\boldsymbol{X}^+)\boldsymbol{b}$.

Proof. $\boldsymbol{c}_1'\mu = \boldsymbol{c}_2'\mu$ iff $(\boldsymbol{c}_1-\boldsymbol{c}_2)'X\beta = 0$ for each β. Hence $\boldsymbol{X}'(\boldsymbol{c}_1-\boldsymbol{c}_2) = \boldsymbol{0}$ i.e., $\boldsymbol{c}_1 - \boldsymbol{c}_2 = (\boldsymbol{I}-\boldsymbol{X}\boldsymbol{X}^+)\boldsymbol{b}$. □

Lemma 6.2.2. (Lehmann and Scheffé) $\boldsymbol{a}'\boldsymbol{y}$ *is the BLUE for* $\boldsymbol{V}\boldsymbol{a}\in\mathcal{L}(\boldsymbol{X})$ *for all* $\boldsymbol{V}\in\Theta$.

Proof. Now $\mathrm{Var}(\boldsymbol{a}'\boldsymbol{y}) = \boldsymbol{a}'\boldsymbol{V}\boldsymbol{a}$ and

$$\mathrm{Var}([\boldsymbol{a}+(\boldsymbol{I}-\boldsymbol{X}\boldsymbol{X}^+)\boldsymbol{b}]'\boldsymbol{y}) = [\boldsymbol{a}+(\boldsymbol{I}-\boldsymbol{X}\boldsymbol{X}^+)\boldsymbol{b}]'\boldsymbol{V}[\boldsymbol{a}+(\boldsymbol{I}-\boldsymbol{X}\boldsymbol{X}^+)\boldsymbol{b}].$$

Hence $\mathrm{Var}(\boldsymbol{a}'\boldsymbol{y})\leqslant \mathrm{Var}([\boldsymbol{a}+(\boldsymbol{I}-\boldsymbol{X}\boldsymbol{X}^+)\boldsymbol{b}]'\boldsymbol{y})$ for each $\boldsymbol{b}$ iff $\boldsymbol{b}'(\boldsymbol{I}-\boldsymbol{X}\boldsymbol{X}^+)\boldsymbol{V}\boldsymbol{a} = 0$ for each $\boldsymbol{b}$, that is $\boldsymbol{V}\boldsymbol{a}\in\mathcal{L}(\boldsymbol{X})$ for each $\boldsymbol{V}\in\Theta$. □

It is easy to prove

Lemma 6.2.3. *Let* $M(\boldsymbol{V}) = \{\boldsymbol{a}|\boldsymbol{V}\boldsymbol{a}\in\mathcal{L}(\boldsymbol{X})\}$. *Then* $M(\boldsymbol{V})$ *is a subspace of* R^n.

Put $\boldsymbol{M}_0 = \bigcap_{\boldsymbol{V}\in\Theta} M(\boldsymbol{V})$. Then $\boldsymbol{M}_0$ is a subspace of $\boldsymbol{R}^n$. Consider the estimate $\boldsymbol{a}'\boldsymbol{y}$ for $\boldsymbol{a}\in\boldsymbol{M}_0$. By Lemma 6.2.2, each $\boldsymbol{a}'\boldsymbol{y}$ is the BLUE of $\boldsymbol{a}'\mu$. Noticing Corollary 1 of Lemma 6.2.1, we obtain

Theorem 6.2.4. $L(y;\ \boldsymbol{X},\Theta)$ *is regular iff there exists a subspace* $\mathcal{L}_0\subset M_0$ *and* $\mathcal{L}_0 + \mathcal{L}(\boldsymbol{X})^\perp = R^n$.

By this theorem we can prove many linear models are regular. These proofs are left as exercises in the last section of this chapter.

6.2.4. Variation of Models

Sometimes we must consider the relations between some specified linear models and look for formulas to express the estimates.

There are three canonical cases:

(1) To add explanatory variables.

Consider models

$$\mathcal{E}(\boldsymbol{Y}) = \boldsymbol{X}\boldsymbol{B}, \qquad \mathcal{D}(\mathrm{vec}\,\boldsymbol{Y}) = \Sigma\otimes\boldsymbol{I}_n, \tag{6.2.10}$$

where $\boldsymbol{Y}$: $n \times p$, $\boldsymbol{X}$: $n \times k$ and $\boldsymbol{B}$: $k \times p$. And

$$\mathcal{E}(\boldsymbol{Y}) = (\boldsymbol{X}\boldsymbol{Z})\begin{pmatrix} \boldsymbol{B}^* \\ \Gamma \end{pmatrix}, \qquad \mathcal{D}(\text{vec}\,\boldsymbol{Y}) = \Sigma \otimes \boldsymbol{I}_n, \tag{6.2.11}$$

where $\boldsymbol{Z}$: $n \times l$. It means that there are l additional explanatory variables $z_1, \ldots, z_l$ in the model (6.2.11). Let $\hat{\boldsymbol{B}}$ and $(\hat{\boldsymbol{B}}_*, \hat{\Gamma})$ be the least square estimates corresponding to models (6.2.10) and (6.2.11) respectively. Then the relations between them are

$$\begin{cases} \hat{\boldsymbol{B}}_* = \hat{\boldsymbol{B}} - (\boldsymbol{X}'\boldsymbol{X})^- \boldsymbol{X}'\boldsymbol{Z}\hat{\Gamma}, \\ \hat{\Gamma} = \boldsymbol{G}^- \boldsymbol{Z}'(\boldsymbol{Y} - \boldsymbol{X}\hat{\boldsymbol{B}}), \text{ where } \boldsymbol{G} = \boldsymbol{Z}'(\boldsymbol{I} - \boldsymbol{X}(\boldsymbol{X}'\boldsymbol{X})^- \boldsymbol{X}')\boldsymbol{Z}. \end{cases} \tag{6.2.12}$$

Using (6.2.2), $\hat{\boldsymbol{B}} = (\boldsymbol{X}'\boldsymbol{X})^- \boldsymbol{X}'\boldsymbol{Y}$ and

$$\begin{pmatrix} \boldsymbol{X}'\boldsymbol{X} & \boldsymbol{X}'\boldsymbol{Z} \\ \boldsymbol{Z}'\boldsymbol{X} & \boldsymbol{Z}'\boldsymbol{Z} \end{pmatrix}\begin{pmatrix} \hat{\boldsymbol{B}}_* \\ \hat{\Gamma} \end{pmatrix} = \begin{pmatrix} \boldsymbol{X}'\boldsymbol{Y} \\ \boldsymbol{Z}'\boldsymbol{Y} \end{pmatrix}.$$

Thus

$$\begin{aligned} \begin{pmatrix} \hat{\boldsymbol{B}}_* \\ \hat{\Gamma} \end{pmatrix} &= \begin{pmatrix} \boldsymbol{X}'\boldsymbol{X} & \boldsymbol{X}'\boldsymbol{Z} \\ \boldsymbol{Z}'\boldsymbol{X} & \boldsymbol{Z}'\boldsymbol{Z} \end{pmatrix}^- \begin{pmatrix} \boldsymbol{X}'\boldsymbol{Y} \\ \boldsymbol{Z}'\boldsymbol{Y} \end{pmatrix} \\ &= \left[\begin{pmatrix} (\boldsymbol{X}'\boldsymbol{X})^- & \boldsymbol{O} \\ \boldsymbol{O} & \boldsymbol{O} \end{pmatrix} + \begin{pmatrix} (\boldsymbol{X}'\boldsymbol{X})^- \boldsymbol{X}'\boldsymbol{Z} \\ -\boldsymbol{I} \end{pmatrix} \boldsymbol{G}^- (\boldsymbol{Z}'\boldsymbol{X}(\boldsymbol{X}'\boldsymbol{X})^- \right. \\ &\qquad \left. - \boldsymbol{I})\right] \begin{pmatrix} \boldsymbol{X}'\boldsymbol{Y} \\ \boldsymbol{Z}'\boldsymbol{Y} \end{pmatrix} \\ &= \begin{pmatrix} \hat{\boldsymbol{B}} - (\boldsymbol{X}'\boldsymbol{X})^- \boldsymbol{X}'\boldsymbol{Z}\hat{\Gamma} \\ \boldsymbol{G}^- \boldsymbol{Z}'(\boldsymbol{I} - \boldsymbol{X}(\boldsymbol{X}'\boldsymbol{X})^- \boldsymbol{X}')\boldsymbol{Y} \end{pmatrix} \end{aligned}$$

where $\boldsymbol{G} = \boldsymbol{Z}'(\boldsymbol{I} - \boldsymbol{X}(\boldsymbol{X}'\boldsymbol{X})^- \boldsymbol{X}')\boldsymbol{Z}$.

Remark. Because the LES of $\boldsymbol{B}_*$ and Γ are not unique, the above equation holds in the sense that there exists a generalized inverse $(\boldsymbol{X}'\boldsymbol{X})^-$ such that the equation is true (cf. Section 6.2.2, Remark).

Let $\boldsymbol{Q}$ and $\boldsymbol{Q}_*$ be the residual square sums, corresponding to original model and varied model respectively. Then

$$\boldsymbol{Q} = (\boldsymbol{Y} - \boldsymbol{X}\hat{\boldsymbol{B}})'(\boldsymbol{Y} - \boldsymbol{X}\hat{\boldsymbol{B}}) = \boldsymbol{Y}'(\boldsymbol{I} - \boldsymbol{X}(\boldsymbol{X}'\boldsymbol{X})^- \boldsymbol{X}')\boldsymbol{Y},$$

and

$$\begin{aligned} \boldsymbol{Q}_* &= \boldsymbol{Y}'(\boldsymbol{I} - (\boldsymbol{X}, \boldsymbol{Z}))\begin{pmatrix} \boldsymbol{X}'\boldsymbol{X} & \boldsymbol{X}'\boldsymbol{Z} \\ \boldsymbol{Z}'\boldsymbol{X} & \boldsymbol{Z}'\boldsymbol{Z} \end{pmatrix}^- \begin{pmatrix} \boldsymbol{X}' \\ \boldsymbol{Z}' \end{pmatrix}\boldsymbol{Y} \\ &= \boldsymbol{Y}'\boldsymbol{Y} - \boldsymbol{Y}'\boldsymbol{X}(\boldsymbol{X}'\boldsymbol{X})^- \boldsymbol{X}'\boldsymbol{Y} - \boldsymbol{Y}'\boldsymbol{X}(\boldsymbol{X}'\boldsymbol{X})^- \boldsymbol{X}'\boldsymbol{Z}\boldsymbol{G}^- \boldsymbol{Z}'\boldsymbol{X}(\boldsymbol{X}'\boldsymbol{X})^- \boldsymbol{X}'\boldsymbol{Y} \\ &\quad - \boldsymbol{Y}'\boldsymbol{X}(\boldsymbol{X}'\boldsymbol{X})^- \boldsymbol{X}'\boldsymbol{Z}\boldsymbol{G}^- \boldsymbol{Z}'\boldsymbol{Y} - \boldsymbol{Y}'\boldsymbol{Z}\boldsymbol{G}^- \boldsymbol{Z}'\boldsymbol{X}(\boldsymbol{X}'\boldsymbol{X})^- \boldsymbol{X}'\boldsymbol{Y} + \boldsymbol{Y}'\boldsymbol{Z}\boldsymbol{G}^- \boldsymbol{Z}'\boldsymbol{Y} \\ &= \boldsymbol{Y}'(\boldsymbol{I} - \boldsymbol{X}(\boldsymbol{X}'\boldsymbol{X})^- \boldsymbol{X}')\boldsymbol{Y} - \boldsymbol{Y}'(\boldsymbol{I} - \boldsymbol{X}(\boldsymbol{X}'\boldsymbol{X})^- \boldsymbol{X}')\boldsymbol{Z}\boldsymbol{G}^- \boldsymbol{Z}'(\boldsymbol{I} - \boldsymbol{X}(\boldsymbol{X}'\boldsymbol{X})^- \boldsymbol{X}')\boldsymbol{Y}. \end{aligned}$$

That is

$$\boldsymbol{Q}_* = \boldsymbol{Q} - \hat{\Gamma}'\boldsymbol{G}\hat{\Gamma}$$

where $\hat{\Gamma}$ and $\boldsymbol{G}$ are given by (6.2.12). When $l=1$, $\boldsymbol{Z}=\boldsymbol{z}$, $\Gamma=\gamma$, putting $d_z = \boldsymbol{z}'(\boldsymbol{I}$

$- X(X'X)^- X')z$, then $Q_* = Q - d_z\hat{\gamma}\hat{\gamma}'$.

(2) To add observations.

Consider models

$$\mathcal{E}(Y) = XB, \qquad \mathcal{D}(\text{vec}\, Y) = \Sigma \otimes I_n \tag{6.2.10}$$

and

$$\mathcal{E}\begin{pmatrix} Y \\ Y_* \end{pmatrix} = \begin{pmatrix} X \\ X_* \end{pmatrix} B_*, \qquad \mathcal{D}\left(\text{vec}\begin{pmatrix} Y \\ Y_* \end{pmatrix}\right) = \Sigma \otimes I_{n+n_*}, \tag{6.2.13}$$

where Y_*: $n_* \times p$.

It means that the matrix Y_* is the additional observations of the response variables $y_1, \ldots, y_p$. Let $\hat{B}$ and $\hat{B}_*$ be the LSE (least square estimate) for B and B_* respectively. Then the relation between them is

$$\hat{B}_* = \hat{B} + (X'X)^{-1}X_*'(I + X_*(X'X)^{-1}X_*')^{-1}(Y_* - X_*\hat{B}) \tag{6.2.14}$$

if $\text{rk}X = k$. This is because

$$\begin{aligned}\hat{B}_* &= (X'X + X_*'X_*)^{-1}(X'Y + X_*'Y_*)\\ &= [I - (X'X)^{-1}X_*'(I + X_*(X'X)^{-1}X_*')^{-1}X_*](X'X)^{-1}(X'Y + X_*'Y_*).\end{aligned}$$

Notice that $\hat{B} = (X'X)^{-1}X'Y$. Then we get (6.2.14).

It means that if we use the estimate $\hat{B}$ for B, then the difference $Y_* - X_*\hat{B}$ gives some additional information about parameters B, and We must adjust the estimate from B to $\hat{B}_*$.

Consider the relationship between Q and Q_*. Note that

$$\begin{aligned}(Y', Y_*')\begin{pmatrix} X \\ X_* \end{pmatrix}\hat{B}_* &= Y'X\hat{B} + Y'X(X'X)^{-1}X_*'(I + X_*(X'X)^{-1}X_*')^{-1}(Y_* - X_*\hat{B})\\ &\quad + Y_*'X_*\hat{B} + Y_*'X_*(X'X)^{-1}X_*'(I + X_*(X'X)^{-1}X_*')^{-1}(Y_* - X_*\hat{B})\\ &= Y'X\hat{B} + \hat{B}'X_*'(I + X_*(X'X)^{-1}X_*')^{-1}(Y_* - X_*\hat{B}) + Y_*'Y_*\\ &\quad - Y_*'(I + X_*(X'X)^{-1}X_*')^{-1}(Y_* - X_*\hat{B}).\end{aligned}$$

We get, by (6.2.13),

$$Q_* = Q + (Y_* - X_*\hat{B})'(I + X_*(X'X)^{-1}X_*')^{-1}(Y_* - X_*\hat{B}),$$

where Q_* depends on the original data only through $(X'X)^{-1}$, $\hat{B}$ and Q. It is convenient to perform the calculation on a computer.

(3) To add response variables.

Consider models

$$\mathcal{E}(Y) = XB, \qquad \mathcal{D}(\text{vec}\, Y) = \Sigma \otimes I_n \tag{6.2.10}$$

and

$$\mathcal{E}(Y, Z) = X(B_*, \Gamma), \qquad \mathcal{D}(\text{vec}(Y, Z)) = \begin{pmatrix} \Sigma & C \\ C' & \Sigma_* \end{pmatrix} \otimes I_n, \tag{6.2.15}$$

where Y: $n \times p$, Z: $n \times q$, B: $p \times p$, Γ: $p \times q$ and X: $n \times p$.

It means that there are q additional response variables $z_1, \ldots, z_q$ and Z is the data of these z_is. Let $\hat{B}$ and $(\hat{B}_*, \hat{\Gamma})$ be the LSE corresponding to B and (B_*, Γ) respectively.

Then the relations between them is

$$\hat{\boldsymbol{B}}_* = \hat{\boldsymbol{B}} \text{ and } \hat{\Gamma} = (\boldsymbol{X}'\boldsymbol{X})^- \boldsymbol{X}'\boldsymbol{Z}. \tag{6.2.16}$$

The relation between the recurrences $\boldsymbol{Q}$ and $\boldsymbol{Q}_*$ is

$$\boldsymbol{Q}_* = \begin{pmatrix} \boldsymbol{Y}' \\ \boldsymbol{Z}' \end{pmatrix} (\boldsymbol{I} - \boldsymbol{X}(\boldsymbol{X}'\boldsymbol{X})^- \boldsymbol{X}')(\boldsymbol{Y},\boldsymbol{Z})$$

$$= \begin{pmatrix} \boldsymbol{Q} & (\boldsymbol{Y} - \boldsymbol{X}\hat{\boldsymbol{B}})'\boldsymbol{Z} \\ \boldsymbol{Z}'(\boldsymbol{Y} - \boldsymbol{X}\hat{\boldsymbol{B}}) & \boldsymbol{Z}'(\boldsymbol{I} - \boldsymbol{X}(\boldsymbol{X}'\boldsymbol{X})^- \boldsymbol{X}')\boldsymbol{Z} \end{pmatrix}.$$

6.3. Variance Components

Let $L(\boldsymbol{y};\ \boldsymbol{X}, \Theta)$ be a linear model with

$$\Theta = \left\{ \sum_{i=1}^{k} \theta_i \boldsymbol{V}_i \,\middle|\, \sum_{i=1}^{k} \theta_i \boldsymbol{V}_i \geqslant 0 \right\} \tag{6.3.1}$$

where $\boldsymbol{V}_1, \ldots, \boldsymbol{V}_k$ are known symmetric matrices. Now we are interested in estimating the linear functions of θ_i's. These θ's are called the *variance components*.

6.3.1. Least Square Method

Definition 6.3.1. Let $L(\boldsymbol{y};\ \boldsymbol{X}, \Theta)$ be a linear model with Θ having the form (6.3.1). The parameter $\boldsymbol{a}'\theta(= \sum_{i=1}^{k} a_i\theta_i)$ is said to be *quadratic estimable or q-estimable*, if there exists a quadratic function $\boldsymbol{y}'\boldsymbol{A}\boldsymbol{y}$ such that $E(\boldsymbol{y}'\boldsymbol{A}\boldsymbol{y}) = \boldsymbol{a}'\theta$, and $\boldsymbol{y}'\boldsymbol{A}\boldsymbol{y}$ is a QUE (*quadratic unbiased estimate*) of $\boldsymbol{a}'\theta$. Similarly we can define the BQUE (minimum variance within all QUE) of each q-estimable $\boldsymbol{a}'\theta$ if it exists.

Notice that $\boldsymbol{y}'\boldsymbol{A}\boldsymbol{y} = \text{tr}\boldsymbol{A}\boldsymbol{y}\boldsymbol{y}'$, and $\text{tr}\boldsymbol{A}'\boldsymbol{B}$ (where $\boldsymbol{A}$ and $\boldsymbol{B}$ are $n \times n$ matrices) is the inner product in the space which is formed by all $n \times n$ matrices. Then each quadratic form $\boldsymbol{y}'\boldsymbol{A}\boldsymbol{y}$ can be rewritten as a linear function of the matrix $\boldsymbol{y}'\boldsymbol{y}$. Now

$$\mathcal{E}(\boldsymbol{y}\boldsymbol{y}') = \mathcal{D}(y) + (\mathcal{E}(y))(\mathcal{E}(y))' \tag{6.3.2}$$

$$= \sum_{i=1}^{k} \theta_i V_i + \boldsymbol{X}\boldsymbol{B}\boldsymbol{B}'\boldsymbol{X}'.$$

Put $\eta = \text{vec}\boldsymbol{B}\boldsymbol{B}'$. Then

$$\mathcal{E}(\text{vec}\,\boldsymbol{y}\boldsymbol{y}') = (\text{vec}\boldsymbol{V}_1, \ldots, \text{vec}\boldsymbol{V}_k)\theta + (\boldsymbol{X} \otimes \boldsymbol{X})\eta \tag{6.3.3}$$

$$\hat{=} \boldsymbol{W}\theta + (\boldsymbol{X} \otimes \boldsymbol{X})\eta$$

$$= (\boldsymbol{W}, \boldsymbol{X} \otimes \boldsymbol{X}) \begin{pmatrix} \theta \\ \eta \end{pmatrix}.$$

It is a linear model $L(\text{vec}\,\boldsymbol{y}\boldsymbol{y}'; (\boldsymbol{W}, \boldsymbol{X} \otimes \boldsymbol{X}), \Omega)$ for $\mathcal{D}(\text{vec}\,\boldsymbol{y}\boldsymbol{y}')$. also, where Ω is a collection of all matrices for $\mathcal{D}(\text{vec}\,\boldsymbol{y}\boldsymbol{y}')$. Each QUE of the q-estimable $\boldsymbol{a}'\theta$ is the LUE of $\boldsymbol{a}'\theta$ in the new model. Hence we can obtain the following results easily by using the results of Section 6.2.

(1) $\boldsymbol{a}'\theta$ is q-estimable iff

(6.3.4) $$W'(\mathrm{vec}A) = a, \qquad X'AX = O$$

or

$$\mathrm{tr}V_iA = a_i, \quad i = 1, 2, \ldots, k; \qquad X'AX = O.$$

From (6.3.4) and Theorem 6.2.2 we obtain all QUE for the q-estimable $a'\theta$, that is, $\{y'Ay \mid A \in Q(a'\theta)\}$, where

(6.3.5) $$Q(a'\theta) = \{PB + B'P \mid B \text{ satisfies } 2\ tr\ V_iPB = a_i, \quad i = 1, 2, \ldots, k\}$$

and

$$P = I - XX^+.$$

(2) For the linear model (6.3.3), the LSE of θ is

(6.3.6) $$\hat{\theta}_* = \hat{\theta} - (W'W)^- W'(X' \otimes X')\hat{\eta}, \qquad \hat{\eta} = G^-(X' \otimes X')(\mathrm{vec}\ yy' - W\hat{\theta}),$$

where $\hat{\theta} = (W'W)^- W'(\mathrm{vec}\ yy')$ and $G = (X' \otimes X')(I - W(W'W)^- W')(X \otimes X)$.

Now the normal equation is

$$\begin{pmatrix} W'W & W'(X \otimes X) \\ (X' \otimes X')W & X'X \otimes X'X \end{pmatrix} \begin{pmatrix} \hat{\theta} \\ \hat{\eta} \end{pmatrix} = \begin{pmatrix} W' \\ X' \otimes X' \end{pmatrix} (\mathrm{vec}\ yy').$$

By a similar method as that used for (6.2.12), we can obtain (6.3.6).

6.3.2. Invariant QUE (IQUE)

Let $L(y;\ X, \Theta)$ be a linear model with (6.3.1). We consider the translation group $\{\alpha \mid \alpha(y) \to y + X\alpha\}$. It is obvious that $\mathscr{D}(y) = \mathscr{D}(\alpha(y))$. Hence the q-estimate of $a'\theta$ must be invariant under this group.

Definition 6.3.2. $y'Ay$ is said to be an *invariant QUE* (*or IQUE*) for estimable $a'\theta$, if $y'Ay = (y + X\alpha)'A(y + X\alpha)$ for all α.

Lemma 6.3.1. *$y'Ay$ is an IQUE iff $y'Ay$ is a QUE and $X'A = O$.*

The proof is left as an exercise (cf. Exercise 6.6).

Thus $y'Ay$ is an IQUE iff $A = PBP$, $B' = B$ and $\mathrm{tr}PV_iPB = a_i$, $i = 1, 2, \ldots, k$, where $P = I - XX^+$.

Let $z = Py$. The linear model $L(y;\ X, \Theta)$ induces a linear model $L(z;\ X, \Theta_*)$, where

$$X_* = PX,$$

$$\Theta_* = \{PVP' \mid V \in \Theta\} = \{PVP \mid V \in \Theta\},$$

and

$$\Theta_* = \left\{ \sum_{i=1}^{k} \theta_i PV_iP \,\middle|\, \sum_{i=1}^{k} \theta_i PV_iP \geq 0 \right\}$$

if Θ is defined by (6.3.1). It is obvious that each QUE of $a'\theta$ for $L(z;\ X_*, \Theta_*)$ is an IQUE of $a'\theta$ for $L(y;\ X, \Theta)$. Hence we can obtain the results about IQUE from that of QUE.

6.3.3. MINQUE

C.R. Rao (1971) suggested a method to estimate the parameter $a'\theta$ for $L(y;\ X, \Theta)$ with (6.3.1). This method is called the *MINQUE method.* We consider the simple case.

Definition 6.3.3. $y'Ay$ is said to be a *MINQUE for estimable* $a'\theta$, if A satisfies

$$(6.3.7)\qquad \begin{cases} X'AX = 0, \quad \mathrm{tr} V_i A = a_i, \quad i = 1, \ldots, k \\ \text{and } \mathrm{tr} AUA(U + 2XX') = \min \end{cases}$$

where $U = \sum_{i=1}^{k} V_i$.

Definition 6.3.4. A MINQUE $y'Ay$ is said to be invariant, if it is an invariant QUE. Then by (6.3.7) we obtain

Lemma 6.3.2. $y'Ay$ *is an IMINQUE (invariant MINQUE) for* $a'\theta$ *iff* A *satisfies*

$$(6.3.8)\qquad \begin{cases} AX = 0, \quad \mathrm{tr} V_i A = a_i, \quad i = 1, 2, \ldots, k \\ \text{and } \mathrm{tr} AUAU = \min \end{cases}$$

where $U = \sum_{i=1}^{k} V_i$.

Theorem 6.3.1. $y'Ay$ *is an IMINQUE for estimable* $a'\theta$ *iff*

$$(6.3.9)\qquad A = \sum_{i=1}^{k} \lambda_i P U_*^+ V_{*i} U_*^+ P + P(H - U_*^+ U_* H U_* U_*^+)P.$$

Here $\lambda_1, \ldots, \lambda_k$ *are the solutions of the equation*

$$\sum_{j=1}^{k} \lambda_j \mathrm{tr}(V_{*i} U_*^+ V_{*j} U_*^+) = a_i + \mathrm{tr}\ V_{*i}(U_*^+ U_* H U_* U_*^+ - H), \qquad i = 1, 2, \ldots, k,$$

$V_{*i} = PV_iP$, $U_* = PUP$, *and* H *is an arbitrary* $n \times n$ *symmetric matrix.*

Proof. (6.3.8) is equivalent to

$$\begin{cases} A = PBP; \quad \mathrm{tr} PV_iPB = a_i,\ i = 1, 2, \ldots, k \\ \text{and } \mathrm{tr} BPUPBPUP = \min. \end{cases}$$

Put

$$U_* = \sum_{i=1}^{k} PV_iP = PUP, \qquad V_{*i} = PV_iP, \qquad i = 1, 2, \ldots, k$$

and

$$g(B) = \mathrm{tr} BU_*BU_* - 2\sum_{i=1}^{k} \lambda_i \mathrm{tr}(V_{*i}B - a_i).$$

$$0 = d\,g(B) = \mathrm{tr}[(dB)U_*BU_* + BU_*(dB)U_* - 2\sum_{i=1}^{k} \lambda_i V_{*i}(dB)]$$

$$= \mathrm{tr}(2U_*BU_* - 2\sum_{i=1}^{k} \lambda_i V_{*i})(dB).$$

Hence B is the solution of

$$(6.3.10)\qquad U_*BU_* = \sum_{i=1}^{k} \lambda_i V_{*i}$$

where $\lambda_1, \ldots, \lambda_k$ are determined by $\mathrm{tr} V_{*i}B = a_i$, $i = 1, 2, \ldots, k$. The general solution for (6.3.10) is

$$B = U_*^+\left(\sum_{i=1}^{k} \lambda_i V_{*i}\right)U_*^+ + H - U_*^+ U_* H U_* U_*^+$$

where H is an arbitrary symmetric $n \times n$ matrix. Hence $\lambda_1, \ldots, \lambda_k$ are determined by

$$\sum_{j=1}^{k} \lambda_j \mathrm{tr}(V_{*i} U_*^+ V_{*j} U_*^+) + \mathrm{tr}(V_{*i} H - V_{*i} U_*^+ U_* H U_* U_*^+) = a_i, \qquad i = 1, \quad \ldots, \quad k,$$

which is equivalent to (6.3.9). □

We can use the same method to get the MINQUE.

6.4. Hypothesis Testing

Let $L(Y; X, \Theta)$ be a linear model. Now we consider the hypothesis testing problem related with the parameter matrix B. Assume that the distribution of Y belongs to the matrix elliptically contoured distributions and it is denoted by ELSL$(Y; X, \Theta)$ (ELS linear), EMSL$(Y; X, \Theta)$ (EMS linear), EVSL $(Y; X, \Theta)$ (EVS linear), or ESSL $(Y; X, \Theta)$ (cf. Chapter III). Throughout this section it is assumed that the density function of the random matrix Y exists. Thus we can use the results related to the distribution of some statistics as those derived in Chapters II and III.

6.4.1. Linear Hypothesis

Suppose ELSL$(Y; X, \Theta)$, $X: n \times k$, is given with parameter matrix B and consider the linear hypothesis

$$H_0 : AB = C, \text{ where } A : l \times k,\ B : k \times p,\ C : k \times p.$$

It is easy to reject H_0 if $AB = C$ is inconsistent. Hence in general we can assume that $AB = C$ is consistent. The general solution of $AB = C$ is

$$B = A^+C + (I - A^+A)\Gamma$$

where Γ is an arbitrary matrix of appropriate size.

Let $Z = Y - XA^+C$. Then $L(Z; X(I - A^+A), \Theta)$ is a linear model of the same structure as ELSL$(Y; X, \Theta)$. Now

$$\begin{aligned}\mathcal{E}(Z) &= X(B - A^+C) = XA^+A(B - A^+C) + X(I - A^+A)(B - A^+C) \\ &\stackrel{\Lambda}{=} XT_1T_1'(B - A^+C) + XT_2T_2'(B - A^+C)\end{aligned}$$

where $T = (T_1\ T_2)$ is a $k \times k$ orthogonal matrix. Put

$$\Gamma = \begin{pmatrix}\Gamma_1 \\ \Gamma_2\end{pmatrix} = T'(B - A^+C) = \begin{pmatrix}T_1' \\ T_2'\end{pmatrix}(B - A^+C)$$

and

$$X_* = (X_{*1} X_{*2}) = (XT_1\ XT_2).$$

Then

$$\mathcal{E}(Z) = X_{*1}\Gamma_1 + X_{*2}\Gamma_2,$$

and

$$H_0 : AB = C$$

is equivalent to

$$H_0 : \Gamma_1 = O.$$

This shows that it is enough to consider the hypothesis $H_0 : \Gamma_1 = O.$ for the ELSL$(Y; X, \Theta)$.

6.4.2. Canonical Form

We use the following lemma to get the canonical form of a linear model. Because the proof is easy, we leave it as an exercise (Exercise 6.7).

Lemma 6.4.1. *Given a matrix* $\boldsymbol{X}$, *there exists an orthogonal matrix* $\boldsymbol{G} \in O(n)$ *and a nonsingular matrix* $\boldsymbol{P} \in GL(p)$ *such that*

$$\boldsymbol{GXP} = \begin{pmatrix} \boldsymbol{I}_r & \boldsymbol{O} \\ \boldsymbol{O} & \boldsymbol{O} \end{pmatrix} \tag{6.4.1}$$

where $r = \text{rk}(\boldsymbol{X})$.

For ELSL($\boldsymbol{Y}$; $\boldsymbol{X}$, Θ), transform $\boldsymbol{Y}$ to $\boldsymbol{Z}$ defined by

$$\boldsymbol{Z} = \boldsymbol{GY}$$

where $\boldsymbol{G}$ is the orthogonal matrix in (6.4.1). Then $\boldsymbol{Z} - \boldsymbol{GXB} \overset{d}{=} \boldsymbol{Y} - \boldsymbol{XB}$ and $\boldsymbol{Z}$ is an ELSL($\boldsymbol{Z}$; $\boldsymbol{GX}$, Θ) also. Let $\eta = \boldsymbol{P}^{-1}\boldsymbol{B}$, where $\boldsymbol{P}$ is the nonsingular matrix $\boldsymbol{P} \in GL(p)$ in (6.4.1). Then

$$\mathcal{E}(\boldsymbol{Z}) = \mathcal{E}(\boldsymbol{GY}) = \boldsymbol{GXB} = \boldsymbol{GXPP}^{-1}\boldsymbol{B}$$

$$= \begin{pmatrix} \boldsymbol{I}_r & \boldsymbol{O} \\ \boldsymbol{O} & \boldsymbol{O} \end{pmatrix}\eta = \begin{pmatrix} \mu \\ \boldsymbol{O} \end{pmatrix},$$

where $r = \text{rk}(\boldsymbol{X})$ and μ is the first r rows of η.

It means that for ELSL($\boldsymbol{Y}$; $\boldsymbol{X}$, Θ), the canonical form is $\boldsymbol{X} = (\boldsymbol{I}_r\ \boldsymbol{O})'$ and

$$\begin{cases} \mathcal{E}(\boldsymbol{Y}) = \begin{pmatrix} \boldsymbol{I}_r \\ \boldsymbol{O} \end{pmatrix}\boldsymbol{B} = \begin{pmatrix} \boldsymbol{B}_1 \\ \boldsymbol{B}_2 \\ \boldsymbol{O} \end{pmatrix} \\ \mathcal{D}(\text{vec}\,\boldsymbol{Y}) = \Sigma \otimes \boldsymbol{I}_n, \qquad \Sigma > 0 \end{cases} \tag{6.4.2}$$

where $\boldsymbol{B}_1$: $s \times p$, $\boldsymbol{B}_2$: $(r-s) \times p$ and $\boldsymbol{Y}$: $n \times p$.

Consider the linear hypothesis H_0: $\boldsymbol{B}_1 = \boldsymbol{O}$. Divide $\boldsymbol{Y}$ into three parts $(\boldsymbol{Y}_1',\ \boldsymbol{Y}_2',\ \boldsymbol{Y}_3')'$ such that

$$\mathcal{E}(\boldsymbol{Y}_1) = \boldsymbol{B}_1, \qquad \mathcal{E}(\boldsymbol{Y}_2) = \boldsymbol{B}_2, \qquad \mathcal{E}(\boldsymbol{Y}_3) = \boldsymbol{O},$$

where $\boldsymbol{Y}_1$: $s \times p$, $\boldsymbol{Y}_2$: $(r-s) \times p$ and $\boldsymbol{Y}_3$: $(n-r) \times p$.

Consider the group

$$G = \{(\boldsymbol{G}_1, \boldsymbol{N}, \boldsymbol{G}_3, \boldsymbol{T}) | \boldsymbol{G}_1 \in O(s),\ G_3 \in O(n-r),\ \boldsymbol{N}: (r-s) \times p,\ \boldsymbol{T} \in GL(p)\},$$

acting as

$$(\boldsymbol{G}_1, \boldsymbol{N}, \boldsymbol{G}_3, \boldsymbol{T})(\boldsymbol{Y}_1, \boldsymbol{Y}_2, \boldsymbol{Y}_3) = (\boldsymbol{G}_1\boldsymbol{Y}_1\boldsymbol{T}, \boldsymbol{Y}_2\boldsymbol{T} + \boldsymbol{N}, \boldsymbol{G}_3\boldsymbol{Y}_3\boldsymbol{T}). \tag{6.4.3}$$

It is obvious that the testing problem is invariant under the group G, From (6.4.2) G is transitive on $\boldsymbol{Y}_2$ so that $\boldsymbol{Y}_2$ can be transformed to **O**. Hence the maximal invariants are functions of $\boldsymbol{Y}_1$ and $\boldsymbol{Y}_3$ only.

Lemma 6.4.2. *The maximal invariants of G are the nonzero latent roots of* $\boldsymbol{Y}_1(\boldsymbol{Y}_3'\boldsymbol{Y}_3)^{-1}\boldsymbol{Y}_1'$, *if* $n - r \geqslant p$.

Proof. $\boldsymbol{G}_1\boldsymbol{Y}_1\boldsymbol{T}(\boldsymbol{T}'\boldsymbol{Y}_3'\boldsymbol{G}_3'\boldsymbol{G}_3\boldsymbol{Y}_3\boldsymbol{T})^{-1}\boldsymbol{T}'\boldsymbol{Y}_1'\boldsymbol{G}_1 = \boldsymbol{G}_1\boldsymbol{Y}_1(\boldsymbol{Y}_3'\boldsymbol{Y}_3)^{-1}\boldsymbol{Y}_1'\boldsymbol{G}'$ has the same nonzero latent roots as $\boldsymbol{Y}_1(\boldsymbol{Y}_3'\boldsymbol{Y}_3)^{-1}\boldsymbol{Y}_1'$. Hence it is invariant.

Suppose $\boldsymbol{Y}_1(\boldsymbol{Y}_3'\boldsymbol{Y}_3)^{-1}\boldsymbol{Y}_1'$ and $\boldsymbol{Z}_1(\boldsymbol{Z}_3'\boldsymbol{Z}_3)^{-1}\boldsymbol{Z}_1'$ have the same nonzero latent roots $f_1, \ldots, f_t$ (say). Then there exist s × s orthogonal matrices $\boldsymbol{H}_1$ and $\boldsymbol{H}_2$ such that

$$\boldsymbol{H}_1\boldsymbol{Y}_1(\boldsymbol{Y}_3'\boldsymbol{Y}_3)^{-1}\boldsymbol{Y}_1'\boldsymbol{H}_1' = \boldsymbol{H}_2\boldsymbol{Z}_1(\boldsymbol{Z}_3'\boldsymbol{Z}_3)^{-1}\boldsymbol{Z}_1'\boldsymbol{H}_2' = \operatorname{diag}(f_1, \ldots, f_t, 0, \ldots, 0).$$

Thus

$$\boldsymbol{Y}_1(\boldsymbol{Y}_3'\boldsymbol{Y}_3)^{-1}\boldsymbol{Y}_1' = \boldsymbol{H}\boldsymbol{Z}_1(\boldsymbol{Z}_3'\boldsymbol{Z}_3)^{-1}\boldsymbol{Z}_1'\boldsymbol{H}'$$

where $\boldsymbol{H} = \boldsymbol{H}_1'\boldsymbol{H}_2$ is an orthogonal matrix. Hence there exists an orthogonal matrix $\boldsymbol{H}_3$ such that

$$(\boldsymbol{Y}_3'\boldsymbol{Y}_3)^{-1/2}\boldsymbol{Y}_1' = \boldsymbol{H}_3(\boldsymbol{Z}_3'\boldsymbol{Z}_3)^{-1/2}\boldsymbol{Z}_1'\boldsymbol{H}_1'$$

i.e.,

$$\boldsymbol{Y}_1 = \boldsymbol{H}\boldsymbol{Z}_1\boldsymbol{T}$$

where $\quad \boldsymbol{T} = (\boldsymbol{Z}_3'\boldsymbol{Z}_3)^{-1/2}\boldsymbol{H}_3'(\boldsymbol{Y}_3'\boldsymbol{Y}_3)^{1/2} \quad$ and $\quad \boldsymbol{T}(\boldsymbol{Y}_3'\boldsymbol{Y}_3)^{-1}\boldsymbol{T}' = (\boldsymbol{Z}_3'\boldsymbol{Z}_3)^{-1} \quad$ or $\boldsymbol{T}'^{-1}(\boldsymbol{Y}_3'\boldsymbol{Y}_3)\boldsymbol{T}^{-1} = \boldsymbol{Z}_3'\boldsymbol{Z}_3$.

Hence there exists an orthogonal matrix H_4 such that

$$\boldsymbol{Y}_3\boldsymbol{T}^{-1} = \boldsymbol{H}_4\boldsymbol{Z}_3$$

i.e.,

$$\boldsymbol{Y}_3 = \boldsymbol{H}_4\boldsymbol{Z}_3\boldsymbol{T}.$$

Hence $(\boldsymbol{Y}_1, \boldsymbol{Y}_3)$ is equivalent to $(\boldsymbol{Z}_1, \boldsymbol{Z}_3)$ under the group G. □

By this lemma any invariant test is a function of the nonzero latent roots $f_1, \ldots, f_t$ of matrix $\boldsymbol{Y}_1(\boldsymbol{Y}_3'\boldsymbol{Y}_3)^{-1}\boldsymbol{Y}_1'$, and the distribution of $f_1, \ldots, f_t$ depends only on the nonzero latent roots of $\boldsymbol{B}_1\Sigma^{-1}\boldsymbol{B}_1'$ (See Exercise 6.9).

By noting that when $r = s = 1$, $\boldsymbol{Y}_1 = \boldsymbol{y}_1'$ and $\boldsymbol{B}_1 = \boldsymbol{\beta}_1'$ are both reduced to $1 \times p$ vector,

$$f_1 = \boldsymbol{y}_1'(\boldsymbol{Y}_3'\boldsymbol{Y}_3)^{-1}\boldsymbol{y}_1$$

is a multiple of Hotelling's T^2. In general, there are many invariant tests and there is no uniformly most powerful invariant test. We list some statistics as follows:

(a) Wilks (1932)

$$W(f_1, \ldots, f_t) = \prod_{i=1}^{t}(1 + f_i)^{-1} = |I + \boldsymbol{Y}_1(\boldsymbol{Y}_3'\boldsymbol{Y}_3)^{-1}\boldsymbol{Y}_1'|^{-1}$$

which can be obtained by the maximum likelihood ratio method (Anderson and Fang, 1982c).

(b) Lawley (1938)

$$V(f_1, \ldots, f_t) = \sum_{i=1}^{t} f_i = \operatorname{tr}(\boldsymbol{Y}_3'\boldsymbol{Y}_3)^{-1}\boldsymbol{Y}_1'\boldsymbol{Y}_1.$$

(c) Pillai (1955)

$$V(f_1, \ldots, f_t) = \sum_{i=1}^{t}(f_i/(1 + f_i)) = \operatorname{tr}\boldsymbol{Y}_1'\boldsymbol{Y}_1(\boldsymbol{Y}_1'\boldsymbol{Y}_1 + \boldsymbol{Y}_3'\boldsymbol{Y}_3)^{-1}.$$

(d) Roy (1957)

$$f_{\max} = \max_{1 \leqslant i \leqslant t} f_i.$$

(e) Anderson (1958)

$$f_{\min} = \min_{1 \leqslant i \leqslant t} f_1.$$

(f) Gnanadeslkan (1965)

$$U = \prod_{i=1}^{t} f_i/(1+f_i).$$

(g) Olson (1974)

$$S = \prod_{i=1}^{t} f_i.$$

(h) Zhang (1977)

$$\rho = \sum_{i=1}^{t} (1+f_i)^{-1} = \operatorname{tr}(\boldsymbol{Y}_1'\boldsymbol{Y}_1 + \boldsymbol{Y}_3'\boldsymbol{Y}_3)^{-1}\boldsymbol{Y}_3'\boldsymbol{Y}_3.$$

We have the random decomposition

$$\begin{pmatrix} \boldsymbol{Y}_1 \\ \boldsymbol{Y}_2 \\ \boldsymbol{Y}_3 \end{pmatrix} \stackrel{d}{=} \begin{pmatrix} \boldsymbol{U}_1 \\ \boldsymbol{U}_2 \\ \boldsymbol{U}_3 \end{pmatrix} \boldsymbol{T} \stackrel{d}{=} \boldsymbol{UT}$$

where $\boldsymbol{U}$: $n \times p$ is distributed uniformly on the Stiefel manifold $\{\boldsymbol{U} \mid \boldsymbol{U}'\boldsymbol{U} = \boldsymbol{I}_p, \boldsymbol{U}: n \times p\}$ and is independent of $\boldsymbol{T}$. Hence

$$\boldsymbol{Y}_1(\boldsymbol{Y}_3'\boldsymbol{Y}_3)^{-1}\boldsymbol{Y}_1' \stackrel{d}{=} \boldsymbol{U}_1\boldsymbol{T}(\boldsymbol{T}'\boldsymbol{U}_3'\boldsymbol{U}_3\boldsymbol{T})^{-1}\boldsymbol{T}'\boldsymbol{U}_1' \stackrel{d}{=} \boldsymbol{U}_1(\boldsymbol{U}_3'\boldsymbol{U}_3)^{-1}\boldsymbol{U}_1'.$$

So the joint distribution of $f_1, \ldots, f_t$ is the same as that in the normal distribution, and the results can be found in most multivariate statistical books. Hence all of the above statistics are distribution-free (See Section 5.1).

The examples of testing linear hypothese will be given in the next section, Section 6.5.)

6.4.3. Pre-Test Estimates and James-Stein Estimates

Now consider the univariate linear model, i.e., $p = 1$, $L(\boldsymbol{y}; \boldsymbol{X}, \Theta)$, where $\boldsymbol{y} \sim EC_n(\boldsymbol{X}\beta, \boldsymbol{I}_n, \phi)$, and

(6.4.4)
$$\begin{cases} \mathcal{E}(\boldsymbol{y}) = \boldsymbol{X}\beta, & \boldsymbol{X}: n \times k, \\ \mathcal{D}(\boldsymbol{y}) = \sigma^2 \boldsymbol{I}_n, & \sigma > 0. \end{cases}$$

For testing the hypothesis

(6.4.5) $\qquad H_0$: $\beta = \mathbf{0}$

the UMP invariant test is F, where

(6.4.6)
$$F = (\hat{\beta}'(\boldsymbol{X}'\boldsymbol{X})\hat{\beta}/S^2)\left(\frac{n-r}{r}\right), \quad r = \operatorname{rk}(\boldsymbol{X})$$

$$S^2 = \boldsymbol{y}'\boldsymbol{y} - \hat{\beta}'(\boldsymbol{X}'\boldsymbol{X})\hat{\beta},$$

which had been proved in the normal case. By 6.4.2, the statistic F **has the same distribution as** $\boldsymbol{y}$ which is normal.

When $\beta = \mathbf{0}$ is accepted, it is obvious that the estimate of β is $\mathbf{0}$; if $\beta = 0$ is not accepted the estimate of β is the least square estimate $\hat{\beta} = (\boldsymbol{X}'\boldsymbol{X})^{-}\boldsymbol{X}'\boldsymbol{y}$. Hence the pre-test estimate of β has the following form:

$$\hat{\beta}_{\text{pre}} = \alpha I_{(F<c)} \cdot \mathbf{0} + (1-\alpha) I_{(F \geqslant c)} \hat{\beta}$$

where I_A is the characteristic function of the region A, c is determined by the significant level and α is the weight of the estimate $\mathbf{0}$.

One may choose α as a constant. But if the value of F is much larger than the critical value, we prefer to use $\hat{\beta}$ rather than $\mathbf{0}$. So the weight α may be dependent on $\boldsymbol{y}$. The natural choice is

$$1-\alpha = \left(1-\frac{c}{F}\right)_+ = \begin{cases} 1-\dfrac{c}{F}, & F \geqslant c \\ 0, & F < c. \end{cases}$$

It means that the weight becomes larger when F is larger, if $F \geqslant c$ is true; and becomes zero, if $F < c$. As F tends to infinity, the weight for $\hat{\beta}$ tends to one. Then

$$\hat{\beta}_{\text{pre}} = \left(1-\frac{c}{F}\right)_+ \hat{\beta}. \tag{6.4.7}$$

Noticing (6.4.6), the $\hat{\beta}_{\text{pre}}$ is indeed a James–Stein estimate of $\hat{\beta}$. If σ^2 is known, and $\boldsymbol{X} = \boldsymbol{I}$, then

$$\left(1-\frac{c}{F}\right)_+ = \left(1-\frac{a}{\boldsymbol{y}'\boldsymbol{y}}\right)_+,$$

and

$$\hat{\beta}_{\text{pre}} = \left(1-\frac{a}{\boldsymbol{y}'\boldsymbol{y}}\right)_+ \boldsymbol{y}$$

is the Stein positive estimate of $\hat{\beta}$, which is better than the Stein estimate $(1-a/\boldsymbol{y}'\boldsymbol{y})\boldsymbol{y}$. The proof is left as an exercise (cf. Exercise 6.8).

6.5. Applications

6.5.1. Double Screening Stepwise Regression (DSSR Method)

There are many methods such as "optimal subsets regression", "the canonical regression" and the "stepwise regression" for obtaining a "best" regression equation relating one predicted variable to several predictors.

When the predicted variables become more and more, there are some relationships among these variables. It is necessary to suggest a method that can aggregate the predicted variables according to their correlations and can select the predictors to get a "best" regression system. The DSSR method was proposed by Zhang and Zhao (1980).

The predictors are denoted by $x_1, \dots, x_m$ and the predicted variables are denoted by $y_1, \dots, y_p$. The raw data are denoted by the following matrices:

$$\boldsymbol{X} = \begin{pmatrix} x_{11} & x_{12} & \cdots & x_{1m} \\ x_{21} & x_{22} & \cdots & x_{2m} \\ x_{n1} & x_{n2} & \cdots & x_{nm} \end{pmatrix}$$

and

$$Y = \begin{pmatrix} y_{11} & y_{12} & \cdots & y_{1p} \\ y_{21} & y_{22} & \cdots & y_{2p} \\ y_{n1} & y_{n2} & \cdots & y_{np} \end{pmatrix}.$$

Assume the model

$$\text{(6.5.1)} \qquad \begin{cases} \mathcal{E}(Y) = (\mathbf{1}_n\, X)\begin{pmatrix} \beta_0 \\ B \end{pmatrix}, \text{ where } Y\text{: } n \times p,\ X\text{: } n \times m, \text{ and } \operatorname{rk}(\mathbf{1}_n\, X) = m + 1 \\ \mathcal{D}(\operatorname{vec} Y) = \Sigma \otimes I_n,\ \Sigma > 0,\ Y \sim \text{ELSL}. \end{cases}$$

where "$Y \sim$ ELSL" means that in the linear model (6.5.1) Y is distributed as ELS(.,.,.,).

Let

$$\bar{x} = \frac{1}{n} X'\mathbf{1}, \qquad \bar{y} = \frac{1}{n} Y'\mathbf{1}$$

and

$$L_{xy} = X'(I - \frac{1}{n} J)\, Y, \text{ where } J = \mathbf{1}\mathbf{1}'$$

and L_{yx}, L_{xx} and L_{yy} are defined similarly. Then we get

$$\text{(6.5.2.)} \qquad \begin{cases} \hat{\beta} = L_{xx}^{-1} L_{xy}, & \hat{\beta}_0 = \bar{y}' - \bar{x}'\hat{B}, \\ \hat{\Sigma} = (n - m - 1)^{-1} Q, \end{cases}$$

where $Q = L_{yy} - L_{yx} L_{xx}^{-1} L_{xy} = L_{yy} - \hat{\beta}' L_{xx} \hat{B}$.

If we intend to add or to eliminate a predictor u, it is enough to consider the relation between the model (6.5.1) and the following expressions (6.5.3).

$$\text{(6.5.3)} \qquad \begin{cases} \mathcal{E}(Y) = (\mathbf{1}\, X\, u)\begin{pmatrix} \beta_0^* \\ B^* \\ \beta_u^* \end{pmatrix}, \operatorname{re}(\mathbf{1}\ X\ u) = m + 2, \\ \mathcal{D}(\operatorname{vec} Y) = \Sigma \otimes I_n,\ \Sigma > 0 \text{ and } Y \sim \text{ELSL}. \end{cases}$$

By (6.2.12), we get

$$\text{(6.5.4)} \qquad \begin{cases} \hat{\beta}_u^* = L_{uu}^{-1}(x) L_{uy}(x) \\ \hat{\beta}_0^* = \bar{y}' - \bar{x}'\hat{\beta}^* - \bar{u}\hat{\beta}_u^* \\ \hat{B}^* = \hat{B} - L_{xx}^{-1} L_{xu} \hat{\beta}_u^{*\prime} \end{cases}$$

where

$$L_{uy}(x) = L_{uy} - L_{ux} L_{xx}^{-} L_{xy},$$
$$L_{uu}(x) = L_{uu} - L_{ux} L_{xx}^{-} L_{xu}$$

and Q_*, which appeared in (6.5.2) for the model (6.5.3), and Q which appeared in (6.5.2) are related by

$$\text{(6.5.5)} \qquad Q_* = Q - L_{yu}(x) L_{uu}^{-1}(x) L_{uy}(x)$$

where

$$L_{yu}(x) = L_{yu} - L_{yx} L_{xx}^{-1} L_{xu}$$

and $L_{yy}(x)$ is defined similarly.

Then the statistics Λ and Λ^* —— the Wilks' statistics for the model (6.5.1) and (6.5.3) respectively —— are used to test the hypothesis: these x_i's are not related with

$y_i's$, i.e., all the regression coefficients are zero. Now

$$\frac{\Lambda^*}{\Lambda} = \frac{|\boldsymbol{Q}^*|/|\boldsymbol{L}_{yy}|}{|\boldsymbol{Q}|/|\boldsymbol{L}_{yy}|} = \frac{|\boldsymbol{L}_{yy}(x) - \boldsymbol{L}_{yu}(x)\boldsymbol{L}_{uu}^{-1}(x)\boldsymbol{L}_{uy}(x)|}{|\boldsymbol{L}_{yy}(x)|}$$
$$= 1 - V_u$$

where

$$V_u = \boldsymbol{L}_{uu}^{-1}(x)\boldsymbol{L}_{uy}(x)\boldsymbol{L}_{yy}^{-1}(x)\boldsymbol{L}_{yu}(x).$$

In fact, V_u is a multiple correlation coefficient, which is the "contribution" of u to y under the condition that we accept these $x_i's$ as predictors.

The statistic

(6.5.6) $$\frac{n-p-m-1}{p}\frac{V_u}{1-V_u} \sim F(p,\ n-p-m-1).$$

It is equivalent to a Hotelling T^2 indeed, and can be used to test H_0: $\beta_u^* = 0$.

If we intend to add or to eliminate a predicted variable, it is enough to consider the relationship between model (6.5.1) and the following expression (6.5.7)

(6.5.7) $$\begin{cases} \mathcal{E}(\boldsymbol{Y}\ \boldsymbol{z}) = (\boldsymbol{1}\ \boldsymbol{X})\begin{pmatrix} \beta_{OX} & \beta_{OZ} \\ \beta_X & \boldsymbol{B}_Z \end{pmatrix}, \\ \mathcal{D}(\text{vec}(\boldsymbol{Y}\ \boldsymbol{z})) = \begin{pmatrix} \Sigma & \Sigma_{yz} \\ \Sigma_{zy} & \Sigma_{zz} \end{pmatrix} \otimes \boldsymbol{I}_n \hat{=} \Sigma_* \otimes \boldsymbol{I}_n, \\ \Sigma_* > 0 \text{ and } (\boldsymbol{Y}\ \boldsymbol{z}) \sim \text{ELSL}. \end{cases}$$

By (6.2.16) we get the relations:

(6.5.8) $$\begin{cases} \hat{\beta}_{OX} = \hat{\beta}_O, \qquad \hat{\boldsymbol{B}}_X = \hat{\boldsymbol{B}}, \\ \hat{\beta}_{OZ} = \bar{\boldsymbol{z}} - \bar{\boldsymbol{x}}'\hat{\boldsymbol{B}}_Z,\ \hat{\boldsymbol{B}}_z = \boldsymbol{L}_{xx}^{-1}\boldsymbol{L}_{xz}, \\ Q_{**} = \begin{pmatrix} \boldsymbol{L}_{yy}(x) & \boldsymbol{L}_{yz}(x) \\ \boldsymbol{L}_{zy}(x) & \boldsymbol{L}_{zz}(x) \end{pmatrix}. \end{cases}$$

The ratio of two Wilks' statistics is

$$\frac{\Lambda^{**}}{\Lambda} = \frac{\begin{vmatrix} \boldsymbol{L}_{yy}(x) & \boldsymbol{L}_{yz}(x) \\ \boldsymbol{L}_{zy}(x) & \boldsymbol{L}_{zz}(x) \end{vmatrix} \Big/ \begin{vmatrix} \boldsymbol{L}_{yy} & \boldsymbol{L}_{yz} \\ \boldsymbol{L}_{zy} & \boldsymbol{L}_{zz} \end{vmatrix}}{|\boldsymbol{L}_{yy}(x)|/|\boldsymbol{L}_{yy}|}$$
$$= \frac{\boldsymbol{L}_{zz}(x) - \boldsymbol{L}_{zy}(x)\boldsymbol{L}_{yy}^{-1}(x)\boldsymbol{L}_{yz}(x)}{\boldsymbol{L}_{zz} - \boldsymbol{L}_{zy}\boldsymbol{L}_{yy}^{-1}\boldsymbol{L}_{yz}}$$
$$= \frac{L_{zz}\binom{x}{y}}{L_{zz}(y)}$$
$$= 1 - V_z^*$$

and the statistic T^2

(6.5.9) $$T^2 = \frac{n-m-p-2}{m}\frac{V_z^*}{1-V_Z^*} \sim F(m, n-m-p-2)$$

can be used to test the hypothesis H_0: $\boldsymbol{B}_z = \boldsymbol{0}$.

It is easy to design a computational procedure, which can be found in Zhang's paper (1980).

6.5.2. Example

In the Yangtze River valley, the forecasting of "Plum Rains" is important. We select the following 30 $\boldsymbol{y}_i's$ as the predicted variables and 47 $\boldsymbol{x}_i's$ as the predictors.

y_1 to y_{27} are the total sums of rainfall at those stations (the rainfall station of the city or the country) during May 1 to August 31 each year.

y_1 Shanghai	y_2 Nantong
y_3 Nanjing	y_4 Hangzhou
y_5 Wuhu	y_6 Hefei
y_7 Jiujang	y_8 Nanchang
y_9 Hankou	y_{10} Yueyang
y_{11} Changsha	y_{12} Yichang
y_{13} Engshi	y_{14} Chongqing
y_{15} Zunyi	y_{16} Chengdu
y_{17} Yibin	y_{18} Xichang
y_{19} Kunming	y_{20} Huili
y_{21} Anqing	y_{22} Jian
y_{23} Hengyang	y_{24} Shaoyang
y_{25} Yuanling	y_{26} Zhijiang
y_{27} Yunxian	

y_{28} the average flow at Hankou flow station (each year)
y_{29} the highest level at Hankou flow station (each year)
y_{30} the average flow at Yichang flow station (each year)

Generally the $x_i's$ are the quantities of the previous year.

x_1 the number of sunspots
x_2 the class of the air temperature in China
x_3 the intensity of the extreme vortex in January
x_4 the location of the East Asian Valley in January
x_5 the intensity at the East Asian Valley in January
x_6 the index of the intensity of semi-high pressure region in January
x_7 the index of the area of semi-high pressure region in January
x_8 the average index of Asia-Enrope atomospheric circulation in January
x_9 the average air temperature of Lasha from October (the previous year) to the next February
x_{10} the average air temperature of Yushu from October (the previous year) to the next February

The following five predictors are the average water temperatures of the sea surface at these location.

x_{11} 20°N, 120°E Dec.
x_{12} 20°N, 125°E Dec.
x_{13} 20°N, 130°E Dec.
x_{14} 25°N, 125°E Dec.

x_{15} 25°N, 130°E Dec.

x_{16}–x_{20} are the same quantities in Nov. of the above five locations respectively.

x_{21}–x_{25} are the same quantities in Jan. of the above five location respectively.

x_{26}–x_{36} are the averages of the 500 mb distance from means of the altitude of the 11 location, which are given in the following, in Jan., and x_{37}—x_{47} are the same quantities in Feb. respectively.

70°N—80°N, 40°E—70°E; 60°N, 70°E;
45°N—50°N, 0°E ; 30°N, 10°E;
45°N—55°N, 80°E—90°E; 30°N—40°N, 60°E—90°E;
50°N—55°N, 110°E—120°E; 40°N, 120°E—130°E;
15°N—30°N, 100°E—130°E; 25°N—30°N, 160°E—170°E;
45°N—50°N, 100°W—60°W.

We used the data from 1954 to 1977 and selected F_x to screen the predictors x and F_y to screen the forecasted variables y respectively. The following results are obtained:

The lower reaches of Yangtze River are divided into two parts:

a) Predicted variables: y_3(Nanjing), y_5(Wuhu), y_6(Hefei).
Predictors: x_7, x_{40}, x_8, x_{15}
$F_x = 4.2$
$F_y = 1.0$
Regressive equations:

$$y_3 = 5644 - 18.0x_7 - 10.3x_8 - 167.8x_{15} - 23.4x_{40}$$
$$y_5 = 9534 - 4.96x_7 - 8.54x_8 - 323.2x_{15} - 23.6x_{40}$$
$$y_6 = 2456 - 2.20x_7 - 9.60x_8 - 48.7x_{15} - 18.9x_{40}$$

b) Predicted variables: y_1(Shanghai), y_4(Hangzhou)
Predictors: x_{35}, x_{47}, x_9, x_{24}
$F_x = 4.0$
$F_y = 1.5$
Regressive equations:

$$y_1 = 1312 + 3.76x_9 - 39.7x_{24} - 22.5x_{35} + 16.2x_{47}$$
$$y_4 = -2810 + 15.1x_9 + 133x_{24} - 41.4x_{35} + 10.8x_{47}$$

The middle reaches of the Yangtze River is divided into three parts:

a) Predicted variables: y_{10}(Yeuyang), y_{12}(Yichang)
Predictors: x_{35}, x_{40}, x_{47}, x_{34}, x_{11}, x_{15}
$F_x = 4.0$
$F_y = 1.5$
Regressive equations:

$$y_{10} = 9724 + 62.4x_{11} - 422x_{15} + 44.9x_{34} - 3.12x_{36} - 11.20x_{40} + 33.9x_{47}$$
$$y_{12} = 3162 - 12.0x_{11} - 84.4x_{15} - 19.1x_{34} + 19.4x_{36} - 16.5x_{40} + 16.7x_{47}$$

b) Predicted variables: y_{21}(Anqing), y_{26}(Zhijiang), y_{28}(flow of Hankou), y_{30}(flow of Yichang)
Predictors: x_4, x_8, x_{25}, x_{35}
$F_x = 4.0$
$F_y = 1.5$
Regressive equations:

$$y_{21} = -6298 + 1.72x_4 + 1.03x_8 + 29.8x_{25} - 56.0x_{35}$$
$$y_{26} = -1098 - 22.5x_4 - 7.7x_8 + 247x_{25} - 7.8x_{35}$$
$$y_{28} = -521 + 3.0x_4 + 0.9x_8 + 11.3x_{25} - 4.0x_{35}$$
$$y_{30} = -231 + 2.5x_4 + 0.5x_8 - 0.98x_{25} - 3.0x_{35}$$

c) Predicted variables: y_9(Hankou), y_{11}(Changsha), y_{13}(Eangshi), y_{24}(Shaoyang), y_8(Nanchang), y_{27}(Yunxian), y_{29}(the highest level of Hankou flow station).
Predictors: x_{25}, x_{26}, x_{40}
$F_x = 4.0$
$F_y = 1.5$
Regressive equations:
$y_9 = -4882 + 245.7x_{25} + 0.59x_{26} - 21.8x_{40}$
$y_{11} = -4889 + 246.6x_{25} + 0.87x_{26} - 14.2x_{40}$
$y_{13} = -1107 + 84.7x_{25} - 6.8x_{26} - 13.6x_{40}$
$y_{24} = -824 + 65.8x_{25} - 3.3x_{26} - 17.0x_{40}$
$y_8 = -3867 + 208.7x_{25} - 4.3x_{26} - 31.4x_{40}$
$y_{27} = 1417 - 44.3x_{25} - 2.8x_{26} - 12.5x_{40}$
$y_{29} = 245 + 102.5x_{25} - 2.0x_{26} - 14.7x_{40}$

The upper reaches of the Yangtze River is divided into two parts:

a) Predicted variables: y_{16}(Chengdu), y_{18}(Xichang),
Predictors: x_7, x_{36}, x_{38}, x_{40}
$F_x = 4.0$
$F_y = 1.5$
Regressive equations:
$y_{16} = 523 + 40.0x_7 - 10.3x_{36} + 12.3x_{38} - 16.5x_{40}$
$y_{18} = 756 - 9.9x_7 + 10.9x_{36} - 6.1x_{38} - 7.9x_{40}$

b) Predicted variables: y_{14}(Chongqing), y_{15}(Zunyi).
Predictors: x_{27}, x_{31}, x_{34}
$F_x = 6.0$
$F_y = 1.5$
Regressive equations:
$y_{14} = 637 + 0.62x_{27} - 4.77x_{31} - 19.1x_{44}$
$y_{15} = 654 - 10.5x_{27} - 17.7x_{31} + 20.0x_{44}$

Thus we obtain 22 regressive equations. The other eight predicted variables y_2, y_7, y_{17}, y_{19}, y_{20}, y_{22}, y_{23}, y_{25} can not be selected in any regressive equation system, and they must be calculated separately.

It is interesting that the regression equation systems, aggregated by this method, correspond to the geographical regions. In the same region, there are common factors influencing the rainfalls, and the factors are different in different regions.

6.5.3. Discriminant Analysis and Regression

In Section 6.1.4, We established a discriminant analysis model in terms of a linear model so that a discriminant problem can be regarded as a regression problem. Now we would like to study it further.

Given k populations $G_1, \cdots, G_k$, we take $n_1, \cdots, n_k$ samples with p variables from these populations respectively. The data matrix of G_i is

$$X_i = \begin{pmatrix} x_{11}^{(i)} & \cdots & x_{1p}^{(i)} \\ \vdots & & \vdots \\ x_{n_i1}^{(i)} & \cdots & x_{n_ip}^{(i)} \end{pmatrix}, \quad i = 1,2,\cdots,k$$

and the total data matrix (say X) is

$$X = \begin{pmatrix} X_1 \\ \vdots \\ X_k \end{pmatrix}.$$

The mean vectors of these populations and the general mean are denoted by $\bar{\boldsymbol{x}}^{(1)}, \cdots, \bar{\boldsymbol{x}}^{(k)}$ and $\bar{\boldsymbol{x}}$ respectively. Without loss of generality, we may suppose $\bar{\boldsymbol{x}} = \boldsymbol{0}$. Consider a dummy variable $\boldsymbol{Y}$, where $Y = y_i$ when the sample is from G_i, i.e.,

$$\boldsymbol{y} = \begin{pmatrix} y_1 \mathbf{1}_{n_1} \\ \vdots \\ y_k \mathbf{1}_{n_k} \end{pmatrix}.$$

Now we use this dummy variable to set up the regression model, i.e., to find the regression coefficients β and $\boldsymbol{y}$ such that

$$\begin{cases} Q = (\boldsymbol{y} - \boldsymbol{X}\beta)'(\boldsymbol{y} - \boldsymbol{X}\beta) \to \min, \\ \boldsymbol{y}'{,}\boldsymbol{y} = 1. \end{cases} \tag{6.5.10}$$

The restriction $\boldsymbol{y}'\boldsymbol{y} = 1$ is natural and necessary; otherwise let $\boldsymbol{y} = \boldsymbol{0}$, $\beta = \boldsymbol{0}$; then we have $Q = 0$.

Denote

$$\boldsymbol{A} = \sum_{i=1}^{k} n_i \bar{\boldsymbol{x}}^{(i)} \bar{\boldsymbol{x}}^{(i)\prime}, \quad \boldsymbol{C} = \sum_{i=1}^{k} \boldsymbol{X}_i' \boldsymbol{X}_i = \boldsymbol{X}'\boldsymbol{X}$$

and

$$\boldsymbol{W} = \boldsymbol{C} - \boldsymbol{A} = \sum_{i=1}^{k} \boldsymbol{X}_i' \left(\boldsymbol{I}_{n_i} - \frac{1}{n_i} \mathbf{1}\mathbf{1}' \right) \boldsymbol{X}_i.$$

Theorem 6.5.1. *Denote the least squares estimates of β and $\boldsymbol{y}$ in the model* (6.5.10) *by $\hat{\beta}$ and $\hat{\boldsymbol{y}}$. Then*

(i) *$\hat{\beta}$ is the characteristic vector of $\boldsymbol{W}$ relative to $\boldsymbol{A}$ corresponding to the least characteristic root λ_1 of $\boldsymbol{W}$ relative to $\boldsymbol{A}$;*

(ii) *$\hat{\boldsymbol{y}} = (y_1 \mathbf{1}'_{n_1}, \ldots, y_k \mathbf{1}'_{n_k})'$ with*

$$y_i = \frac{1}{1 - \lambda_1} \bar{\boldsymbol{x}}^{(i)\prime} \beta. \tag{6.5.11}$$

Proof. Using the Lagrangian multipliers, we must find $\hat{y}_i$'s and $\hat{\beta}$ such that

$$\begin{aligned} Q^* &= (\hat{\boldsymbol{y}} - \boldsymbol{X}\hat{\beta})'(\hat{\boldsymbol{y}} - \boldsymbol{X}\hat{\beta}) - \lambda(\boldsymbol{y}'\boldsymbol{y} - 1) \\ &= \sum_{i=1}^{\bar{k}} (n_i \hat{y}_i^2 - 2 n_i \hat{y}_i \hat{\beta}' \bar{\boldsymbol{x}}^{(i)} + \hat{\beta}' \boldsymbol{X}_i' \boldsymbol{X}_i \hat{\beta}) - \lambda \left(\sum_{i=1}^{k} n_i y_i^2 - 1 \right) \end{aligned}$$

is a mimimum. Let the partial derivative of Q^* with respect to $\hat{y}_i$ equal zero. Then we obtain

$$\hat{y}_i = (1/1 - \lambda) \hat{\beta}' \bar{\boldsymbol{x}}^{(i)}.$$

Let the partial derivative of Q^* with respect to $\hat{\beta}$ equal zero. Then we get

$$\left(\sum_{i=1}^{k} \boldsymbol{X}_i' \boldsymbol{X}_i \right) \hat{\beta} = \sum_{i=1}^{k} n_i \hat{y}_i \bar{\boldsymbol{x}}^{(i)}. \tag{6.5.12}$$

Noting (6.5.11), the above equation becomes

$$C\hat{\beta} = \sum_{i=1}^{k} n_i(1/1-\lambda)\bar{\boldsymbol{x}}^{(i)}\bar{\boldsymbol{x}}^{(i)\prime}\hat{\beta} = (1/1-\lambda)\boldsymbol{A}\hat{\beta},$$

or

$$\left(\boldsymbol{C} - \frac{1}{1-\lambda}\boldsymbol{A}\right)\hat{\beta} = \boldsymbol{0}, \tag{6.5.13}$$

or

$$\left(\boldsymbol{W} - \frac{\lambda}{1-\lambda}\boldsymbol{A}\right)\hat{\beta} = \boldsymbol{0}; \tag{6.5.14}$$

Hence $\hat{\beta}$ is a characteristic vector of $\boldsymbol{W}$ relative to $\boldsymbol{A}$. In order to minimize Q, substituting (6.5.11) into Q and using (6.5.12), we have

$$\begin{aligned}
Q &= \sum_{i=1}^{k} (n_i y_i^2 - 2n_i \bar{x}^{(i)\prime} y_i \hat{\beta}) + \hat{\beta}'\boldsymbol{C}\hat{\beta} \\
&= \sum_{i=1}^{k} (n_i y_i^2 - n_i y_i \bar{\boldsymbol{x}}^{(i)\prime}\hat{\beta}) \\
&= (1-\lambda)^{-2}\hat{\beta}'\Big(\sum_{i=1}^{k} n_i \bar{\boldsymbol{x}}^{(i)}\bar{\boldsymbol{x}}^{(i)\prime}\Big)\hat{\beta} - (1-\lambda)^{-1}\left(\sum_{i=1}^{k} n_i \hat{\beta}\bar{\boldsymbol{x}}^{(i)}\bar{\boldsymbol{x}}^{(i)\prime}\hat{\beta}\right) \\
&= (1-\lambda)^{-2}\hat{\beta}'\boldsymbol{A}\hat{\beta} - (1-\lambda)^{-1}\hat{\beta}'\boldsymbol{A}\hat{\beta} = \lambda(1-\lambda)^{-2}\hat{\beta}'\boldsymbol{A}\hat{\beta}.
\end{aligned}$$

By $\hat{\boldsymbol{y}}'\hat{\boldsymbol{y}} = 1$, we have $(1-\lambda)^{-2}\hat{\beta}'\boldsymbol{A}\hat{\beta} = 1$ and $Q = \lambda$. To minimize Q is equivalent to minimize λ. It is obvious that $0 \leqslant \lambda < 1$ by (6.5.13) and (6.5.14). As $\lambda(1-\lambda)^{-1}$ as a monotone increasing function of λ in $(0, 1)$, to minimize λ is equivalent to minimize the characteristic roots of $\boldsymbol{W}$ relative to $\boldsymbol{A}$ and the theorem follows. □

We can obviously generalize the preceding conclusion for more than one dummy variable. Let the dummy variables be

$$\boldsymbol{y}_i = \begin{pmatrix} y_1^{(i)}\boldsymbol{1}_{n_1} \\ \vdots \\ y_k^{(i)}\boldsymbol{1}_{n_k} \end{pmatrix}, \quad i = 1, \dots, l;\ 1 \leqslant l \leqslant \min(p, k-1),\ \boldsymbol{Y} = (\boldsymbol{y}_1, \dots, \boldsymbol{y}_l) \tag{6.5.15}$$

and the corresponding coefficients of regression be $\boldsymbol{B} = (\beta_1, \dots, \beta_l)$. The regression model with these dummy variables is

$$\begin{cases} Q = \operatorname{tr}(\boldsymbol{Y} - \boldsymbol{X}\boldsymbol{B})'(\boldsymbol{Y} - \boldsymbol{X}\boldsymbol{B}) \to \min \\ \boldsymbol{y}_i'\boldsymbol{y}_i = \delta_{ij} = \begin{cases} 1, \text{ if } i = j, \\ 0, \text{ if } i = j, \end{cases} \quad i, j = 1, \dots, l, \end{cases} \tag{6.5.16}$$

where $\boldsymbol{Y}$ and the regression coefficient matrix $\boldsymbol{B}$ are unknown variables.

Theorem 6.5.2. *Denote the least square estimates of* $\boldsymbol{B}$ *and* $\boldsymbol{Y}$ *in the model* (6.5.16) *by* $\hat{\boldsymbol{B}} = (\hat{\beta}_1, \dots, \hat{\beta}_l)$ *and* $\hat{\boldsymbol{Y}} = (\hat{\boldsymbol{y}}_1, \dots, \hat{\boldsymbol{y}}_l)$. *Then*

(1) $\hat{\beta}_i$ *is the characteristic vector of* $\boldsymbol{W}$ *relative to* $\boldsymbol{A}$ *corresponding to the i-th least characteristic root* λ_i *of* $\boldsymbol{W}$ *relative to* $\boldsymbol{A}$;

(2) $\hat{\boldsymbol{y}}_i = (\hat{y}_1^{(i)}\boldsymbol{1}_{n_1}', \dots, y_k^{(i)}\boldsymbol{1}_{n_k}')'$ *with* $\hat{y}_j^{(i)} = (1-\lambda_i)^{-1}\hat{\beta}_i'\bar{\boldsymbol{x}}^{(j)}$,
$i = 1, \dots; j = 1, \dots k.$

Proof. Since

$$Q=\sum_{i=1}^{l}(\boldsymbol{y}_i-\boldsymbol{X}\hat{\beta}_i)'(\boldsymbol{y}_i-\boldsymbol{X}\hat{\beta}_i),$$

by a method similar to that in Theorem 6.5.1, the proof is straightforward. Hence we only point out

$$\begin{aligned}\hat{\boldsymbol{y}}_i'\hat{\boldsymbol{y}}_j &= \sum_{t=1}^{k} y_t^{(i)}y_t^{(j)}n_t=\sum_{t=1}^{k}\frac{n_t}{(1-\lambda_i)(1-\lambda_j)}\hat{\beta}_i'\bar{\boldsymbol{x}}^{(t)}\bar{\boldsymbol{x}}^{(t)\prime}\hat{\beta}_j\\ &=\frac{1}{(1-\lambda_i)(1-\lambda_j)}\hat{\beta}_i'\left(\sum_{i=1}^{k}n_t\bar{\boldsymbol{x}}^{(t)}\bar{\boldsymbol{x}}^{(t)\prime}\right)\hat{\beta}_j\\ &=\frac{1}{(1-\lambda_i)(1-\lambda_j)}\hat{\beta}_i'\boldsymbol{A}\hat{\beta}_j=\frac{1}{\lambda_j(1-\lambda_i)}\hat{\beta}_i'\boldsymbol{W}\hat{\beta}_j=0.\end{aligned}$$

The last step above is from the property of relative characteristic roots. □

Remark. Usually, the LSE $\hat{\boldsymbol{B}}$ of $\boldsymbol{B}$ can be obtained immediately by the Fisher discriminant criterion. Fisher's suggestion was to look for the linear function $\boldsymbol{a}'\boldsymbol{x}$ which maximizes the ratio of the between-groups sum of squares to the with-groups sum of squares. That is, let

$$\boldsymbol{Z}=\boldsymbol{X}\boldsymbol{a}=\begin{pmatrix}\boldsymbol{X}_1\boldsymbol{a}\\ \vdots\\ \boldsymbol{X}_k\boldsymbol{a}\end{pmatrix}=\begin{pmatrix}\boldsymbol{z}_1\\ \vdots\\ \boldsymbol{z}_k\end{pmatrix}$$

be a linear combination of the columns of $\boldsymbol{X}$. As $\bar{\boldsymbol{x}}=0$, then $\boldsymbol{z}$ has a total sum of squares

$$\boldsymbol{z}'\boldsymbol{z}=\boldsymbol{a}'\boldsymbol{X}'\boldsymbol{X}\boldsymbol{a}=\boldsymbol{a}'\boldsymbol{C}\boldsymbol{a},$$

which can be partitioned as a sum of the with-groups sum of squares,

$$\sum_{i=1}^{k}\boldsymbol{z}_i'\left(\boldsymbol{I}_{n_i}-\frac{1}{n_i}\boldsymbol{1}\boldsymbol{1}'\right)\boldsymbol{z}_i=\boldsymbol{a}'\left(\sum_{i=1}^{k}\boldsymbol{X}_i'\left(\boldsymbol{I}-\frac{1}{n_i}\boldsymbol{1}\boldsymbol{1}'\right)\boldsymbol{X}_i\boldsymbol{a}=\boldsymbol{a}'\boldsymbol{W}\boldsymbol{a},\right.$$

plus the between-groups sum of squares,

$$\sum_{i=1}^{k}n_i\bar{z}_i^2=\sum_{i=1}^{k}n_i\boldsymbol{a}'\bar{\boldsymbol{x}}^{(i)}\bar{\boldsymbol{x}}^{(i)\prime}\boldsymbol{a}=\boldsymbol{a}'\boldsymbol{A}\boldsymbol{a},$$

where $\bar{z}_i$ is the mean of z_i.

Fisher's criterion is to maximize the ratio

$$\frac{\boldsymbol{a}'\boldsymbol{A}\boldsymbol{a}}{\boldsymbol{a}'\boldsymbol{W}\boldsymbol{a}}$$

with respect to $\boldsymbol{a}$. It can be shown that the vector $\boldsymbol{a}$ in Fisher discriminant model is the characteristic vector of $\boldsymbol{W}^{-1}\boldsymbol{A}$ corresponding to the largest characteristic root. This means that the solution $\boldsymbol{a}$ is just $\hat{\beta}$ in Theorem 6.5.1. The fact that the equivalence between the Fisher discriminant model and regression model is first established by Fang (1982). If the dummy variable is 0 or 1, Glahn (1968) discussed the relationship between the above two models.

References

Anderson (1984), Anderson and Fang (1982c), Fang (1982), Glahn (1968), Gnot, Klonecki and Zmyslony (1980), Lawley (1938), Lehmann and Scheffé (1950), Pillai (1955), Rao (1971), Roy

(1957), Wilks (1932), Zhang (1978), Zhang and Zhao (1980).

Exercises 6

6.1 Given a linear model $L(y;\ X, \Theta)$ and a subspace $H_0 \subset R^n$, the parameter $\alpha'\beta$ is said to be H_0–estimable if there is a vector $\boldsymbol{b} \in H_0$ such that $\boldsymbol{b}'\boldsymbol{y}$ is the LUE of the $\boldsymbol{a}'\beta$.
Put the results in the Sections 6.2.2 and 6.2.3 to the H_0–*estimablity*, H_0–BLUE and H_0–regularity.

6.2 Prove Theorem 6.2.3 by the results of Theorem 6.2.4.

6.3 Using Theorem 6.2.4, obtain the conditions of the regularity of the linear models $L(y;\ X, \Theta)$, where
(1) $\Theta = \{\sigma^2 \boldsymbol{I} \mid \sigma > 0\}$;
(2) $\Theta = \{\sigma^2 \boldsymbol{I} + \rho(\boldsymbol{J} - \boldsymbol{I}) \mid \sigma > 0\}$;
(3) $\boldsymbol{y} = \mathrm{vec}\boldsymbol{Y},\ \boldsymbol{Y}\colon n \times p,\ \Theta = \{\Sigma_1 \otimes \Sigma_2 | \Sigma_i \geqslant 0,\ i = 1, 2,\ \Sigma_1\colon p \times p,\ \Sigma_2\colon n \times n\}$.

6.4 Given a linear model $L(y;\ X, \Theta)$ with regularity, when is the $\boldsymbol{a}\hat{\beta}$ the BLUE for each estimable $\boldsymbol{a}'\beta$, where $\hat{\beta}$ is the LSE?

6.5 Prove that for symmetric $\boldsymbol{A}$, $\boldsymbol{X}'\boldsymbol{A}\boldsymbol{X} = \boldsymbol{O}$ iff $\boldsymbol{A} = \boldsymbol{P}\boldsymbol{B} + \boldsymbol{B}'\boldsymbol{P}$ where $\boldsymbol{P} = \boldsymbol{I} - \boldsymbol{X}\boldsymbol{X}^+$ and $\boldsymbol{B}$ is an $n \times n$ matrix.

6.6 Prove that: $\boldsymbol{y}'\boldsymbol{A}\boldsymbol{y} = (\boldsymbol{y} + \boldsymbol{X}\alpha)'\boldsymbol{A}(\boldsymbol{y} + \boldsymbol{X}\alpha)$ for all α and $\boldsymbol{y}$ iff $\boldsymbol{X}'\boldsymbol{A} = 0$ or $\boldsymbol{A}\boldsymbol{X} = \boldsymbol{O}$.

6.7 Prove Lemma 6.4.1.

6.8 Prove that:

$$E(\hat{\theta}_+ - \theta)'(\hat{\theta}_+ - \theta) \leqslant E(\hat{\theta}_s - \theta)'(\hat{\theta}_s - \theta)$$

where $\boldsymbol{y} \sim N(\theta, \boldsymbol{I}_p)$ with $p > 2$, and

$$\hat{\theta}_s = \left(1 - \frac{a}{\boldsymbol{y}'\boldsymbol{y}}\right)\boldsymbol{y},\ \hat{\theta}_+ = \left(1 - \frac{a}{\boldsymbol{y}'\boldsymbol{y}}\right)_+ \boldsymbol{y}.$$

6.9 Let $\boldsymbol{Y} = (\boldsymbol{Y}_1', \boldsymbol{Y}_2', \boldsymbol{Y}_3') \sim \mathrm{ELSL}(\boldsymbol{Y};\ \boldsymbol{X}, \Theta)$ and G, a transformation group defined in Lemma 6.4.2. Show that the distribution of latent roots of the matrix $\boldsymbol{Y}_1(\boldsymbol{Y}_3'\boldsymbol{Y}_3)^{-1}\boldsymbol{Y}_1'$ depends only on the latent roots of $\boldsymbol{B}_1\Sigma^{-1}\boldsymbol{B}_1$.

REFERENCES

[1] Anderson,T.W. (1958, 1984), *An Introdution to Multivariate Statistical Analysis*, First and Second Edition, Wiley, New York.

[2] Anderson, T.W., and Fang,K.T. (1982a), Distributions of quadratic forms and Cochran's Theorem for elliptically contoured distributions and their applications, *Technical Report* No. 53, ONR Contract N00014-75-C-0442, Department of Statistics, Stanford University, Califonia.

[3] Anderson, T.W., and Fang,K.T. (1982b), On the theory of multivariate elliptically contoured distributions and their applications, *Technical Report* No.54, Contract as the above.

[4] Anderson, T.W., and Fang,K.T. (1982c), Maximum likelihood estimators and likelihood ratio criteria for multivariate elliptically contoured distributions, *Technical Report* No. 1, ARO Contract DAAG 29-82-K-0156, Department of Statistics, Stanford University, California.

[5] Anderson,T.W., Fang,K.T., and Hsu, H. (1986), Maximum-likelihood estimates and likelihood-ratio criteria for multivariate elliptically contoured distributions, *The Canad. J. Statist.*, **14**, 55—59.

[6] Anderson,T.W., and Fang,K.T. (1984), Cochran's theorem for elliptically contoured distributions, *Contributed Papers*, China-Japan Symposium on Statistics, 4—7, appear in *Sankhyá*, 1987.

[7] Anderson, T.W., and Styan, G.P.H. (1982), Cochran's theorem, rank additivity, and tripotent matrices, *Statistics and Probability*: *Essays in Honor of C.R.Rao.*, G. Kallianpur, P.R. Krishnaiah, and J.K. Ghosh eds, North-Holland, 1—23.

[8] Bai,Z., Su,C., Fang,K.T., and Chen,P. (1980), A problem on independence of random variables, *Kexue Tongbao*, a special volume on math., phys. and chem., 90—92 (in Chinese).

[9] Bariloche, F. (1976), Robust M-estimators of multivariate location and scatter, *Ann.Statist.*, **4**, 51—67.

[10] Bellman, R. (1970), *Introduction to Matrix Analysis* (2nd ed), McGraw-Hill, New York.

[11] Beran,R. (1979), Testing for elliptical symmetry of a multivariate density, *Ann.Statist.* **7**, 150—162.

[12] Berger, J. (1975), Minimax estimation of location vectors for a wide class of densities, *Ann. Statist.* **3**, 1318—1328.

[13] Bian, G., Wang,J., and Zhang, Y. (1984), The uniqueness conditions for the distribution of a statistic in the class of left O(n)-invariant distributions, *Contributed Papers*, China-Japan Symposium on Statistics. 17—19.

[14] Bian,G. and Zhang, Y. (1984), Estimators and tests of the functional relationship with the left O(n)-invariant errors, *Contributed Papers*, China-Japan Symposium on Statistics. 20—22.

[15] Blackwell, D. (1956), On a class of probability spaces, *Proc. Berkeley Symp. Math. Statist, Prob., 3rd*, University of California Press, Berkeley, California.

[16] Blyth, C.R. (1951), On minimax statistical decision procedures and their admissibility, *Ann. Math. Statist.*, **22**, 22—42.

[17] Box, G.E.P. (1949), A general distribution theory for a class of likelihood ratio criteria, *Biometrika*, **36**, 317—346.

[18] Brandwein, A.C. (1979), Minimax estimation of the mean of spherically symmetric distributions under general quadratic loss, *J.Mult. Anal.*, **9**, 579—588.

[19] Brandwein, A.R.C., and Strawderman,W.E. (1978), Minimax estimation of location

parameter for spherically symmetric unimodal distributions under quadratic loss, *Ann. Statist.*, **6**, 377—416.

[20] Cacoullos, T., and Koutras, M. (1985), Minimum-distance discrimination for spherical distributions, *Statistical Theory and Data Analysis*, K. Matusita ed., Elsevier Science Publishers, North-Holland, 91—102.

[21] Cacoullos. T., and Koutras, M. (1984), Quadratic forms in spherical random variables generalized noncentral x^2-distribution,*Naval Res. Logist. Quart.*, **31**, 447—461.

[22] Cambanis, S., Huang,S., and Simons, G. (1981), On the theory of elliptically contoured distributions, *J. Mult. Anal.*, **11**, 368—385.

[23] Cambanis, S., Keener, R., and Simons, G. (1983), On α-symmetric multivariate distributions, *J. Mult. Anal.*, **13**, 213—233.

[24] Chen, X. (1964), A minimax estimate of transition vector, *Acta Math. Sinica*, **14**, 276—290 (in Chinese).

[25] Chmielewski, M.A. (1980), Invariant scale matrix hypothesis tests under elliptical symmetry, *J. Mult. Anal.*, **10**, 343—350.

[26] Chmielewski, M.A. (1981), Elliptically symmetric distributions: A review and bibliography, *Inter. Statist. Review*, **49**, 67—74.

[27] Chow,Y.S. (1978), *Probability Theory*, Springer-Verlag, Berlin.

[28] Cohen, A. (1966), All admissible linear estimates of the mean Vector, *Ann. Math. Statist.*, **37**, 458—463.

[29] Davis, A.W. (1971), Percentile approximations for a class of likelihood ratio criteria, *Biometrika*, **27**, 349—356.

[30] Dawid, A.P. (1977), Spherical matrix distributions and a multivariate model, *J.R. Statist. Soc.*, **39**, 254—261.

[31] Dawid, A.P. (1978), Extendability of spherical matrix distributions, *J.Mult. Anal.*, **8**, 559—566.

[32] Dawid, A.P. (1985), Invariance and independence in multivariate distribution theory, *J. Mult. Anal.*, **17**, 304—315.

[33] Deng, W. (1984), Testing for ellipsoidal symmetry, *Contributed Papers*, China-Japan Symposium on Statistics. 55—58.

[34] Dykstra,R.L. (1970), Establishing the positive definiteness of the sample covariance matrix, *Ann. Math. Statist.*, **41**, 2153—2154.

[35] Esseen, C.(1945), Fourier analysis of distribution function, mathemetical study of the Laplace-Gaussian law, *Acta Math.*, 1—125.

[36] Eaton, M.L. (1981), On the projections of isotropic distributions, *Ann. Stat.*, **9**, 391—400.

[37] Eaton, M.L. (1983), *Multivariate Statistics*-A Vector Space Approach, Wiley, New York.

[38] Eaton, M.L., and Kariya, T. (1984), A condition for null robustness, *J.Mult. Anal.*, **14**, 155—168.

[39] Fan, J. (1984), Generalized non-central *t*-, *F*- and T^2-distributions, *J. Graduate School*, **2**, 134—148 (in Chinese).

[40] Fan,J. (1986), Shrinkage estimators and ridge regression estimators for elliptically contoured distributions, *Acta Appl. Math. Sinica*, **9**, 237—250 (in Chinese).

[41] Fan, J. (1986), Non-central Cochran's theorem for elliptically contoured distributions, to appear in *Acta Math. Sinica* (New Ser.), 2. 185—198.

[42] Fan,J., and Fang,K.T. (1985), Inadmissibility of sample mean and sample regression parameters for elliptically contoured distributions, *Northeastern Math. J.*, **1**, 68—81 (in Chinese).

[43] Fan, J., and Fang, K.T. (1985), Minimax estimators and Stein two-stage estimators of location parameters for elliptically contoured distributions, *Chinese J. Appl. Prob. Stat.*, **1**, 103—114 (in Chinese).

[44] Fan.J., and Fang, K.T. (1986), Inadmissibility of the usual estimator for the location parameters of spherically symmetric distributions, to appear in *Kexue Tongbao*.

[45] Fan,J., and Fang,K.T. (1986), Maximum likelihood characterization of distributions, to appear in *Acta Appl. Math. Sinica* (English Ser.).

[46] Fang,K.T. (1982), Equivalence between Fisher discriminant model and regression model, *Kexue Tongbao*, **27**, 803—806.

[47] Fang, K.T. (1987), A review: on the theory of elliptically contoured distributions, *Math. Advance Sinica* 16, 1—15 (in Chinese).

[48] Fang, K.T., and Chen, H.F. (1984), Relationships among classes of spherical matrix distributions, *Acta Math. Appl. Sinica* (English Ser.), **1**, 139—147.

[49] Fang, K.T., and Chen, H.F.(1986), On the spectral decompositions and some of their subclasses, *J. Math. Res. & Exposition*, No.14, 147—156.

[50] Fang,K.T., and Wu, Y.H. (1984), Distributions of quadratic forms and generalized Cochran's theorem, *Math. in Economics*, **1**, 29—48. (in Chinese).

[51] Fang,K.T., and Xu,J.L. (1985), Likelihood ratio criteria testing hypotheses about parameters of elliptically contoured distributions, *Math. in Economics*,, **2**, 1—19 (in Chinese).

[52] Fang,K.T., and Xu,J.L. (1986), The direct operations of symmetric and lower-triangular matrices with their applications, *Northeastern Math. J.*, **2**, 4—16.

[53] Feller, W.(1971), *An Introduction to Probability Theory and Its Applications*, Vol. II, Wiley, New York.

[54] Fraser,D.A. and Ng,K.W. (1980), Multivariate regression analysis with spherical error, in Multivariate Analysis V, (Krishnaiah, P.R.ed.), North-Holland, 369—386.

[55] Giri, N. (1977), *Multivariate Statistical Inference*, Academic Press, New York.

[56] Glahn,H.R. (1968), Canonical correlation and its relationship to discriminant analysis and multiple regression, *J. Atmospheric Sci.*, **25**, 23—31.

[57] Gnot, S., Klonecki,W. and Zmyslony,R (1980), Best unbiased linear estimation, a coordinate free approach, *Prob. and Math. Statist.*, **1**, 1—13.

[58] Graham,A.(1981), *Kronecker Products and Matrix Calculus: with Applications*, Ellis. Hovwood Limited.

[59] Haff, L.R. (1980), Empirical Bayes estimation of the multivariate normal covariance matrix, *Ann. Statist.*, **8**, 586—597.

[60] Halmos, P.L. and Savage,L.J. (1949), Application of Radon-Nikodym Theorem of the theory of sufficient statistics, *Ann. Math. Statist.*, **20**, 225—241.

[61] Hsu,P.L. (1940a), On generalized analysis of variance (I), *Biometrika*, **31**, 221—237.

[62] Hsu,P.L.(1940b), An algebraic derivation of the distribution of rectangular coordinate, *Proc. Edin. Math. Soc. 6*, **2**, 185—189.

[63] Hsu,P.L. (1954, 1983), On characteristic functions which coincide in the neighbourhood of zero, *Acta Math.Sinica*, **4**, 21—31, and in *Pao-Lu Xsu, Collected Works*, Springer-Verlag.

[64] Huber,P.J. (1964), Robust estimation of the location parameter, *Ann. Math. Statist.*, **35**, 73—101.

[65] James,W. and Stein, C. (1961), Estimation with quadratic loss, *Proc. Fourth Berkeley Symp. Math. Statist. and Prob.*, **1**, 361—379.

[66] Jensen, D.R. and Good, I.J. (1981), Invariant distributions associated with matrix laws under structural symmetry, *J. Royal Statist. Soc., B.* **43**, 327—332.

[67] Johnson, N.J. and Kotz,S. (1972), *Distributions in Statistics: Continuous Multivariate Distributions*, Wiley, New York.

[68] Kariya, T. (1981), Robustness of multivariate tests, *Ann. Statist.*, **9**, 1267—1275.

[69] Kariya,T. and Eaton, M.L. (1977), Robust tests for spherical symmetry, *Ann. Statist.*, **5**, 206—215.

[70] Kelker, D, (1970), Distribution theory of spherical distributions and a location-scale

parameter generalization, *Sankhyā, A*, **32**, 419—430.

[71] Khatri, C.G. (1970), A note on Mitra's paper "A density-free approach to the matrix variate beta distribution", *Sankhyā, A*, **32**, 311—318.

[72] Khatri, C.G. and Srivastava, M.S. (1971), On exact non-null distributions of likelihood ratio criteria for sphericity test and equality of two covariance matrices, *Sankhyā, A*, **33**, 201—206.

[73] Kiefer, J. (1957), Invariance, minimax sequential estimation, and continuous time processes, *Ann. Math. Statist.*, **28**, 537—601.

[74] King, M.L. (1980), Robust tests for spherical symmetry and their application to least squares regression, *Ann. Statist.*, **8**, 1165—1271.

[75] Kingman, J.F.C. (1972), On random sequences with spherical symmetry, *Biometrika*, **59**, 492—494.

[76] Kotz,S. (1975), Multivariate distributions at a cross-road, in *Statistical Distributions in Scientific Work, Vol.I*, (G.P. Patil, S.Kotz and J.K.Ord eds), D.Reidel Publ.Co.

[77] Kozial,J.A. (1982), A class of invariant procedures for assessing multivariate normality, to be submitted.

[78] Kres, H. (1975), *Statistische Tafeln zur Multivariate Analysis*, Springer-Verlag.

[79] Kshirsagar, A.M. (1961), The noncentral multivariate beta distribution, *Ann. Math. Statist.*, **32**, 104—111.

[80] Kshirsagar, A.M. (1972), *Multivariate Analysis*, Dekker, New York.

[81] Laurent,A.G. (1974), The intersection of random sphere and the noncentral radial error distribution for spherical models, *Ann. Statist.*, **30**, 180—187.

[82] Lawley,D.N. (1938), A generalization of Fisher's Z-test, *Biometrika*, **30**, 180—187.

[83] Lehmann,E.L. and Scheffé, H. (1950), Completeness, similar regions and unbiased estimation Part 1, *Sankhyā*, **10**, 305—340.

[84] Lehmann,E.L. and Stein,C. (1948), Most powerful tests of composite hypothesis, I. Normal distributions, *Ann.Math. Statist.*, **10**, 495—516.

[85] Li,G. (1984), Moments of random vector and its quadratic forms, *Contributed Papers*, China-Japan Symposium on Statistics, 134—135.

[86] Loève, M. (1960, 1977), *Probability Theory*, (2nd and 4th edition), Springer-Verlag, Berlin.

[87] Lukacs, E. (1956), Characterization of populations by properties of suitable statistics, *Proc. 3rd Berkeley Symp. Math. Statist. and Prob.*, **2**, University of California Press, Los Angeles & Berkeley.

[88] Marcinkiewicz, J. (1938), Sur les functions independantes III, *Fundamenta Mathematicae*, **31**, 66—102.

[89] Maronna,R.A. (1976), Robust M-estimators of multivariate location and scatter, *Ann. Statist.*, **4**, 51—67.

[90] Mauchly,J.W. (1940), Significance test for sphericity of a normal n-variate distribution, *Ann. Math. Statist.*, **11**, 204—209.

[91] Muirhead,R.J. (1980), The effects of elliptical distributions on some standard procedures involving correlation coefficients: A review, in Multivariate Statistical Analysis (R.P. Gupta ed.), North-Holland.

[92] Muirhead,R.J. (1982), *Aspects of Multivariate Statistical Theory*, Wiley, New York.

[93] Olkin, I. and Selliah,J.B. (1977), Estimating covariances in multivariate normal distribution, in *Statistical Decision Theory and Related Topics*, Vol. 1, (S.S. Gupta and D.S.Moore eds), Academic Press, New York, 313—326.

[94] Pillai,K.C.S. (1955), Some new test criteria in multivariate analysis, *Ann. Math., Statist.*, **26**, 117—121.

[95] Puri, L.P. and Sen, P.K. (1971), *Nonparametric Methods in Multivariate Analysis*, Wiley, New York.

[96] Quan, H. (1987), Some optima parameter tests for elliptically contoured distribution

class, *Acta Math. Sinica* (English Ser.), **3**, 1—14.

[97] Quan, H. and Fang,K.T. (1987), Unbiasedness of some testing hypotheses in elliptically contoured population, *Acta Math. Appl. Sinica*, **10**, 211—220.

[98] Quan, H., Fang,K.T. and Teng,C.Y. (1985), The applications of information function for spherical distributions, to be submitted to *J. Graduate School.*

[99] Rao,C.R. (1973), *Linear Statistical Inference and Its Applications*, Second Edi., Wiley, New York.

[100] Rao,C.R. (1971), Estimation on variance and covariance components ——MINQUE theory, *J. Mult. Anal.*, **3**, 257—275.

[101] Roy, S. N. (1957), *Some Aspects of Multivariate Analysis*, Wiley, New York.

[102] Schatzoff,M. (1966), Exact distribution of Wilks' likelihood ratio criterion, *Biometrika*, **53**, 347—358.

[103] Schoenberg,I.J. (1938), Metric spaces and completely monotone functions, *Ann. Math.*, **39**, 811—841.

[104] Srivastava,M.S. and Khatri,C.G. (1979), *An Introduction to Multivariate Statistics*, North-Holland, New York.

[105] Stein,C. (1956), Inadmissibility of the usual estimator of the mean of a multivariate normal distribution, *Proc. Third Berkeley Symposium Math. Statist. Prob.*, **1**, 197—206.

[106] Strawderman,W,E. (1971), Proper Bayes minimax estimators of the multivariate normal mean, *Ann. Math. Statist.*, **42**, 385—388.

[107] Strawderman, W.E. (1972), On the existence of proper Bayes minimax estimators of the mean of a multivariate normal distribution, *Proc. 6th Berkeley Symp. Math. Statist. Prob.*, Vol. I, 51—55.

[108] Strawderman,W.E. (1974), Minimax estimation of location parameters for certain spherically symmetric distribution, *J. Mult. Anal.*, **4**, 255—264.

[109] Wijsman,R.A. (1967), Cross-sections of orbits and their application to densities of maximal invariants, *Proc. Fifth Berkeley Symp. Math. Statist. Prob.*, **I**, 389—400

[110] Wilks, S.S. (1932), Certain generalizations of analysis of variance, *Biometrika*, **24**, 471—494.

[111] Xu, J.L. (1984), Inverse Dirichlet distribution and its applications, *Contributed Papers*, China-Japan Symposium on Statistics, 367—370.

[112] Zhang,H.Q. and Fang,K.T. (1987), Some properties of left-spherical and right-spherical matrix distributions, *Chinese J. Appl. Prob. Stat.* **3**, 1—9 (in Chinese).

[113] Zhang,H.Q. and Fang,K.T. (1987), Some characteristics of normal matrix variate distribution, *J. Graduate School*, **4**, 22—30 (in Chinese).

[114] Zhang,Y. (1978), Generalized correlation coefficients and their applications, *Acta Appl. Math. Sinica*, **4**, 312—320, (in Chinese).

[115] Zhang,Y. (1983), The representations of random with spherical distribution and their applications, to be submitted.

[116] Zhang,Y. and Fang,K.T. (1982), *An Introduction to Multivariate Analysis*, Science Press, Beijing, (in Chinese).

[117] Zhang,Y., Fang,K.T. and Chen,H.F. (1985), On matrix elliptically contoured distributions, *Acta Math. Scientia*, **5**, 341—353 (in Chinese).

[118] Zhang, Y. and Zhao,Z. (1980), A two-way step-wise multiple regression procedure, *Acta Appl. Math, Sinica*, **3**, 161—165, (in Chinese).

[119] Zygmund,A. (1951), *Proc. Second Berkeley Symposium*, 369—372.

INDEX